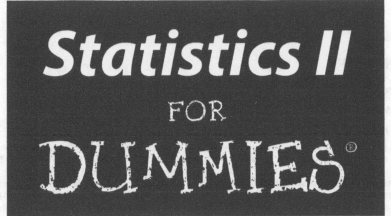

# Statistics II
## FOR
## DUMMIES®

## by Deborah Rumsey, PhD

WILEY

Wiley Publishing, Inc.

**Statistics II For Dummies®**

Published by
**Wiley Publishing, Inc.**
111 River St.
Hoboken, NJ 07030-5774
www.wiley.com

For general information on our other products and services, please contact our Customer Care Department within the U.S. at 877-762-2974, outside the U.S. at 317-572-3993, or fax 317-572-4002.

For technical support, please visit www.wiley.com/techsupport.

Wiley also publishes its books in a variety of electronic formats. Some content that appears in print may not be available in electronic books.

Library of Congress Control Number: 2009928737

ISBN: 978-0-470-46646-9

Manufactured in the United States of America

C10005940_110718

WILEY

# Dedication

To my husband Eric: My sun rises and sets with you. To my son Clint: I love you up to the moon and back.

# About the Author

**Deborah Rumsey** has a PhD in Statistics from The Ohio State University (1993), where she's a Statistics Education Specialist/Auxiliary Faculty Member for the Department of Statistics. Dr. Rumsey has been given the distinction of being named a Fellow of the American Statistical Association. She has also won the Presidential Teaching Award from Kansas State University. She's the author of *Statistics For Dummies, Statistics Workbook For Dummies*, and *Probability For Dummies* and has published numerous papers and given many professional presentations on the subject of statistics education. Her passions include being with her family, bird watching, getting more seat time on her Kubota tractor, and cheering the Ohio State Buckeyes on to another National Championship.

# Author's Acknowledgments

Thanks again to Lindsay Lefevere and Kathy Cox for giving me the opportunity to write this book; to Natalie Harris and Chrissy Guthrie for their unwavering support and perfect chiseling and molding of my words and ideas; to Kim Gilbert, University of Georgia, for a thorough technical view; and to Elizabeth Rea and Sarah Westfall for great copy-editing. Special thanks to Elizabeth Stasny for guidance and support from day one; and to Joan Garfield for constant inspiration and encouragement.

## Publisher's Acknowledgments

We're proud of this book; please send us your comments through our Dummies online registration form located at http://dummies.custhelp.com. For other comments, please contact our Customer Care Department within the U.S. at 877-762-2974, outside the U.S. at 317-572-3993, or fax 317-572-4002.

Some of the people who helped bring this book to market include the following:

**Acquisitions, Editorial, and Media Development**

**Project Editors:** Natalie Faye Harris, Chrissy Guthrie

**Acquisitions Editors:** Lindsay Lefevere, Kathy Cox

**Copy Editors:** Elizabeth Rea, Sarah Westfall

**Assistant Editor:** Erin Calligan Mooney

**Editorial Program Coordinator:** Joe Niesen

**Technical Editor:** Kim Gilbert

**Editorial Manager:** Christine Meloy Beck

**Editorial Assistants:** Jennette ElNaggar, David Lutton

**Cover Photos:** iStock

**Cartoons:** Rich Tennant (www.the5thwave.com)

**Composition Services**

**Project Coordinator:** Lynsey Stanford

**Layout and Graphics:** Carl Byers, Carrie Cesavice, Julie Trippetti, Christin Swinford, Christine Williams

**Proofreaders:** Melissa D. Buddendeck, Caitie Copple

**Indexer:** Potomac Indexing, LLC

---

**Publishing and Editorial for Consumer Dummies**

    **Diane Graves Steele,** Vice President and Publisher, Consumer Dummies

    **Kristin Ferguson-Wagstaffe,** Product Development Director, Consumer Dummies

    **Ensley Eikenburg,** Associate Publisher, Travel

    **Kelly Regan,** Editorial Director, Travel

**Publishing for Technology Dummies**

    **Andy Cummings,** Vice President and Publisher, Dummies Technology/General User

**Composition Services**

    **Debbie Stailey,** Director of Composition Services

# Contents at a Glance

# Table of Contents

# Introduction

● ● ● ● ● ● ● ● ● ● ● ● ● ● ● ● ● ● ● ● ● ● ● ● ● ● ● ● ● ● ● ● ● ● ● ● ● ● ● ● ● ● ● ● ● ● ● ● ● ●

**S**o you've gone through some of the basics of statistics. Means, medians, and standard deviations all ring a bell. You know about surveys and experiments and the basic ideas of correlation and simple regression. You've studied probability, margin of error, and a few hypothesis tests and confidence intervals. Are you ready to load your statistical toolbox with a new level of tools? *Statistics II For Dummies* picks up right where *Statistics For Dummies* (Wiley) leaves off and keeps you moving along the road of statistical ideas and techniques in a positive, step-by-step way.

The focus of *Statistics II For Dummies* is on finding more ways of analyzing data. I provide step-by-step instructions for using techniques such as multiple regression, nonlinear regression, one-way and two-way analysis of variance (ANOVA), Chi-square tests, and nonparametric statistics. Using these new techniques, you estimate, investigate, correlate, and congregate even more variables based on the information at hand.

## About This Book

This book is designed for those who have completed the basic concepts of statistics through confidence intervals and hypothesis testing (found in *Statistics For Dummies*) and are ready to plow ahead to get through the final part of Stats I, or to tackle Stats II. However, I do pepper in some brief overviews of Stats I as needed, just to remind you of what was covered and make sure you're up to speed. For each new technique, you get an overview of when and why it's used, how to know when you need it, step-by-step directions on how to do it, and tips and tricks from a seasoned data analyst (yours truly). Because it's very important to be able to know which method to use when, I emphasize what makes each technique distinct and what the results say. You also see many applications of the techniques used in real life.

I also include interpretation of computer output for data analysis purposes. I show you how to use the software to get the results, but I focus more on how to interpret the results found in the output, because you're more likely to be interpreting this kind of information rather than doing the programming specifically. And because the equations and calculations can get too involved by hand, you often use a computer to get your results. I include instructions for using Minitab to conduct many of the calculations in this book. Most statistics teachers who cover these topics hold this philosophy as well. (What a relief!)

This book is different from the other Stats II books in many ways. Notably, this book features

- ✔ **Full explanations of Stats II concepts.** Many statistics textbooks squeeze all the Stats II topics at the very end of Stats I coverage; as a result, these topics tend to get condensed and presented as if they're optional. But no worries; I take the time to clearly and fully explain all the information you need to survive and thrive.

- ✔ **Dissection of computer output.** Throughout the book, I present many examples that use statistical software to analyze the data. In each case, I present the computer output and explain how I got it and what it means.

- ✔ **An extensive number of examples.** I include plenty of examples to cover the many different types of problems you'll face.

- ✔ **Lots of tips, strategies, and warnings.** I share with you some trade secrets, based on my experience teaching and supporting students and grading their papers.

- ✔ **Understandable language.** I try to keep things conversational to help you understand, remember, and put into practice statistical definitions, techniques, and processes.

- ✔ **Clear and concise step-by-step procedures.** In most chapters, you can find steps that intuitively explain how to work through Stats II problems — and remember how to do it on your own later on.

# Conventions Used in This Book

Throughout this book, I've used several conventions that I want you to be aware of:

- ✔ I indicate multiplication by using a times sign, indicated by a lowered asterisk, *.

- ✔ I indicate the null and alternative hypotheses as Ho (for the null hypothesis) and Ha (for the alternative hypothesis).

- ✔ The statistical software package I use and display throughout the book is Minitab 14, but I simply refer to it as Minitab.

- ✔ Whenever I introduce a new term, I italicize it.

- ✔ Keywords and numbered steps appear in **boldface.**

- ✔ Web sites and e-mail addresses appear in monofont.

# What You're Not to Read

At times I get into some of the more technical details of formulas and procedures for those individuals who may need to know about them — or just really want to get the full story. These minutiae are marked with a Technical Stuff icon. I also include sidebars as an aside to the essential text, usually in the form of a real-life statistics example or some bonus info you may find interesting. You can feel free to skip those icons and sidebars because you won't miss any of the main information you need (but by reading them, you may just be able to impress your stats professor with your above-and-beyond knowledge of Stats II!).

# Foolish Assumptions

Because this book deals with Stats II, I assume you have one previous course in introductory statistics under your belt (or at least have read *Statistics For Dummies*), with topics taking you up through the Central Limit Theorem and perhaps an introduction to confidence intervals and hypothesis tests (although I review these concepts briefly in Chapter 3). Prior experience with simple linear regression isn't necessary. Only college algebra is needed for the mathematics details. And, some experience using statistical software is a plus but not required.

As a student, you may be covering these topics in one of two ways: either at the tail end of your Stats I course (perhaps in a hurried way, but in some way nonetheless); or through a two-course sequence in statistics in which the topics in this book are the focus of the second course. If so, this book provides you the information you need to do well in those courses.

You may simply be interested in Stats II from an everyday point of view, or perhaps you want to add to your understanding of studies and statistical results presented in the media. If this sounds like you, you can find plenty of real-world examples and applications of these statistical techniques in action as well as cautions for interpreting them.

# How This Book Is Organized

This book is organized into five major parts that explore the main topic areas in Stats II, along with one bonus part that offers a series of quick top-ten references for you to use. Each part contains chapters that break down the part's major objective into understandable pieces. The nonlinear setup of this book allows you to skip around and still have easy access to and understanding of any given topic.

# Part I: Tackling Data Analysis and Model-Building Basics

This part goes over the big ideas of descriptive and inferential statistics and simple linear regression in the context of model-building and decision-making. Some material from Stats I receives a quick review. I also present you with the typical jargon of Stats II.

# Part II: Using Different Types of Regression to Make Predictions

In this part, you can review and extend the ideas of simple linear regression to the process of using more than one predictor variable. This part presents techniques for dealing with data that follows a curve (nonlinear models) and models for yes or no data used to make predictions about whether or not an event will happen (logistic regression). It includes all you need to know about conditions, diagnostics, model-building, data-analysis techniques, and interpreting results.

# Part III: Analyzing Variance with ANOVA

You may want to compare the means of more than two populations, and that requires that you use analysis of variance (ANOVA). This part discusses the basic conditions required, the F-test, one-way and two-way ANOVA, and multiple comparisons. The final goal of these analyses is to show whether the means of the given populations are different and if so, which ones are higher or lower than the rest.

## Part IV: Building Strong Connections with Chi-Square Tests

This part deals with the Chi-square distribution and how you can use it to model and test categorical (qualitative) data. You find out how to test for independence of two categorical variables using a Chi-square test. (No more making speculations just by looking at the data in a two-way table!) You also see how to use a Chi-square to test how well a model for categorical data fits.

## Part V: Nonparametric Statistics: Rebels without a Distribution

This part helps you with techniques used in situations where you can't (or don't want to) assume your data comes from a population with a certain distribution, such as when your population isn't normal (the condition required by most other methods in Stats II).

## Part VI: The Part of Tens

Reading this part can give you an edge in a major area beyond the formulas and techniques of Stats II: ending the problem right (knowing what kinds of conclusions you can and can't make). You also get to know Stats II in the real world, namely how it can help you stand out in a crowd.

You also can find an appendix at the back of this book that contains all the tables you need to understand and complete the calculations in this book.

# Icons Used in This Book

I use icons in this book to draw your attention to certain text features that occur on a regular basis. Think of the icons as road signs that you encounter on a trip. Some signs tell you about shortcuts, and others offer more information that you may need; some signs alert you to possible warnings, while others leave you with something to remember.

When you see this icon, it means I'm explaining how to carry out that particular data analysis using Minitab. I also explain the information you get in the computer output so you can interpret your results.

I use this icon to reinforce certain ideas that are critical for success in Stats II, such as things I think are important to review as you prepare for an exam.

When you see this icon, you can skip over the information if you don't want to get into the nitty-gritty details. They exist mainly for people who have a special interest or obligation to know more about the more technical aspects of certain statistical issues.

This icon points to helpful hints, ideas, or shortcuts that you can use to save time; it also includes alternative ways to think about a particular concept.

I use warning icons to help you stay away from common misconceptions and pitfalls you may face when dealing with ideas and techniques related to Stats II.

# Where to Go from Here

This book is written in a nonlinear way, so you can start anywhere and still understand what's happening. However, I can make some recommendations if you want some direction on where to start.

If you're thoroughly familiar with the ideas of hypothesis testing and simple linear regression, start with Chapter 5 (multiple regression). Use Chapter 1 if you need a reference for the jargon that statisticians use in Stats II.

If you've covered all topics up through the various types of regression (simple, multiple, nonlinear, and logistic) or a subset of those as your professor deemed important, proceed to Chapter 9, the basics of analysis of variance (ANOVA).

Chapter 14 is the place to begin if you want to tackle categorical (qualitative) variables before hitting the quantitative stuff. You can work with the Chi-square test there.

Nonparametric statistics are presented starting with Chapter 16. This area is a hot topic in today's statistics courses, yet it's also one that doesn't seem to get as much space in textbooks as it should. Start here if you want the full details on the most common nonparametric procedures.

# Part I
# Tackling Data Analysis and Model-Building Basics

The 5th Wave                    By Rich Tennant

"I ran an evaluation of our last pie chart. Apparently it's boysenberry."

## In this part . . .

To get you up and moving from the foundational concepts of statistics (covered in your Stats I textbook as well as *Statistics For Dummies*) to the new and exciting methods presented in this book, I first go over the basics of data analysis, important terminology, main goals and concepts of model-building, and tips for choosing appropriate statistics to fit the job. I refresh your memory of the most heavily referred to items from Stats I, and you also get a head start on making and looking at some basic computer output.

# Chapter 1

# Beyond Number Crunching: The Art and Science of Data Analysis

*In This Chapter*

▶ Realizing your role as a data analyst

▶ Avoiding statistical faux pas

▶ Delving into the jargon of Stats II

*B*ecause you're reading this book, you're likely familiar with the basics of statistics and you're ready to take it up a notch. That next level involves using what you know, picking up a few more tools and techniques, and finally putting it all to use to help you answer more realistic questions by using real data. In statistical terms, you're ready to enter the world of the *data analyst*.

In this chapter, you review the terms involved in statistics as they pertain to data analysis at the Stats II level. You get a glimpse of the impact that your results can have by seeing what these analysis techniques can do. You also gain insight into some of the common misuses of data analysis and their effects.

## Data Analysis: Looking before You Crunch

It used to be that statisticians were the only ones who really analyzed data because the only computer programs available were very complicated to use, requiring a great deal of knowledge about statistics to set up and carry out analyses. The calculations were tedious and at times unpredictable, and they required a thorough understanding of the theories and methods behind the calculations to get correct and reliable answers.

Today, anyone who wants to analyze data can do it easily. Many user-friendly statistical software packages are made expressly for that purpose — Microsoft Excel, Minitab, SAS, and SPSS are just a few. Free online programs are available, too, such as Stat Crunch, to help you do just what it says — crunch your numbers and get an answer.

Each software package has its own pros and cons (and its own users and protesters). My software of choice and the one I reference throughout this book is Minitab, because it's very easy to use, the results are precise, and the software's loaded with all the data-analysis techniques used in Stats II. Although a site license for Minitab isn't cheap, the student version is available for rent for only a few bucks a semester.

The most important idea when applying statistical techniques to analyze data is to know what's going on behind the number crunching so you (not the computer) are in control of the analysis. That's why knowledge of Stats II is so critical.

Many people don't realize that statistical software can't tell you when to use and not to use a certain statistical technique. You have to determine that on your own. As a result, people think they're doing their analyses correctly, but they can end up making all kinds of mistakes. In the following sections, I give examples of some situations in which innocent data analyses can go wrong and why it's important to spot and avoid these mistakes before you start crunching numbers.

Bottom line: Today's software packages are too good to be true if you don't have a clear and thorough understanding of the Stats II that's underneath them.

---

## Remembering the old days

In the old days, in order to determine whether different methods gave different results, you had to write a computer program using code that you had to take a class to learn. You had to type in your data in a specific way that the computer program demanded, and you had to submit your program to the computer and wait for the results. This method was time consuming and a general all-around pain.

The good news is that statistical software packages have undergone an incredible evolution in the last 10 to 15 years, to the point where you can now enter your data quickly and easily in almost any format. Moreover, the choices for data analysis are well organized and listed in pull-down menus. The results come instantly and successfully, and you can cut and paste them into a word-processing document without blinking an eye.

# Nothing (not even a straight line) lasts forever

Bill Prediction is a statistics student studying the effect of study time on exam score. Bill collects data on statistics students and uses his trusty software package to predict exam score using study time. His computer comes up with the equation $y = 10x + 30$, where $y$ represents the test score you get if you study a certain number of hours $(x)$. Notice that this model is the equation of a straight line with a $y$-intercept of 30 and a slope of 10.

So Bill predicts, using this model, that if you don't study at all, you'll get a 30 on the exam (plugging $x = 0$ into the equation and solving for $y$; this point represents the $y$-intercept of the line). And he predicts, using this model, that if you study for 5 hours, you'll get an exam score of $y = (10 * 5) + 30 = 80$. So, the point (5, 80) is also on this line.

But then Bill goes a little crazy and wonders what would happen if you studied for 40 hours (since it always seems that long when he's studying). The computer tells him that if he studies for 40 hours, his test score is predicted to be $(10 * 40) + 30 = 430$ points. Wow, that's a lot of points! Problem is, the exam only goes up to a total of 100 points. Bill wonders where his computer went wrong.

But Bill puts the blame in the wrong place. He needs to remember that there are limits on the values of $x$ that make sense in this equation. For example, because $x$ is the amount of study time, $x$ can never be a number less than zero. If you plug a negative number in for $x$, say $x = -10$, you get $y = (10 * -10) + 30 = -70$, which makes no sense. However, the equation itself doesn't know that, nor does the computer that found it. The computer simply graphs the line you give it, assuming it'll go on forever in both the positive and negative directions.

After you get a statistical equation or model, you need to specify for what values the equation applies. Equations don't know when they work and when they don't; it's up to the data analyst to determine that. This idea is the same for applying the results of any data analysis that you do.

# Data snooping isn't cool

Statisticians have come up with a saying that you may have heard: "Figures don't lie. Liars figure." Make sure that you find out about all the analyses that were performed on a data set, not just the ones reported as being statistically significant.

Suppose Bill Prediction (from the previous section) decides to try to predict scores on a biology exam based on study time, but this time his model doesn't fit. Not one to give in, Bill insists there must be some other factors that predict biology exam scores besides study time, and he sets out to find them.

Bill measures everything from soup to nuts. His set of 20 possible variables includes study time, GPA, previous experience in statistics, math grades in high school, and whether you chew gum during the exam. After his multitude of various correlation analyses, the variables that Bill found to be related to exam score were study time, math grades in high school, GPA, and gum chewing during the exam. It turns out that this particular model fits pretty well (by criteria I discuss in Chapter 5 on multiple linear regression models).

But here's the problem: By looking at all possible correlations between his 20 variables and exam score, Bill is actually doing 20 separate statistical analyses. Under typical conditions that I describe in Chapter 3, each statistical analysis has a 5 percent chance of being wrong just by chance. I bet you can guess which one of Bill's correlations likely came out wrong in this case. And hopefully he didn't rely on a stick of gum to boost his grade in biology.

Looking at data until you find something in it is called *data snooping*. Data snooping results in giving the researcher his five minutes of fame but then leads him to lose all credibility because no one can repeat his results.

## *No (data) fishing allowed*

Some folks just don't take no for an answer, and when it comes to analyzing data, that can lead to trouble.

Sue Gonnafindit is a determined researcher. She believes that her horse can count by stomping his foot. (For example, she says "2" and her horse stomps twice.) Sue collects data on her horse for four weeks, recording the percentage of time the horse gets the counting right. She runs the appropriate statistical analysis on her data and is shocked to find no significant difference between her horse's results and those you would get simply by guessing.

Determined to prove her results are real, Sue looks for other types of analyses that exist and plugs her data into anything and everything she can find (never mind that those analyses are inappropriate to use in her situation). Using the famous hunt-and-peck method, at some point she eventually stumbles upon a significant result. However, the result is bogus because she tried so many analyses that weren't appropriate and ignored the results of the appropriate analysis because it didn't tell her what she wanted to hear.

Funny thing, too. When Sue went on a late night TV program to show the world her incredible horse, someone in the audience noticed that whenever the horse got to the correct number of stomps, Sue would interrupt him and say "Good job!" and the horse quit stomping. He didn't know how to count; all he knew to do was to quit stomping when she said "Good job!"

Redoing analyses in different ways in order to try to get the results you want is called *data fishing*, and folks in the stats biz consider it to be a major no-no. (However, people unfortunately do it all too often to verify their strongly held beliefs.) By using the wrong data analysis for the sake of getting the results you desire, you mislead your audience into thinking that your hypothesis is actually correct when it may not be.

# *Getting the Big Picture: An Overview of Stats II*

Stats II is an extension of Stats I (introductory statistics), so the jargon follows suit and the techniques build on what you already know. In this section, you get an introduction to the terminology you use in Stats II along with a broad overview of the techniques that statisticians use to analyze data and find the story behind it. (If you're still unsure about some of the terms from Stats I, you can consult your Stats I textbook or see my other book, *Statistics For Dummies* (Wiley), for a complete rundown.)

## *Population parameter*

A *parameter* is a number that summarizes the *population*, which is the entire group you're interested in investigating. Examples of parameters include the mean of a population, the median of a population, or the proportion of the population that falls into a certain category.

Suppose you want to determine the average length of a cellphone call among teenagers (ages 13–18). You're not interested in making any comparisons; you just want to make a good guesstimate of the average time. So you want to estimate a population parameter (such as the mean or average). The population is all cellphone users between the ages of 13 and 18 years old. The parameter is the average length of a phone call this population makes.

## Sample statistic

Typically you can't determine population parameters exactly; you can only estimate them. But all is not lost; by taking a *sample* (a subset of individuals) from the population and studying it, you can come up with a good estimate of the population parameter. A *sample statistic* is a single number that summarizes that subset.

For example, in the cellphone scenario from the previous section, you select a sample of teenagers and measure the duration of their cellphone calls over a period of time (or look at their cellphone records if you can gain access legally). You take the average of the cellphone call duration. For example, the average duration of 100 cellphone calls may be 12.2 minutes — this average is a statistic. This particular statistic is called the *sample mean* because it's the average value from your sample data.

Many different statistics are available to study different characteristics of a sample, such as the proportion, the median, and standard deviation.

## Confidence interval

A *confidence interval* is a range of likely values for a population parameter. A confidence interval is based on a sample and the statistics that come from that sample. The main reason you want to provide a range of likely values rather than a single number is that sample results vary.

For example, suppose you want to estimate the percentage of people who eat chocolate. According to the Simmons Research Bureau, 78 percent of adults reported eating chocolate, and of those, 18 percent admitted eating sweets frequently. What's missing in these results? These numbers are only from a single sample of people, and those sample results are guaranteed to vary from sample to sample. You need some measure of how much you can expect those results to move if you were to repeat the study.

This expected variation in your statistic from sample to sample is measured by the *margin of error*, which reflects a certain number of standard deviations of your statistic you add and subtract to have a certain confidence in your results (see Chapter 3 for more on margin of error). If the chocolate-eater results were based on 1,000 people, the margin of error would be approximately 3 percent. This means the actual percentage of people who eat chocolate in the entire population is expected to be 78 percent, ± 3 percent (that is, between 75 percent and 81 percent).

## Hypothesis test

A *hypothesis test* is a statistical procedure that you use to test an existing claim about the population, using your data. The claim is noted by Ho (the null hypothesis). If your data support the claim, you fail to reject Ho. If your data don't support the claim, you reject Ho and conclude an alternative hypothesis, Ha. The reason most people conduct a hypothesis test is not to merely show that their data support an existing claim, but rather to show that the existing claim is false, in favor of the alternative hypothesis.

The Pew Research Center studied the percentage of people who turn to ESPN for their sports news. Its statistics, based on a survey of about 1,000 people, found that in 2000, 23 percent of people said they go to ESPN; in 2004, only 20 percent reported going to ESPN. The question is this: Does this 3 percent reduction in viewers from 2000 to 2004 represent a significant trend that ESPN should worry about?

To test these differences formally, you can set up a hypothesis test. You set up your null hypothesis as the result you have to believe without your study, Ho = No difference exists between 2000 and 2004 data for ESPN viewership. Your alternative hypothesis (Ha) is that a difference is there. To run a hypothesis test, you look at the difference between your statistic from your data and the claim that has been already made about the population (in Ho), and you measure how far apart they are in units of standard deviations.

With respect to the example, using the techniques from Chapter 3, the hypothesis test shows that 23 percent and 20 percent aren't far enough apart in terms of standard deviations to dispute the claim (Ho). You can't say the percentage of viewers of ESPN in the entire population changed from 2000 to 2004.

As with any statistical analysis, your conclusions can be wrong just by chance, because your results are based on sample data, and sample results vary. In Chapter 3 I discuss the types of errors that can be made in conclusions from a hypothesis test.

# Analysis of variance (ANOVA)

ANOVA is the acronym for *analysis of variance*. You use ANOVA in situations where you want to compare the means of more than two populations. For example, you want to compare the lifetimes of four brands of tires in number of miles. You take a random sample of 50 tires from each group, for a total of 200 tires, and set up an experiment to compare the lifetime of each tire, and record it. You have four means and four standard deviations now, one for each data set.

Then, to test for differences in average lifetime for the four brands of tires, you basically compare the variability between the four data sets to the variability within the entire data set, using a ratio. This ratio is called the *F-statistic*. If this ratio is large, the variability between the brands is more than the variability within the brands, giving evidence that not all the means are the same for the different tire brands. If the *F*-statistic is small, not enough difference exists between the treatment means compared to the general variability within the treatments themselves. In this case, you can't say that the means are different for the groups. (I give you the full scoop on ANOVA plus all the jargon, formulas, and computer output in Chapters 9 and 10.)

## Multiple comparisons

Suppose you conduct ANOVA, and you find a difference in the average lifetimes of the four brands of tire (see the preceding section). Your next questions would probably be, "Which brands are different?" and "How different are they?" To answer these questions, use multiple-comparison procedures.

A *multiple-comparison procedure* is a statistical technique that compares means to each other and finds out which ones are different and which ones aren't. With this information, you're able to put the groups in order from those with the largest mean to those with the smallest mean, realizing that sometimes two or more groups were too close to tell and are placed together in a group.

Many different multiple-comparison procedures exist to compare individual means and come up with an ordering in the event that your *F*-statistic does find that some difference exists. Some of the multiple-comparison procedures include Tukey's test, LSD, and pairwise *t*-tests. Some procedures are better than others, depending on the conditions and your goal as a data analyst. I discuss multiple-comparison procedures in detail in Chapter 11.

 Never take that second step to compare the means of the groups if the ANOVA procedure doesn't find any significant results during the first step. Computer software will never stop you from doing a follow-up analysis, even if it's wrong to do so.

## Interaction effects

An *interaction effect* in statistics operates the same way that it does in the world of medicine. Sometimes if you take two different medicines at the same time, the combined effect is much different than if you were to take the two individual medications separately.

Interaction effects can come up in statistical models that use two or more variables to explain or compare outcomes. In this case you can't automatically study the effect of each variable separately; you have to first examine whether or not an interaction effect is present.

For example, suppose medical researchers are studying a new drug for depression and want to know how this drug affects the change in blood pressure for a low dose versus a high dose. They also compare the effects for children versus adults. It could also be that dosage level affects the blood pressure of adults differently than the blood pressure of children. This type of model is called a *two-way ANOVA model,* with a possible interaction effect between the two factors (age group and dosage level). Chapter 11 covers this subject in depth.

# Correlation

The term *correlation* is often misused. Statistically speaking, the correlation measures the strength and direction of the linear relationship between two *quantitative variables* (variables that represent counts or measurements only).

You aren't supposed to use correlation to talk about relationships unless the variables are quantitative. For example, it's wrong to say that a correlation exists between eye color and hair color. (In Chapter 14, you explore associations between two categorical variables.)

Correlation is a number between –1.0 and +1.0. A correlation of +1 indicates a perfect positive relationship; as you increase one variable, the other one increases in perfect sync. A correlation of –1.0 indicates a perfect negative relationship between the variables; as one variable increases, the other one decreases in perfect sync. A correlation of zero means you found no linear relationship at all between the variables. Most correlations in the real world fall somewhere in between –1.0 and +1.0; the closer to –1.0 or +1.0, the stronger the relationship is; the closer to 0, the weaker the relationship is.

Figure 1-1 shows a plot of the number of coffees sold at football games in Buffalo, New York, as well as the air temperature (in degrees Fahrenheit) at each game. This data set seems to follow a downhill straight line fairly well, indicating a negative correlation. The correlation turns out to be –0.741; number of coffees sold has a fairly strong negative relationship with the temperature of the football game. This makes sense because on days when the temperature is low, people get cold and want more coffee. I discuss correlation further, as it applies to model building, in Chapter 4.

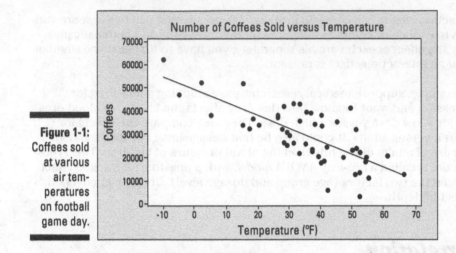

**Figure 1-1:** Coffees sold at various air temperatures on football game day.

# Linear regression

After you've found a correlation and determined that two variables have a fairly strong linear relationship, you may want to try to make predictions for one variable based on the value of the other variable. For example, if you know that a fairly strong negative linear relationship exists between coffees sold and the air temperature at a football game (see the previous section), you may want to use this information to predict how much coffee is needed for a game, based on the temperature. This method of finding the best-fitting line is called *linear regression*.

Many different types of regression analyses exist, depending on your situation. When you use only one variable to predict the response, the method of regression is called *simple linear regression* (see Chapter 4). Simple linear regression is the best known of all the regression analyses and is a staple in the Stats I course sequence.

However, you use other flavors of regression for other situations.

✔ If you want to use more than one variable to predict a response, you use *multiple linear regression* (see Chapter 5).

✔ If you want to make predictions about a variable that has only two outcomes, yes or no, you use *logistic regression* (see Chapter 8).

✔ For relationships that don't follow a straight line, you have a technique called (no surprise) *nonlinear regression* (see Chapter 7).

## Chi-square tests

Correlation and regression techniques all assume that the variable being studied in most detail (the response variable) is quantitative — that is, the variable measures or counts something. You can also run into situations where the data being studied isn't quantitative, but rather categorical — that is, the data represent categories, not measurements or counts. To study relationships in categorical data, you use a Chi-square test for independence. If the variables are found to be unrelated, they're declared independent. If they're found to be related, they're declared dependent.

Suppose you want to explore the relationship between gender and eating breakfast. Because each of these variables is categorical, or qualitative, you use a Chi-square test for independence. You survey 70 males and 70 females and find that 25 men eat breakfast and 45 do not; for the females, 35 do eat breakfast and 35 do not. Table 1-1 organizes this data and sets you up for the Chi-square test for this scenario.

### Table 1-1    Table Setup for the Breakfast and Gender Question

|        | Do Eat Breakfast | Don't Eat Breakfast | Total |
|--------|------------------|---------------------|-------|
| Male   | 25               | 45                  | 70    |
| Female | 35               | 35                  | 70    |

A Chi-square test first calculates what you expect to see in each cell of the table if the variables are independent (these values are brilliantly called the *expected cell counts*). The Chi-square test then compares these expected cell counts to what you observed in the data (called the *observed cell counts*) and compares them using a Chi-square statistic.

In the breakfast gender comparison, fewer males than females eat breakfast (25 ÷ 70 = 35.7 percent compared to 35 ÷ 70 = 50 percent). Even though you know results will vary from sample to sample, this difference turns out to be enough to declare a relationship between gender and eating breakfast, according to the Chi-square test of independence. Chapter 14 reveals all the details of doing a Chi-square test.

You can also use the Chi-square test to see whether your theory about what percent of each group falls into a certain category is true or not. For example, can you guess what percentage of M&M'S fall into each color category? You can find more on these Chi-square variations, as well as the M&M'S question, in Chapter 15.

# *Nonparametrics*

*Nonparametrics* is an entire area of statistics that provides analysis techniques to use when the conditions for the more traditional and commonly used methods aren't met. However, people sometimes forget or don't bother to check those conditions, and if the conditions are actually not met, the entire analysis goes out the window, and the conclusions go along with it!

Suppose you're trying to test a hypothesis about a population mean. The most common approach to use in this situation is a *t*-test. However, to use a *t*-test, the data needs to be collected from a population that has a normal distribution (that is, it has to have a bell-shaped curve). You collect data and graph it, and you find that it doesn't have a normal distribution; it has a skewed distribution. You're stuck — you can't use the common hypothesis test procedures you know and love (at least, you shouldn't use them).

This is where nonparametric procedures come in. Nonparametric procedures don't require nearly as many conditions be met as the regular parametric procedures do. In this situation of skewed data, it makes sense to run a hypothesis test for the median rather than the mean anyway, and plenty of nonparametric procedures exist for doing so.

If the conditions aren't met for a data-analysis procedure that you want to do, chances are that an equivalent nonparametric procedure is waiting in the wings. Most statistical software packages can do them just as easily as the regular (parametric) procedures.

Before doing a data analysis, statistical software packages don't automatically check conditions. It's up to you to check any and all appropriate conditions and, if they're seriously violated, to take another course of action. Many times a nonparametric procedure is just the ticket. For much more information on different nonparametric procedures, see Chapters 16 through 19.

# Chapter 2

# Finding the Right Analysis for the Job

### In This Chapter
▶ Deciphering the difference between categorical and quantitative variables
▶ Choosing appropriate statistical techniques for the task at hand
▶ Evaluating bias and precision levels
▶ Interpreting the results properly

One of the most critical elements of statistics and data analysis is the ability to choose the right statistical technique for each job. Carpenters and mechanics know the importance of having the right tool when they need it and the problems that can occur if they use the wrong tool. They also know that the right tool helps to increase their odds of getting the results they want the first time around, using the "work smarter, not harder" approach.

In this chapter, you look at the some of the major statistical analysis techniques from the point of view of the carpenters and mechanics — knowing what each statistical tool is meant to do, how to use it, and when to use it. You also zoom in on mistakes some number crunchers make in applying the wrong analysis or doing too many analyses.

Knowing how to spot these problems can help you avoid making the same mistakes, but it also helps you to steer through the ocean of statistics that may await you in your job and in everyday life.

If many of the ideas you find in this chapter seem like a foreign language to you and you need more background information, don't fret. Before continuing on in this chapter, head to your nearest Stats I book or check out another one of my books, *Statistics For Dummies* (Wiley).

# Categorical versus Quantitative Variables

After you've collected all the data you need from your sample, you want to organize it, summarize it, and analyze it. Before plunging right into all the number crunching though, you need to first identify the type of data you're dealing with. The type of data you have points you to the proper types of graphs, statistics, and analyses you're able to use.

Before I begin, here's an important piece of jargon: Statisticians call any quantity or characteristic you measure on an individual a *variable;* the data collected on a variable is expected to vary from person to person (hence the creative name).

The two major types of variables are the following:

- **Categorical:** A *categorical variable,* also known as a *qualitative variable,* classifies the individual based on categories. For example, political affiliation may be classified into four categories: Democrat, Republican, Independent, and Other; gender as a variable takes on two possible categories: male and female. Categorical variables can take on numerical values only as placeholders.

- **Quantitative:** A *quantitative variable* measures or counts a quantifiable characteristic, such as height, weight, number of children you have, your GPA in college, or the number of hours of sleep you got last night. The quantitative variable value represents a quantity (count) or a measurement and has numerical meaning. That is, you can add, subtract, multiply, or divide the values of a quantitative variable, and the results make sense as numbers.

Because the two types of variables represent such different types of data, it makes sense that each type has its own set of statistics. Categorical variables, such as gender, are somewhat limited in terms of the statistics that can be performed on them.

For example, suppose you have a sample of 500 classmates classified by gender — 180 are male and 320 are female. How can you summarize this information? You already have the total number in each category (this statistic is called the *frequency*). You're off to a good start, but frequencies are hard to interpret because you find yourself trying to compare them to a total in your mind in order to get a proper comparison. For example, in this case you may be thinking, "One hundred and eighty males out of what? Let's see, it's out of 500. Hmmm . . . what percentage is that?"

The next step is to find a means to relate these numbers to each other in an easy way. You can do this by using the *relative frequency,* which is the percentage of data that falls into a specific category of a categorical variable. You can find a category's relative frequency by dividing the frequency by the sample total and then multiplying by 100. In this case, you have

$\frac{180}{500} = 0.36 * 100 = 36$ percent males and $\frac{320}{500} = 0.64 * 100 = 64$ percent females.

You can also express the relative frequency as a proportion in each group by leaving the result in decimal form and not multiplying by 100. This statistic is called the *sample proportion.* In this example, the sample proportion of males is 0.36, and the sample proportion of females is 0.64.

You mainly summarize categorical variables by using two statistics — the number in each category (frequency) and the percentage (relative frequency) in each category.

# Statistics for Categorical Variables

The types of statistics done on categorical data may seem limited; however, the wide variety of analyses you can perform using frequencies and relative frequencies offers answers to an extensive range of possible questions you may want to explore.

In this section, you see that the proportion in each group is the number-one statistic for summarizing categorical data. Beyond that, you see how you can use proportions to estimate, compare, and look for relationships between the groups that comprise the categorical data.

## Estimating a proportion

You can use relative frequencies to make estimates about a single population proportion. (Refer to the earlier section "Categorical versus Quantitative Variables" for an explanation of relative frequencies.)

Suppose you want to know what proportion of females in the United States are Democrats. According to a sample of 29,839 female voters in the U.S. conducted by the Pew Research Foundation in 2003, the percentage of female Democrats was 36. Now, because the Pew researchers based these results on only a sample of the population and not on the entire population, their results will vary if they take another sample. This variation in sample results is cleverly called — you guessed it — sampling variability.

The sampling variability is measured by the *margin of error* (the amount that you add and subtract from your sample statistic), which for this sample is only about 0.5 percent. (To find out how to calculate margin of error, turn to Chapter 3.) That means that the estimated percentage of female Democrats in the U.S. voting population is somewhere between 35.5 percent and 36.5 percent.

The margin of error, combined with the sample proportion, forms what statisticians call a confidence interval for the population proportion. Recall from Stats I that a *confidence interval* is a range of likely values for a population parameter, formed by taking the sample statistic plus or minus the margin of error. (For more on confidence intervals, see Chapter 3.)

## Comparing proportions

Researchers, the media, and even everyday folk like you and me love to compare groups (whether you like to admit it or not). For example, what proportion of Democrats support oil drilling in Alaska, compared to Republicans? What percentage of women watch college football, compared to men? What proportion of readers of *Statistics II For Dummies* pass their stats exams with flying colors, compared to nonreaders?

To answer these questions, you need to compare the sample proportions using a hypothesis test for two proportions (see Chapter 3 or your Stats I textbook).

Suppose you've collected data on a random sample of 1,000 voters in the U.S. and you want to compare the proportion of female voters to the proportion of male voters and find out whether they're equal. Suppose in your sample you find that the proportion of females is 0.53, and the proportion of males is 0.47. So for this sample of 1,000 people, you have a higher proportion of females than males.

But here's the big question: Are these sample proportions different enough to say that the entire population of American voters has more females in it than males? After all, sample results vary from sample to sample. The answer to this question requires comparing the sample proportions by using a hypothesis test for two proportions. I demonstrate and expand on this technique in Chapter 3.

# Looking for relationships between categorical variables

Suppose you want to know whether two categorical variables are related; for example, is gender related to political affiliation? Answering this question requires putting the sample data into a two-way table (using rows and columns to represent the two variables) and analyzing the data by using a Chi-square test (see Chapter 14).

By following this process, you can determine if two categorical variables are independent (unrelated) or if a relationship exists between them. If you find a relationship, you can use percentages to describe it.

Table 2-1 shows an example of data organized in a two-way table. The data was collected by the Pew Research Foundation.

| Table 2-1 | Gender and Political Affiliation of 56,735 U.S. Voters | | |
|---|---|---|---|
| *Gender* | *Republican* | *Democrat* | *Other* |
| *Males* | 32% | 27% | 41% |
| *Females* | 29% | 36% | 35% |

Notice that the percentage of male Republicans in the sample is 32 and the percentage of female Republicans in the sample is 29. These percentages are quite close in relative terms. However, the percentage of female Democrats seems much higher than the percentage of male Democrats (36 percent versus 27 percent); also, the percentage of males in the "Other" category is quite a bit higher than the percentage of females in the same category (41 percent versus 35 percent).

These large differences in the percentages indicate that gender and political affiliation are related in the sample. But do these trends carry over to the population of all American voters? This question requires a hypothesis test to answer. Because gender and political affiliation are both categorical variables, the particular hypothesis test you need in this situation is a Chi-square test. (I discuss Chi-square tests in detail in Chapter 14.)

To make a two-way table from a data set by using Minitab, first enter the data in two columns, where column one is the row variable (in this case, gender) and column two is the column variable (in this case, political affiliation). For example, suppose the first person is a male Democrat. In row one of Minitab, enter *M* (for male) in column one and *D* (Democrat) in column two. Then go to Stat>Tables>Cross Tabulation and Chi-square. Highlight column one and click Select to enter this variable in the For Rows line. Highlight column two and click Select to enter this variable in the For Columns line. Click OK.

People often use the word *correlation* to discuss relationships between variables, but in the world of statistics, correlation only relates to the relationship between two quantitative (numerical) variables, not two categorical variables. *Correlation* measures how closely the relationship between two quantitative variables, such as height and weight, follows a straight line and tells you the direction of that line as well. In total, for any two quantitative variables, *x* and *y*, the correlation measures the strength and direction of their linear relationship. As one increases, what does the other one do?

Because categorical variables don't have a numerical order to them, they don't increase or decrease in value. For example, just because male = 1 and female = 2 doesn't mean that a female is worth twice as much as a male (although some women may want to disagree). Therefore, you can't use the word *correlation* to describe the relationship between, say, gender and political affiliation. (Chapter 4 covers correlation.)

The appropriate term to describe the relationships of categorical variables is *association*. You can say that political affiliation is associated with gender and then explain how. (For full details on association, see Chapter 13.)

## *Building models to make predictions*

You can build models to predict the value of a categorical variable based on other related information. In this case, building models is more than a lot of little plastic pieces and some irritatingly sticky glue.

When you build a statistical model, you look for variables that help explain, estimate, or predict some response you're interested in; the variables that do this are called *explanatory variables*. You sort through the explanatory variables and figure out which ones do the best job of predicting the response. Then you put them together into a type of equation like $y = 2x + 4$ where $x$ = shoe size and $y$ = estimated calf length. That equation is a *model*.

For example, suppose you want to know which factors or variables can help you predict someone's political affiliation. Is a woman without children more likely to be a Republican or a Democrat? What about a middle-aged man who proclaims Hinduism as his religion?

In order for you to compare these complex relationships, you must build a model to evaluate each group's impact on political affiliation (or some other categorical variable). This kind of model-building is explored in-depth in Chapter 8, where I discuss the topic of logistic regression.

*Logistic regression* builds models to predict the outcome of a categorical variable, such as political affiliation. If you want to make predictions about a quantitative variable, such as income, you need to use the standard type of regression (check out Chapters 4 and 5).

# Statistics for Quantitative Variables

Quantitative variables, unlike categorical variables, have a wider range of statistics that you can do, depending on what questions you want to ask. The main reason for this wider range is that *quantitative data* are numbers that represent measurements or counts, so it makes sense that you can order, add or subtract, and multiply or divide them — and the results all have numerical meaning. In this section, I present the major data-analysis techniques for quantitative data. I expand on each technique in later chapters of this book.

## Making estimates

Quantitative variables take on numerical values that involve counts or measurements, so they have means, medians, standard deviations, and all those good things that categorical variables don't have. Researchers often want to know what the average or median value is for a population (these are called parameters). To do this requires taking a sample and making a good guess, also known as an estimate, of that parameter.

To find an estimate for any population parameter requires a confidence interval. For categorical variables, you would find a confidence interval to estimate the population mean, median, or standard deviation, but by far the most common parameter of interest is the population mean.

A confidence interval for the population mean is the sample mean plus or minus a margin of error. (To calculate the margin of error in this case, see Chapter 3.) The result will be a range of likely values you have produced for the real population mean. Because the variable is quantitative, the confidence interval will take on the same units as the variable does. For example, household incomes will be in thousands of dollars.

There is no rule of thumb regarding how large or small the margin of error should be for a quantitative variable; it depends on what the variable is counting or measuring. For example, if you want average household income for the state of New York, a margin of error of plus or minus $5,000 is not unreasonable. If the variable is the average number of steps from the first floor to the second floor of a two-story home in the U.S., the margin of error will be much smaller. Estimates of categorical variables, on the other hand, are percentages; most people want those confidence intervals to be within plus or minus 2 to 3 percent.

## Making comparisons

Suppose you want to look at income (a quantitative variable) and how it relates to a categorical variable, such as gender or region of the country. Your first question may be: Do males still make more money than females? In this case, you can compare the mean incomes of two populations — males and females. This assessment requires a hypothesis test of two means (often called a *t*-test for independent samples). I present more information on this technique in Chapter 3.

When comparing the means of *more* than two groups, don't simply look at all the possible *t*-tests that you can do on the pairs of means because you have to control for an overall error rate in your analysis. Too many analyses can result in errors — adding up to disaster. For example, if you conduct 100 hypothesis tests, each one with a 5 percent error rate, then 5 of those 100 tests will come out statistically significant on average, just by chance, even if no real relationship exists.

If you want to compare the average wage in different regions of the country (the East, the Midwest, the South, and the West, for example), this comparison requires a more sophisticated analysis because you're looking at four groups rather than just two. The procedure for comparing more than two means is called *analysis of variance* (ANOVA, for short), and I discuss this method in detail in Chapters 9 and 10.

## Exploring relationships

One of the most common reasons data is collected is to look for relationships between variables. With quantitative variables, the most common type of relationship people look for is a linear relationship; that is, as one variable increases, does the other increase/decrease along with it in a similar way? Relationships between any variables are examined using specialized plots and statistics. Since a linear relationship is so common, it has its own special statistic called correlation. You find out how statisticians make graphs and statistics to explore relationships in this section, paying particular attention to linear relationships.

Suppose you're an avid golfer and you want to figure out how much time you should spend on your putting game. The question is this: Is the number of putts related to your total score? If the answer is yes, then spending time on your putting game makes sense. If not, then you can slack off on it a bit. Both of these variables are quantitative variables, and you're looking for a connection between them. You collect data on 100 rounds of golf played by golfers at your favorite course over a weekend. Following are the first few lines of your data set.

| Round | Number of Putts | Total Score |
|---|---|---|
| 1 | 23 | 76 |
| 2 | 27 | 80 |
| 3 | 28 | 80 |
| 4 | 29 | 80 |
| 5 | 30 | 80 |
| 6 | 29 | 82 |
| 7 | 30 | 83 |
| 8 | 31 | 83 |
| 9 | 33 | 83 |
| 10 | 26 | 84 |

The first step in looking for a connection between putts and total scores (or any other quantitative variables) is to make a scatterplot of the data. A *scatterplot* graphs your data set in two-dimensional space by using an X,Y plane. You can take a look at the scatterplot of the golf data in Figure 2-1. Here, *x* represents the number of putts, and *y* represents the total score. For example, the point in the lower-left corner of the graph represents someone who had only 23 putts and a total score of 75. (For instructions on making a scatterplot by using Minitab, see Chapter 4.)

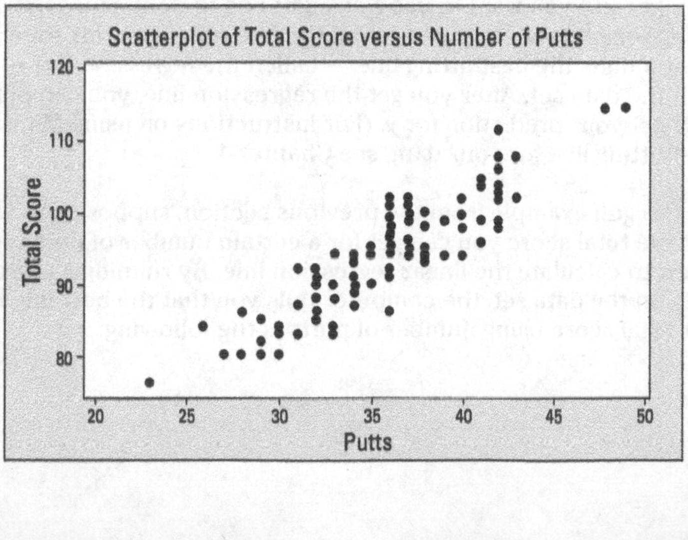

**Figure 2-1:** The two-dimensional scatterplot helps you look for relationships in data.

Scatterplot of Total Score versus Number of Putts

According to Figure 2-1, it appears that as the number of putts increases, so does the golfer's total score. It also shows that the variables increase in a linear way; that is, the data form a pattern that resembles a straight line. The relationship seems pretty strong — the number of putts plays a big part in determining the total score.

Now you need a measure of how strong the relationship is between $x$ and $y$ and whether it goes uphill or downhill. Different measures are used for different types of patterns seen in a scatterplot. Because the relationship we see in this case resembles a straight line, the correlation is the measure that we use to quantify the relationship. Correlation is the number that measures how close the points follow a straight line. Correlation is always between –1.0 and +1.0, and the more closely the points follow a straight line, the closer the correlation is to –1.0 or +1.0.

- **A positive correlation means that as $x$ increases on the $x$-axis, $y$ also increases on the $y$-axis.** Statisticians call this type of relationship an *uphill relationship.*

- **A negative correlation means that as $x$ increases on the $x$-axis, $y$ goes down.** Statisticians call this type of relationship — you guessed it — a *downhill relationship.*

For the golf data set, the correlation is 0.896 = 0.90, which is extremely high as correlations go. The sign of the correlation is positive, so as you increase number of putts, your total score increases (an uphill relationship). For instructions on calculating a correlation in Minitab, see Chapter 4.

## Predicting y using x

If you want to predict some response variable *(y)* using one explanatory variable *(x)* and you want to use a straight line to do it, you can use *simple linear regression* (see Chapter 4 for all the fine points on this topic). Linear regression finds the best-fitting line — called the *regression line* — that cuts through the data set. After you get the regression line, you can plug in a value of $x$ and get your prediction for $y$. (For instructions on using Minitab to find the best-fitting line for your data, see Chapter 4.)

To use the golf example from the previous section, suppose you want to predict the total score you can get for a certain number of putts. In this case, you want to calculate the linear regression line. By running a regression analysis on the data set, the computer tells you that the best line to use to predict total score using number of putts is the following:

Total score = 39.6 + 1.52 * Number of putts

So if you have 35 putts in an 18-hole golf course, your total score is predicted to be about 39.6 + 1.52 * 35 = 92.8, or 93. (Not bad for 18 holes!)

Don't try to predict *y* for *x*-values that fall outside the range of where the data was collected; you have no guarantee that the line still works outside of that range or that it will even make sense. For the golf example, you can't say that if *x* (the number of putts) = 1 the total score would be 39.6 + 1.52 * 1 = 41.12 (unless you just call it good after your ball hits the green). This mistake is called *extrapolation*.

You can discover more about simple linear regression, and expansions on it, in Chapters 4 and 5.

# *Avoiding Bias*

Bias is the bane of a statistician's existence; it's easy to create and very hard (if not impossible) to deal with in most situations. The statistical definition of *bias* is the systematic overestimation or underestimation of the actual value. In language the rest of us can understand, it means that the results are always off by a certain amount in a certain direction.

For example, a bathroom scale may always report a weight that's five pounds more than it should be (I'm convinced this is true of the scale at my doctor's office).

Bias can show up in a data set in a variety of different ways. Here are some of the most common ways bias can creep into your data:

✔ **Selecting the sample from the population:** Bias occurs when you either leave some groups out of the process that should have been included, or give certain groups too much weight.

For example, TV surveys that ask viewers to phone in their opinion are biased because no one has selected a prior sample of people to represent the population — viewers who want to be involved select themselves to participate by calling in on their own. Statisticians have found that folks who decide to participate in "call-in" or Web site polls are very likely to have stronger opinions than those who have been randomly selected but choose not to get involved in such polls. Such samples are called *self-selected samples* and are typically very biased.

✔ **Designing the data-collection instrument:** Poorly designed instruments, including surveys and their questions, can result in inconsistent or even incorrect data. A survey question's wording plays a large role in whether or not results are biased. A leading question can make people feel like they should answer a certain way. For example, "Don't you think that the president should be allowed to have a line-item veto to prevent government spending waste?" Who would feel they should say *no* to that?

✔ **Collecting the data:** In this case, bias can infiltrate the results if someone makes errors in recording the data or if interviewers deviate from the script.

✔ **Deciding how and when the data is collected:** The time and place you collect data can affect whether your results are biased. For example, if you conduct a telephone survey during the middle of the day, people who work from 9 to 5 aren't able to participate. Depending on the issue, the timing of this survey could lead to biased results.

The best way to deal with bias is to avoid it in the first place, but you also can try to minimize it by

✔ **Using a random process to select the sample from the population.** The only way a sample is truly random is if every single member of the population has an equal chance of being selected. Self-selected samples aren't random.

✔ **Making sure the data is collected in a fair and consistent way.** Be sure to use neutral question wording and time the survey properly.

---

# Don't put all your data in one basket!

An animal science researcher came to me one time with a data set he was so proud of. He was studying cows and the variables involved in helping determine their longevity. His super-mega data set contained over 100,000 observations. He was thinking, "Wow, this is gonna be great! I've been collecting this data for years and years, and I can finally have it analyzed. There's got to be loads of information I can get out of this. The papers I'll write, the talks I'll be invited to give . . . the raise I'll get!" He turned his precious data over to me with an expectant smile and sparkling eyes.

But after looking at his data for a few minutes I made a terrible realization — all his data came from exactly one cow. With no other cows to compare with and a sample size of just one, he had no way to even measure how much those results would vary if he wanted to apply them to another cow. His results were so biased toward that one animal that I couldn't do anything with the data. After I summoned the courage to tell him so, it took a while to peel him off the floor. The moral of the story, I suppose, is to run your big plans by a statistician before you go down a cow path like this guy did.

# *Measuring Precision with Margin of Error*

*Precision* is the amount of movement you expect to have in your sample results if you repeat your entire study again with a new sample. Precision comes in two forms:

- ✔ **Low precision** means that you expect your sample results to move a lot (not a good thing).

- ✔ **High precision** means you expect your sample results to remain fairly close in the repeated samples (a good thing).

In this section, you find out what precision does and doesn't measure and you see how to measure the precision of a statistic in general terms.

Before you report or try to interpret any statistical results, you need to have some measurement of how much those results are expected to vary from sample to sample. This measurement is called the *margin of error.* You always hope, and may even assume, that statistical results shouldn't change much with another sample, but that's not always the case.

## Up close and personal: Survey results

The Gallup Organization states its survey results in a universal, statistically correct format. Using a specific example from a recent survey it conducted, here's the language it uses to report its results:

"These results are based on telephone interviews with a randomly selected national sample of 1,002 adults, aged 18 years and older, conducted June 9–11, 2006. For results based on this sample, one can say with 95 percent confidence that the maximum error attributable to sampling and other random effects is ± 3 percentage points. In addition to sampling error, question wording and practical difficulties in conducting surveys can introduce error or bias into the findings of public opinion polls."

The first sentence of the quote refers to how the Gallup Organization collected the data, as well as the size of the sample. As you can guess, precision is related to the sample size, as seen in the section "Measuring Precision with Margin of Error."

The second sentence of the quote refers to the precision measurement: How much did Gallup expect these sample results to vary? The fact that Gallup is 95 percent confident means that if this process were repeated a large number of times, in 5 percent of the cases the results would be wrong, just by chance. This inconsistency occurs if the sample selected for the analysis doesn't represent the population — not due to biased reasons, but due to chance alone. Check out the section "Avoiding Bias" to get the info on why the third sentence is included in this quote.

The margin of error is affected by two elements:

- The sample size
- The amount of diversity in the population (also known as the population standard deviation)

You can read more about these elements in Chapter 3, but here's the big picture: As your sample size increases, you have more data to work with, and your results become more precise. As a result, the margin of error goes down.

On the other hand, a high amount of diversity in your population reduces your level of precision because the diversity makes it harder to get a handle on what's going on. As a result, the margin of error increases. (To offset this problem, just increase the sample size to get your precision back.)

To interpret the margin of error, just think of it as the amount of play you allow in your results to cover most of the other samples you could have taken.

Suppose you're trying to estimate the proportion of people in the population who support a certain issue, and you want to be 95 percent confident in your results. You sample 1,002 individuals and find that 65 percent support the issue. The margin of error for this survey turns out to be plus or minus 3 percentage points (you can find the details of this calculation in Chapter 3). That result means that you could expect the sample proportion of 65 percent to change by as much as 3 percentage points either way if you were to take a different sample of 1,002 individuals. In other words, you believe the actual population proportion is somewhere between 65 – 3 = 62 percent and 65 + 3 = 68 percent. That's the best you can say.

Any reported margin of error is calculated on the basis of having zero bias in the data. However, this assumption is rarely true. Before interpreting any margin of error, check first to be sure that the sampling process and the data-collection process don't contain any obvious sources of bias. Ignore results that are based on biased data, or at least take them with a great deal of skepticism.

For more details on how to calculate margin of error in various statistical techniques, turn to Chapter 3.

## Knowing Your Limitations

The most important goal of any data analyst is to remain focused on the big picture — the question that you or someone else is asking — and make sure that the data analysis used is appropriate and comprehensive enough to answer that question correctly and fairly.

Here are some tips for analyzing data and interpreting the results, in terms of the statistical procedures and techniques that you may use — at school, in your job, and in everyday life. These tips are implemented and reinforced throughout this book:

✔ **Be sure that the research question being asked is clear and definitive.** Some researchers don't want to be pinned down on any particular set of questions because they have the intent of mining the data — looking for any relationship they can find and then stating their results after the fact. This practice can lead to overanalyzing the data, making the results subject to skepticism by statisticians.

✔ **Double-check that you clearly understand the type of data being collected.** Is the data categorical or quantitative? The type of data used drives the approach that you take in the analysis.

✔ **Make sure that the statistical technique you use is designed to answer the research question.** If you want to make comparisons between two groups and your data is quantitative, use a hypothesis test for two means. If you want to compare five groups, use analysis of variance (ANOVA). Use this book as a resource to help you determine the technique you need.

✔ **Look for the limitations of the data analysis.** For example, if the researcher wants to know whether negative political ads affect the population of voters and she bases her study on a group of college students, you can find severe limitations here. For starters, student reactions to negative ads don't necessarily carry over to all voters in the population. In this case, it's best to limit the conclusions to college students in that class (which no researcher would ever want to do). Better to take a sample that represents the intended population of all voters in the first place (a much more difficult task, but well worth it).

# Chapter 3

# Reviewing Confidence Intervals and Hypothesis Tests

---

## In This Chapter

▶ Utilizing confidence intervals to estimate parameters

▶ Testing models by using hypothesis tests

▶ Finding the probability of getting it right and getting it wrong

▶ Discovering power in a large sample size

---

*O*ne of the major goals in statistics is to use the information you collect from a sample to get a better idea of what's going on in the entire population you're studying (because populations are generally large and exact info is often unknown). Unknown values that summarize the population are called *population parameters*. Researchers typically want to either get a handle on what those parameters are or test a hypothesis about the population parameters.

In Stats I, you probably went over confidence intervals and hypothesis tests for one and two population means and one and two population proportions. Your instructor hopefully emphasized that no matter which parameters you're trying to estimate or test, the general process is the same. If not, don't worry; this chapter drives that point home.

This chapter reviews the basic concepts of confidence intervals and hypothesis tests, including the probabilities of making errors by chance. I also discuss how statisticians measure the ability of a statistical procedure to do a good job — of detecting a real difference in the populations, for example.

# Estimating Parameters by Using Confidence Intervals

*Confidence intervals* are a statistician's way of covering his you-know-what when it comes to estimating a population parameter. For example, instead of just giving a one-number guess as to what the average household income is in the United States, a statistician gives a range of likely values for this number. He does this because

✔ All good statisticians know sample results vary from sample to sample, so a one-number estimate isn't any good.

✔ Statisticians have developed some awfully nice formulas to give a range of likely values, so why not use them?

In this section, you get the general formula for a confidence interval, including the margin of error, and a good look at the common approach to building confidence intervals. I also discuss interpretation and the chance of making an error.

## Getting the basics: The general form of a confidence interval

The big idea of a confidence interval is coming up with a range of likely values for a population parameter. The *confidence level* represents the chance that if you were to repeat your sample-taking over and over, you'd get a range of likely values that actually contains the actual population parameter. In other words, the confidence level is the long-term chance of being correct.

The general formula for a confidence interval is

Confidence interval = Sample statistic ± Margin of error

The confidence interval has a certain level of precision (measured by the margin of error). Precision measures how close you expect your results to be to the truth.

For example, suppose you want to know the average amount of time a student at The Ohio State University spends listening to music on an MP3 player per day. The average time for the entire population of OSU students who are MP3-player users is the parameter you're looking for. You take a random sample of 1,000 students and find that the average time a student uses an MP3 player per day to listen to music is 2.5 hours, and the standard deviation is 0.5 hours. Is it right to say that the population of all OSU-student,

MP3-player owners use their players an average of 2.5 hours per day for music listening? You hope and may assume that the average for the whole population is close to 2.5, but it probably isn't exact.

What's the solution to this problem? The solution is to not only report the average from your sample but along with it report some measure of how much you expect that sample average to vary from one sample to the next, with a certain level of confidence. The number that you use to represent this level of precision in your results is called the *margin of error.*

# Finding the confidence interval for a population mean

The sample statistic part of the confidence-interval formula is fairly straightforward.

✔ **To estimate the population mean,** you use the sample mean plus or minus a margin of error, which is based on standard error. The sample mean has a standard error of $\frac{\sigma}{\sqrt{n}}$. In this formula, you can see the population standard deviation (σ) and the sample size (*n*).

✔ **To estimate the population proportion,** you use the sample proportion plus or minus a margin of error.

In many cases, the standard deviation of the population, σ, is not known. To estimate the population mean by using a confidence interval when σ is unknown, you use the formula $\bar{x} \pm t_{n-1}\left(\frac{s}{\sqrt{n}}\right)$. This formula contains the sample standard deviation *(s)*, the sample size *(n)*, and a *t*-value representing how many standard errors you want to add and subtract to get the confidence you need. To get the margin of error for the mean, you see the standard error, $\frac{s}{\sqrt{n}}$, is being multiplied by a factor of *t*. Notice that *t* has $n-1$ as a subscript to indicate which of the myriad *t*-distributions you use for your confidence interval. The $n-1$ is called *degrees of freedom*.

The value of *t* in this case represents the number of standard errors you add and subtract to or from the sample mean to get the confidence you want. If you want to be 95 percent confident, for example, you add and subtract 1.96 of those standard errors. If you want to be 99.7 percent confident, you add or subtract about three of them. (See Table A-1 in the appendix to find *t*-values for various confidence levels; use $\left(\frac{1-\text{confidence level}}{2}\right)$ for the area to the right and find the *t*-value that goes with it.)

If you know the population standard deviation, you should certainly use it. In that case, you use the corresponding number from the Z-distribution (standard normal distribution) in the confidence interval formula. (The Z-distribution from your Stats I textbook can give you the numbers you need.) But I would be remiss in saying that while textbooks and teachers always include problems where σ is known, rarely is σ known in the real world. Why teach it this way? This issue is up for debate; for now just go with it, and I can keep you posted.

For the MP3 player example from the preceding section, a random sample of 1,000 all OSU students spends an average of 2.5 hours using their MP3 players to listen to music. The standard deviation is 0.5 hours. Plugging this information into the formula for a confidence interval, you get $2.5 \pm 1.96\left(\dfrac{0.5}{\sqrt{1,000}}\right)$. You conclude that *all* OSU-student MP3-owners spend an average of between 2.47 and 2.53 hours listening to music on their players.

# What changes the margin of error?

What do you need to know in order to come up with a margin of error? Margin of error, in general, depends on three elements:

- ✔ The standard deviation of the population, σ (or an estimate of it, denoted by *s*, the sample standard deviation)
- ✔ The sample size, *n*
- ✔ The level of confidence you need

You can see these elements in action in the formula for margin of error of the sample mean: $\pm t_{n-1} * \dfrac{s}{\sqrt{n}}$. Here I assume that σ isn't known; $t_{n-1}$ represents the value on the *t*-distribution table (see Table A-1 in the appendix) with $n-1$ degrees of freedom.

Each of these three elements has a major role in determining how large the margin of error will be when you estimate the mean of a population. In the following sections, I show how each of the elements of the margin of error formula work separately and together to affect the size of the margin of error.

## Population standard deviation

The standard deviation of the population is typically combined with the sample size in the margin of error formula, with the population standard deviation on top of the fraction and *n* on the bottom. (In this case, the standard error of the population, σ, is estimated by the standard deviation of the sample, *s*, because σ is typically unknown.)

This combination of standard deviation of the population and sample size is known as the *standard error* of your statistic. It measures how much the sample statistic deviates from its mean in the long term.

How does the standard deviation of the population ($\sigma$) affect margin of error? As it gets larger, the margin of error increases, so your range of likely values is wider.

Suppose you have two gas stations, one on a busy corner (gas station #1) and one farther off the main drag (gas station #2). You want to estimate the average time between customers at each station. At the busy gas station #1, customers are constantly using the gas pumps, so you basically have no downtime between customers. At gas station #2, customers sometimes come all at once, and sometimes you don't see a single person for an hour or more. So the time between customers varies quite a bit.

For which gas station would it be easier to estimate the overall average time between customers as a whole? Gas station #1 has much more consistency, which represents a smaller standard deviation of time between customers. Gas station #2 has much more variability in time between customers. That means $\sigma$ for gas station #1 is smaller than $\sigma$ for gas station #2. So the average time between customers is easier to estimate at gas station #1.

### Sample size

Sample size affects margin of error in a very intuitive way. Suppose you're trying to estimate the average number of pets per household in your city. Which sample size would give you better information: 10 homes or 100 homes? I hope you'd agree that 100 homes would give more precise information (as long as the data on those 100 homes was collected properly).

If you have more data to base your conclusions on and that data is collected properly, your results will be more precise. Precision is measured by margin of error, so as the sample size increases, the margin of error of your estimate goes down.

Bigger is only better in terms of sample size if the data is collected properly — that is, with minimal bias. If the quality of the data can't be maintained with a larger sample size, it does no good to have it.

### Confidence level

For each problem at hand, you have to address how confident you need to be in your results over the long term, and, of course, more confidence comes with a price in the margin of error formula. This level of confidence in your results over the long term is reflected in a number called the *confidence level*, which you report as a percentage. In general, more confidence requires a wider range of likely values. So, as the confidence level increases, so does the margin of error.

Every margin of error is interpreted as plus or minus a certain number of standard errors. The number of standard errors added and subtracted is determined by the confidence level. If you need more confidence, you add and subtract more standard errors. If you need less confidence, you add and subtract fewer standard errors. The number that represents how many standard errors to add and subtract is different from situation to situation. For one population mean, you use a value on the $t$-distribution, represented by $t_{n-1}$, where $n$ is the sample size (see Table A-1 in the appendix).

Suppose you have a sample size of 20, and you want to estimate the mean of a population with 90 percent confidence. The number of standard errors you add and subtract is represented by $t_{n-1}$, which in this case is $t_{19} = 1.73$. (To find these values of $t$, see Table A-1 in the appendix, with $n - 1$ degrees of freedom for the row, and $\dfrac{\left(1 - \text{confidence level}\right)}{2}$ for the column.)

Now suppose you want to be 95 percent confident in your results, with the same sample size of $n = 20$. The degrees of freedom are $20 - 1 = 19$ (row) and the column is for $\dfrac{\left(1 - .95\right)}{2} = .025$. The $t$-table gives you the value of $t_{19} = 2.09$.

Notice that this value of $t$ is larger than the value of $t$ for 90 percent confidence, because in order to be more confident, you need to go out more standard deviations on the $t$-distribution table to cover more possible results.

### Large confidence, narrow intervals — just the right size

A narrow confidence interval is much more desirable than a wide one. For example, claiming that the average cost of a new home is $150,000 plus or minus $100,000 isn't helpful at all because your estimate is anywhere between $50,000 and $250,000. (Who has an extra $100,000 to throw around?) But you *do* want a high confidence level, so your statistician has to add and subtract more standard errors to get there, which makes the interval that much wider (a downer).

Wait, don't panic — you can have your cake and eat it too! If you know you want to have a high level of confidence but you don't want a wide confidence interval, just increase your sample size to meet that level of confidence.

Suppose the standard deviation of the house prices from a previous study is $s = \$15,000$, and you want to be 95 percent confident in your estimate of average house price. Using a large sample size, your value of $t$ (from Table A-1 in the appendix) is 1.96.

With a sample of 100 homes, your margin of error is $\pm 1.96 * \dfrac{15,000}{\sqrt{100}} = \$2,940$.

If this is too large for you but you still want 95 percent confidence, crank up your value of *n*. If you sample 500 homes, the margin of error decreases to $\pm 1.96 * \frac{15,000}{\sqrt{500}}$, which brings you down to $1,314.81.

You can use a formula to find the sample size you need to meet a desired margin of error. That formula is $n = \left( \frac{t_{n-1}s}{MOE} \right)^2$, where *MOE* is the desired margin of error (as a proportion), *s* is the sample standard deviation, and *t* is the value on the *t*-distribution that corresponds with the confidence level you want. (For large sample sizes, the *t*-distribution is approximately equal to the *Z*-distribution; you can use the last line of Table A-1 in the appendix for the appropriate *t*-values, or use a *Z*-table from your Stats I textbook.)

## Interpreting a confidence interval

Interpreting a confidence interval involves a couple of subtle but important issues. The big idea is that a *confidence interval* presents a range of likely values for the population parameter, based on your sample. However, you interpret it not in terms of your own sample, but in terms of an infinite number of other samples out there that could have been selected, yours just being one of them. For example, suppose 1,000 people each took a sample and they each formed a 95 percent confidence interval for the mean. The "95 percent confidence" part means that of those 1,000 confidence intervals, about 950 of them can be expected to be correct on average. (Correct means the confidence interval actually contains the true value of the parameter.)

A 95 percent confidence interval doesn't mean that your particular confidence interval has a 95 percent chance of capturing the actual value of the parameter; after the sample has been taken, the parameter is either in the interval or it isn't. A confidence interval represents the chances of capturing the actual value of the population parameter over many different samples.

Suppose a polling organization wants to estimate the percentage of people in the United States who drive a car with more than 100,000 miles on it, and it wants to be 95 percent confident in its results. The organization takes a random sample of 1,200 people and finds that 420 of them (35 percent) drive a car with that minimum mileage; the margin of error turns out to be plus or minus 3 percent. (See your Stats I text for determining margin of error for percentages.)

The meaty part of the interpretation lies in the confidence level — in this case, the 95 percent. Because the organization took a sample of 1,200 people in the U.S., asked each of them whether his or her car has more than 100,000 miles on it, and made a confidence interval out of the results, the polling organization is, in essence, accounting for all the other samples out there that it could have gotten by building in the margin of error (± 3 percent). The organization wants to cover its bases on 95 percent of those other situations, and ± 3 percent satisfies that.

Another way of thinking about the confidence interval is to say that if the organization sampled 1,200 people over and over again and made a confidence interval from its results each time, 95 percent of those confidence intervals would be right. (You just have to hope that yours is one of those right results.)

Using stat notation, you can write confidence levels as $(1 - \alpha)\%$. So if you want 95 percent confidence, you write it as $1 - 0.05$. Here, $\alpha$ represents the chance that your confidence interval is one of the wrong ones. This number, $\alpha$, is also related to the random chance of making a certain kind of error with a hypothesis test, which I explain in the later section "False alarms and missed opportunities: Type I and II errors."

# What's the Hype about Hypothesis Tests?

Suppose a shipping company claims that its packages are on time 92 percent of the time, or a campus official claims that 75 percent of students live off campus. If you're questioning these claims, how can you use statistics to investigate?

In this section, you see the big ideas of hypothesis testing that are the basis for the data-analysis techniques in this book. You review and expand on the concepts involved in a hypothesis test, including the hypotheses, the test statistic, and the $p$-value.

## What Ho and Ha really represent

You use a hypothesis test in situations where you have a certain model in mind and want to see whether that model fits your data. Your model may be one that just revolves around the population mean (testing whether that mean is equal to ten, for example). Your model may be testing the slope of a regression line (whether or not it's zero, for example, with zero meaning you find no relationship between $x$ and $y$). You may be trying to use several different variables to predict the marketability of a product, and you believe a model using customer age, price, and shelf location can help predict it, so you need to run one or more hypothesis tests to see whether that model works. (This particular process is called multiple regression, and you can find more info on it in Chapter 5.)

A hypothesis test is made up of two hypotheses:

- **The null hypothesis, Ho:** Ho symbolizes the current situation — the one that everyone assumed was true until you got involved.

- **The alternative hypothesis, Ha:** Ha represents the alternative model that you want to consider. It stands for the researcher's hypothesis, and the burden of proof lies on the researcher.

Ho is the model that's on trial. If you get enough evidence against it, you conclude Ha, which is the model you're claiming is the right one. If you don't get enough evidence against Ho, then you can't say that your model (Ha) is the right one.

# Gathering your evidence into a test statistic

A *test statistic* is the statistic from your sample, standardized so you can look it up on a table, basically. Although each hypothesis test is a little different, the main thought is the same. Take your statistic and standardize it in the appropriate way so you can use the corresponding table for it. Then look up your test statistic on a table to see where it stands. That table may be the *t*-table (Table A-1 in the appendix), the Chi-square table (Table A-3 in the appendix), or a different table. The type of test you need to use on your data dictates which table you use.

In the case of testing a hypothesis for a population mean, $\mu$, you use the sample mean, $\bar{x}$, as your statistic. To standardize it, you take $\bar{x}$ and convert it to a value of $t$ by using the formula $t_{n-1} = \dfrac{\bar{x} - \mu_0}{\frac{s}{\sqrt{n}}}$, where $\mu_0$ is the value in Ho.

This value is your test statistic, which you compare to the *t*-distribution.

# Determining strength of evidence with a p-value

If you want to know whether your data has the brawn to stand up against Ho, you need to figure out the *p*-value and compare it to a predetermined cutoff, $\alpha$ (typically 0.05). The *p-value* is a measure of the strength of your evidence against Ho. You calculate the *p*-value through these steps:

1. **Calculate the test statistic (refer to the preceding section for more info on this).**

2. **Look up the test statistic on the appropriate table (such as the *t*-table, Table A-1 in the appendix).**

3. **Find the percentage of values on the table that fall beyond your test statistic. This percentage is the *p*-value.**

4. **If your Ha is "not equal to," double the percentage that you got in step three because your test statistic could have gone either way before the data was collected. (See your Stats I textbook or *Statistics For Dummies* for full details on obtaining *p*-values for hypothesis tests.)**

Your friend $\alpha$ is the cutoff for your *p*-value. ($\alpha$ is typically set at 0.05, but sometimes it's 0.10.) If your *p*-value is less than your predetermined value of $\alpha$, reject Ho because you have sufficient evidence against it. If your *p*-value is greater than or equal to $\alpha$, you can't reject Ho.

For example, if your *p*-value is 0.002, your test statistic is so far away from Ho that the chance of getting this result by chance is only 2 out of 1,000. So, you conclude that Ho is very likely to be false. If your *p*-value turns out to be 0.30, this same result is expected to happen 30 percent of the time anyway, so you see no red flags there, and you can't reject Ho. You don't have enough evidence against it. If your *p*-value is close to the cutoff line, say $p = 0.049$ or 0.51, you say the result is marginal and let the reader make her own conclusions. That's the main advantage of the *p*-value: It lets other folks determine whether your evidence is strong enough to reject Ho in their minds.

# False alarms and missed opportunities: Type 1 and 11 errors

Any technique you use in statistics to make a conclusion about a population based on a sample of data has the chance of making an error. The errors I am talking about, Type I and Type II errors, are due to random chance.

The way you set up your test can help to reduce these kinds of errors, but they're always out there. As a data analyst, you need to know how to measure and understand the impact of the errors that can occur with a hypothesis test and what you can do to possibly make those errors smaller. In the following sections, I show you how you can do just that.

## Making false alarms with Type 1 errors

A *Type I error* is the conditional probability of rejecting Ho, given that Ho is true. I think of a Type I error as a false alarm: You blew the whistle when you shouldn't have.

The chance of making a Type I error is equal to $\alpha$, which is predetermined before you begin collecting your data. This $\alpha$ is the same $\alpha$ that represents the chance of missing the boat in a confidence interval. It makes some sense that these two probabilities are both equal because the probability of rejecting Ho when you shouldn't (a Type I error) is the same as the chance that the true population parameter falls out of the range of likely values when it shouldn't. That chance is $\alpha$.

Suppose someone claims that the mean time to deliver packages for a company is 3.0 days on average (so Ho is $\mu = 3.0$), but you believe it's not equal to that (so Ha is $\mu \neq 3.0$). Your $\alpha$ level is 0.05, and because you have a two-sided test, you have 0.025 on each side. Your sample of 100 packages has a mean of 3.5 days with a standard deviation of 1.5 days. The test statistic equals $\frac{3.5 - 3.0}{\frac{1.5}{\sqrt{100}}}$ = 3.33, which is greater than 1.96 (the value on the last row and the 0.025 column of the *t*-distribution table — see Table A-1 in the appendix). So 3.0 is not a likely value for the mean time of delivery for all packages, and you reject Ho.

But suppose that just by chance, your sample contained some longer than normal delivery times and that, in reality, the company's claim is right. You just made a Type I error. You made a false alarm about the company's claim.

To reduce the chance of a Type I error, reduce your value of $\alpha$. However I don't recommend reducing it too far. On the positive side, this reduction makes it harder to reject Ho because you need more evidence in your data to do so. On the negative side, by reducing your chance of a false alarm (Type I error) you increase the chance of a missed opportunity (a Type II error.)

### Missing an opportunity with a Type II error

A *Type II error* is the conditional probability of not rejecting Ho, given that Ho is false. I call it a missed opportunity because you were supposed to be able to find a problem with Ho and reject it, but you didn't. You didn't blow the whistle when you should have.

The chance of making a Type II error depends on a couple of things:

- ✔ **Sample size:** If you have more data, you're less likely to miss something that's going on. For example, if a coin actually is unfair, flipping the coin only ten times may not reveal the problem. But if you flip the coin 1,000 times, you have a good chance of seeing a pattern that favors heads over tails, or vice versa.

- ✔ **Actual value of the parameter:** A Type II error is also related to how big the problem is that you're trying to uncover. For example, suppose a company claims that the average delivery time for packages is 3.5 days. If the actual average delivery time is 5.0 days, you won't have a very hard time detecting that with your sample (even a small sample). But if the actual average delivery time is 4.0 days, you have to do more work to actually detect the problem.

To reduce the chance of a Type II error, take a larger sample size. A greater sample size makes it easier to reject Ho but increases the chance of a Type I error.

Type I and Type II errors sit on opposite ends of a seesaw — as one goes up, the other goes down. Try to meet in the middle by choosing a large sample size (the bigger, the better; see Figures 3-1 and 3-2) and a small α level (0.05 or less) for your hypothesis test.

# The power of a hypothesis test

Type II errors, which I explain in the preceding section, show the downside of a hypothesis test. But statisticians, despite what many may think, actually try to look on the bright side once in a while; so instead of looking at the chance of *missing* a difference from Ho that actually is there, they look at the chance of *detecting* a difference that really *is* there. This detection is called the *power of a hypothesis test*.

The power of a hypothesis test is 1 – the probability of making a Type II error. So *power* is a number between 0 and 1 that represents the chance that you rejected Ho when Ho was false. (You can even sing about it: "If Ho is false and you know it, clap your hands. . . .") Remember that power (just like Type II errors) depends on two elements: the sample size and the actual value of the parameter (see the preceding section for a description of these elements).

In the following sections, you discover what power means in statistics (not being one of the bigwigs, mind you); you also find out how to quantify power by using a power curve.

### Throwing a power curve

The specific calculations for the power of a hypothesis test are beyond the scope of this book (so you can take a sigh of relief), but computer programs and graphs are available online to show you what the power is for different hypothesis tests and various sample sizes (just type "power curve for the [blah blah blah] test" into an Internet search engine).

These graphs are called *power curves* for a hypothesis test. A power curve is a special kind of graph that gives you an idea of how much of a difference from Ho you can detect with the sample size that you have. Because the precision of your test statistic increases as your sample size increases, sample size is directly related to power. But it also depends on how much of a difference from Ho you're trying to detect. For example, if a package delivery company claims that its packages arrive in 2 days or less, do you want to blow the whistle if it's actually 2.1 days? Or wait until it's 3 days? You need a much larger sample size to detect the 2.1-days situation versus the 3-days situation just because of the precision level needed.

In Figure 3-1, you can see the power curve for a particular test of Ho: $\mu = 0$ versus Ha: $\mu > 0$. You can assume that $\sigma$ (the standard deviation of the population) is equal to 2 (I give you this value in each problem) and doesn't change. I set the sample size at 10 throughout.

The horizontal $(x)$ axis on the power curve shows a range of actual values of $\mu$. For example, you hypothesize that $\mu$ is equal to 0, but it may actually be 0.5, 1.0, 2.0, 3.0, or any other possible value. If $\mu$ equals 0, then Ho is true, and the chance of detecting this (and therefore rejecting Ho) is equal to 0.05, the set value of $\alpha$. You work from that baseline. (Notice the low power in this situation makes sense because there's nothing to detect for values of $\mu$ that are close to 0.) So, on the graph in Figure 3-1, when $x = 0$, you get a $y$-value of 0.05.

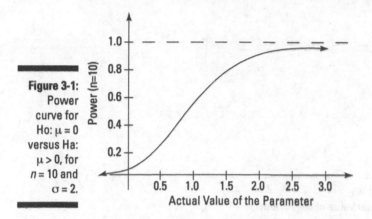

**Figure 3-1:** Power curve for Ho: $\mu = 0$ versus Ha: $\mu > 0$, for $n = 10$ and $\sigma = 2$.

Suppose that $\mu$ is actually 0.5, not 0, as you hypothesized. A computer tells you that the chance of rejecting Ho (what you're supposed to do here) is $0.197 = 0.20$, which is the power. So, you have about a 20-percent chance of detecting this difference with a sample size of 10. As you move to the right, away from 0 on the horizontal $(x)$ axis, you can see that the power goes up and the $y$-values get closer and closer to 1.0.

For example, if the actual value of $\mu$ is 1.0, the difference from 0 is easier to detect than if it's 0.50. In fact, the power at 1.0 is equal to $0.475 = 0.48$, so you have almost a 50 percent chance of catching the difference from Ho in this case. And as the values of the mean increase, the power gets closer and closer to 1.0. Power never reaches 1.0 because statistics can never prove anything with 100 percent accuracy, but you can get close to 1.0 if the actual value is far enough from your hypothesis.

### Controlling the sample size

How can you increase the power of your hypothesis test? You don't have any control over the actual value of the parameter, because that number is unknown. So what do you have control over? The sample size. As the sample size increases, it becomes easier to detect a real difference from Ho.

Figure 3-2 shows the power curve with the same numbers as Figure 3-1, except for the sample size *(n)*, which is 100 instead of 10. Notice that the curve increases much more quickly and approaches 1.0 when the actual mean is 1.0, compared to your hypothesis of 0. You want to see this kind of curve that moves up quickly toward the value of 1.0, while the actual values of the parameter increase on the *x*-axis.

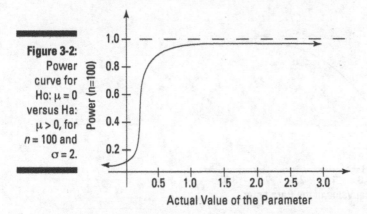

**Figure 3-2:**
Power curve for Ho: $\mu = 0$ versus Ha: $\mu > 0$, for $n = 100$ and $\sigma = 2$.

If you compare the power of your test when $\mu$ is 1.0 for the $n = 10$ situation (in Figure 3-1) versus the $n = 100$ situation (in Figure 3-2), you see that the power increases from 0.475 to more than 0.999. Table 3-1 shows the different values of power for the $n = 10$ case versus the $n = 100$ case, when you test Ho: $\mu = 0$ versus Ha: $\mu > 0$, assuming a value of $\sigma = 2$.

| Table 3-1 | Comparing the Values of Power for $n = 10$ Versus $n = 100$ (Ho is $\mu = 0$) | |
|---|---|---|
| *Actual Value of $\mu$* | *Power When n = 10* | *Power When n = 100* |
| 0.00 | 0.050 = 0.05 | 0.050 = 0.05 |
| 0.50 | 0.197 = 0.20 | 0.804 = 0.81 |
| 1.00 | 0.475 = 0.48 | approx. 1.0 |
| 1.50 | 0.766 = 0.77 | approx. 1.0 |

| Actual Value of $\mu$ | Power When $n = 10$ | Power When $n = 100$ |
| --- | --- | --- |
| 2.00 | 0.935 = 0.94 | approx. 1.0 |
| 3.00 | 0.999 = approx. 1.0 | approx. 1.0 |

You can find power curves for a variety of hypothesis tests under many different scenarios. Each has the same general look and feel to it: starting at the value of $\alpha$ when Ho is true, increasing in an *S*-shape as you move from left to right on the *x*-axis, and finally approaching the value of 1.0 at some point. Power curves with large sample sizes approach 1.0 faster than power curves with low sample sizes.

It's possible to have too much power. For example, if you make the power curve for $n$ = 10,000 and compare it to Figures 3-1 and 3-2, you find that it's practically at 1.0 already for any number other than 0.0 for the mean. In other words, the actual mean could be 0.05 and with your hypothesis Ho: $\mu$ = 0.00, you would reject Ho because of your huge sample size. Unless a researcher really wants to detect very small differences from Ho (such as in medical studies or quality control situations), inflated values of $n$ are usually suspect. People sometimes increase $n$ just to be able to say they've found a difference, no matter how small, so watch for that. If you zoom in enough, you can always detect something, even if that something makes no practical difference. Beware of surveys and experiments with an excessive sample size, such as one in the tens of thousands. Their results are guaranteed to be inflated.

## Power in manufacturing and medicine

The power of a test plays a role in the manufacturing process. Manufacturers often have very strict specifications regarding the size, weight, and/or quality of their products. During the manufacturing process, manufacturers want to be able to detect deviations from these specifications, even small ones, so they must determine how much of a difference from Ho they want to detect, and then figure out the sample size needed in order to detect that difference when it appears. For example, if the candy bar is supposed to weigh 2.0 ounces, the manufacturer may want to blow the whistle if the actual average weight shifts to 2.2 ounces. Statisticians can work backward in calculating the power and find the sample size they need to know to stop the process.

Medical scientists also think about power when they set up their studies (called *clinical trials*). Suppose they're checking to see whether an antidepressant adversely affects blood pressure (as a side effect of taking the drug). Scientists need to be able to detect small differences in blood pressure, because for some patients, any change in blood pressure is important to note and treat.

# Part II

# Using Different Types of Regression to Make Predictions

The 5th Wave      By Rich Tennant

Jimmy spent all day concentrating on his upcoming Statistics II test. This proved an excellent study habit up until he tried milking his father's prize bull.

## In this part . . .

**T**his part takes you beyond using one variable to predict another variable using a straight line (that's what simple linear regression is about). Instead, you find ways to predict one variable using many, and you also discover ways to make predictions using curves. Finally, you make predictions for probabilities, not just average values. It's a one-stop-shopping place for all things regression. Because these methods allow you to solve more complex problems, they lend themselves nicely to many real-world applications.

Exploring Relationships with Scatterplots
and Correlations

# Chapter 4

# Getting in Line with Simple Linear Regression

. . . . . . . . . . . . . . . . . . . . . . . . . . . . . . . . . . . . .

## *In This Chapter*

▶ Using scatterplots and correlation coefficients to examine relationships

▶ Building a simple linear regression model to estimate *y* from *x*

▶ Testing how well the model fits

▶ Interpreting the results and making good predictions

. . . . . . . . . . . . . . . . . . . . . . . . . . . . . . . . . . . . .

*L*ooking for relationships and making predictions is one of the staples of data analysis. Everyone wants to answer questions like, "Can I predict how many units I'll sell if I spend *x* amount of advertising dollars?"; or "Does drinking more diet cola really relate to more weight gain?"; or "Do children's backpacks seem to get heavier with each year of school, or is it just me?"

*Linear regression* tries to find relationships between two or more variables and comes up with a model that tries to describe that relationship, much like the way the line *y* = 2*x* + 3 explains the relationship between *x* and *y*. But unlike in math, where functions like *y* = 2*x* + 3 tell the entire story about the two variables, in statistics things don't come out that perfectly; some variability and error is involved (that's what makes it fun!).

This chapter is partly a review of the concepts of simple linear regression presented in a typical Stats I textbook. But the fun doesn't stop there. I expand on the ideas about regression that you picked up in your Stats I course and set you up for some of the other types of regression models you see in Chapters 5 through 8.

In this chapter, you see how to build a simple linear regression model that examines the relationship between two variables. You also see how simple linear regression works from a model-building standpoint.

# Exploring Relationships with Scatterplots and Correlations

Before looking ahead to predicting a value of *y* from *x* using a line, you need to

✔ Establish that you have a legitimate reason to do so by using a straight line.

✔ Feel confident that using a line to make that prediction will actually work well.

In order to accomplish both of these important steps, you need to first plot the data in a pairwise fashion so you can visually look for a relationship; then you need to somehow quantify that relationship in terms of how well those points follow a line. In this section, you do just that, using scatterplots and correlations.

Here's a perfect example of a situation where simple linear regression is useful: In 2004, the California State Board of Education wrote a report entitled "Textbook Weight in California: Analysis and Recommendations." This report discussed the great concern over the weight of the textbooks in students' backpacks and the problems it presents for students. The board conducted a study where it weighed a variety of textbooks from each of four core areas studied in grades 1–12 (reading, math, science, and history — where's statistics?) over a range of textbook brands and found the average total weight for all four books for each grade.

The board consulted pediatricians and chiropractors, who recommended that the weight of a student's backpack should not exceed 15 percent of his or her body weight. From there, the board hypothesized that the total weight of the textbooks in these four areas increases for each grade level and wanted to see whether it could find a relationship between the average child's weight in each grade and the average weight of his or her books. So along with the average weight of the four core-area textbooks for each grade, researchers also recorded the average weight for the students in that grade. The results are shown in Table 4-1.

| Table 4-1 | Average Textbook Weight and Student Weight (Grades 1–12) | |
|---|---|---|
| Grade | Average Student Weight (In Pounds) | Average Textbook Weight (In Pounds) |
| 1 | 48.50 | 8.00 |
| 2 | 54.50 | 9.44 |
| 3 | 61.25 | 10.08 |
| 4 | 69.00 | 11.81 |
| 5 | 74.50 | 12.28 |
| 6 | 85.00 | 13.61 |
| 7 | 89.00 | 15.13 |
| 8 | 99.00 | 15.47 |
| 9 | 112.00 | 17.36 |
| 10 | 123.00 | 18.07 |
| 11 | 134.00 | 20.79 |
| 12 | 142.00 | 16.06 |

In this section, you begin exploring whether or not a relationship exists between these two quantitative variables. You start by displaying the pairs of data using a two-dimensional scatterplot to look for a possible pattern, and you quantify the strength and direction of that pattern using the correlation coefficient.

## Using scatterplots to explore relationships

In order to explore a possible relationship between two variables, such as textbook weight and student weight, you first plot the data in a special graph called a *scatterplot*. A scatterplot is a two-dimensional graph that displays pairs of data, one pair per observation in the *(x, y)* format. Figure 4-1 shows a scatterplot of the textbook-weight data from Table 4-1.

You can see that the relationship appears to follow the straight line that's included on the graph, except possibly for the last point, where textbook weight is 16.06 pounds and student weight is 142 pounds (for grade 12). This point appears to be an *outlier* — it's the only point that doesn't fall into the pattern. So overall, an uphill, or *positive* linear relationship appears to exist between textbook weight and student weight; as student weight increases, so does textbook weight.

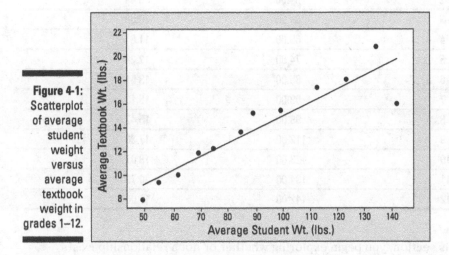

**Figure 4-1:** Scatterplot of average student weight versus average textbook weight in grades 1–12.

To make a scatterplot in Minitab, enter the data in columns one and two of the spreadsheet. Go to Graphs>Scatterplot. Click Simple and then OK. Highlight the response variable *(y)* in the left-hand box, and click Select. This variable shows up as the *y* variable in the scatterplot. Click on the explanatory *(x)* variable in the left-hand box, and click Select. It shows up in the *x* variable box. Click OK, and you get the scatterplot.

## Collating the information by using the correlation coefficient

After you've displayed the data using a scatterplot (see the preceding section), the next step is to find a statistic that quantifies the relationship somehow. The *correlation coefficient* (also known as *Pearson's correlation coefficient*, especially in statistical software packages) measures the strength and direction of the linear relationship between two quantitative variables *x* and *y*. It's a number between –1 and +1 that's *unit-free*, which means that if you change from pounds to ounces, the correlation coefficient doesn't change. (What a messed-up world it would be if this wasn't the case!)

If the relationship between *x* and *y* is uphill, or positive (as *x* increases, so does *y*), the correlation is a positive number. If the relationship is downhill, or negative (as *x* increases, *y* gets smaller), then the correlation is negative. The following list translates different correlation values:

- ✔ **A correlation value of zero means that you can find no linear relationship between *x* and *y*.** (It may be that a different relationship exists, such as a curve; see Chapter 7 for more on this.)

- ✔ **A correlation value of +1 or –1 indicates that the points fall in a perfect, straight line.** (Negative values indicate a downhill relationship; positive values indicate an uphill relationship.)

- ✔ **A correlation value close to +1 or –1 signifies a strong relationship.** A general rule of thumb is that correlations close to or beyond 0.7 or –0.7 are considered to be strong.

- ✔ **A correlation closer to +0.5 or –0.5 shows a moderate relationship.**

You can calculate the correlation coefficient by using a formula involving the standard deviation of *x*, the standard deviation of *y*, and the covariance of *x* and *y*, which measures how *x* and *y* move together in relation to their means. However, the formula isn't the focus here (you can find it in your Stats I textbook or in my other book, *Statistics For Dummies,* published by Wiley); it's the concept that's important. Any computer package can calculate the correlation coefficient for you with a simple click of the mouse.

To have Minitab calculate a correlation for you, go to Stat>Basic Statistics> Correlation. Highlight the variables you want correlations for, and click Select. Then click OK.

The correlation for the textbook-weight example is (can you guess before looking at it?) 0.926, which is very close to 1.0. This correlation means that a very strong linear relationship is present between average textbook weight and average student weight for grades 1–12, and that relationship is positive and linear (it follows a straight line). This correlation is confirmed by the scatterplot shown in Figure 4-1.

Data analysts should never make any conclusions about a relationship between *x* and *y* based solely on either the correlation or the scatterplot alone; the two elements need to be examined together. It's possible (but of course not a good idea) to manipulate graphs to look better or worse than they really are just by changing the scales on the axes. Because of this, statisticians never go with the scatterplot alone to determine whether or not a linear relationship exists between *x* and *y*. A correlation without a scatterplot is dangerous, too, because the relationship between *x* and *y* may be very strong but just not linear.

# Building a Simple Linear Regression Model

After you have a handle on which *x* variables may be related to *y* in a linear way, you go about the business of finding that straight line that best fits the data. You find the slope and *y*-intercept, put them together to make a line, and you use the equation of that line to make predictions for *y*. All this is part of building a simple linear regression model.

In this section, you set the foundation for regression models in general (including those you can find in Chapters 5 through 8). You plot the data, come up with a model that you think makes sense, assess how well it fits, and use it to guesstimate the value of *y* given another variable, *x*.

## Finding the best-fitting line to model your data

After you've established that *x* and *y* have a strong linear relationship, as evidenced by both the scatterplot and the correlation coefficient (close to or beyond 0.7 and –0.7; see the previous sections), you're ready to build a model that estimates *y* using *x*. In the textbook-weight case, you want to estimate average textbook weight using average student weight.

The most basic of all the regression models in the *simple linear regression model* that comes in the general form of $y = \alpha + \beta x + \varepsilon$. Here, $\alpha$ represents the *y*-intercept of the line, $\beta$ represents the slope, and $\varepsilon$ represents the error in the model due to chance.

A straight line that's used in simple linear regression is just one of an entire family of models (or functions) that statisticians use to express relationships between variables. A *model* is just a general name for a function that you can use to describe what outcome will occur based on some given information about one or more related variables.

Note that you will never know the true model that describes the relationship perfectly. The best you can do is estimate it based on data.

To find the right model for your data, the idea is to scour all possible lines and choose the one that fits the data best. Thankfully, you have an algorithm that does this for you (computers use it in their calculations). Formulas also exist for finding the slope and y-intercept of the best-fitting line by hand. The best-fitting line based on your data is $y = a + bx$, where $a$ estimates $\alpha$ and $b$ estimates $\beta$ from the true model. (You can find those formulas in your Stats I text or in *Statistics For Dummies*.)

To run a linear regression analysis in Minitab, go to Stat>Regression> Regression. Highlight the response *(y)* variable in the left-hand box, and click Select. The variable shows up in the Response Variable box. Then highlight your explanatory *(x)* variable, and click Select. This variable shows up in the Predictor Variable box. Click OK.

The equation of the line that best describes the relationship between average textbook weight and average student weight is $y = 3.69 + 0.113x$, where $x$ is the average student weight for that grade, and $y$ is the average textbook weight. Figure 4-2 shows the Minitab output of this analysis.

**Figure 4-2:**
Simple linear regression analysis for the textbook-weight example.

```
The regression equation is
textbook wt = 3.69 + 0.113 student wt


Predictor        Coef   SE Coef      T      P
Constant        3.694     1.395   2.65  0.024
student wt    0.11337   0.01456   7.78  0.000


S = 1.51341     R-Sq = 85.8%     R-Sq(adj) = 84.4%
```

By writing $y = 3.69 + 0.113x$, you mean that this equation represents your estimated value of $y$, given the value of $x$ that you observe with your data. Statisticians technically write this equation by using a caret (or *hat* as statisticians call it), like $\hat{y}$, so everyone can know it's an estimate, not the actual value of $y$. This y-hat is your estimate of the average value of $y$ over the long term, based on the observed values of $x$. However, in many Stats I texts, the hat is left off because statisticians have an unwritten understanding as to what $y$ represents. This issue comes up again in Chapters 5 through 8. (By the way, if you think y-hat is a funny term here, it's even funnier in Mexico, where statisticians call it *y-sombrero* — no kidding!)

## The y-intercept of the regression line

Selected parts of that Minitab output shown in Figure 4-2 are of importance to you at this point. First, you can see that under the Coef column you have the numerical values on the right side of the equation of the line — in other words, the slope and y-intercept. The number 3.69 represents the coefficient of "Constant," which is a fancy way of saying that's the y-intercept (because the *y*-intercept is just a constant — it never changes). The *y*-intercept is the point where the line crosses the *y*-axis; in other words, it's the value of *y* when *x* equals zero.

The *y*-intercept of a regression line may or may not have a practical meaning depending on the situation. To determine whether the *y*-intercept of a regression line has practical meaning, look at the following:

- ✔ Does the *y*-intercept fall within the actual values in the data set? If yes, it has practical meaning.

- ✔ Does the *y*-intercept fall into negative territory where negative *y*-values aren't possible? For example, if the *y*-values are weights, they can't be negative. Then the *y*-intercept has no practical meaning. The *y*-intercept is still needed in the equation though, because it just happens to be the place where the line, if extended to the *y*-axis, crosses the *y*-axis.

- ✔ Does the value *x* = 0 have practical meaning? For example, if *x* is temperature at a football game in Green Bay, then *x* = 0 is a value that's relevant to examine. If *x* = 0 has practical meaning, then the *y*-intercept does too, because it represents the value of *y* when *x* = 0. If the value of *x* = 0 doesn't have practical meaning in its own right (such as when *x* represents height of a toddler), then the *y*-intercept doesn't either.

In the textbook example, the *y*-intercept doesn't really have a practical meaning because students don't weigh zero pounds, so you don't really care what the estimated textbook weight is for that situation. But you do need to find a line that fits the data you do have (where average student weights go from 48.5 to 142 pounds). That best-fitting line must include a *y*-intercept, and for this problem, that *y*-intercept happens to be 3.69 pounds.

## The slope of the regression line

The value 0.113 from Figure 4-2 indicates the coefficient (or number in front) of the student-weight variable. This number is also known as the *slope*. It represents the change in *y* (textbook weight) is associated with a one-unit increase in *x* (student weight). As student weight increases by 1 pound, textbook weight increases by about 0.113 pounds, on average. To make this relationship more meaningful, you can multiply both quantities by 10 to say that as student weight increases by 10 pounds, the textbook weight goes up by about 1.13 pounds on average.

Whenever you get a number for the slope, take that number and put it over 1 to help you get started on a proper interpretation of slope. For example, a slope of 0.113 is rewritten as $\frac{0.113}{1}$. Using the idea that slope equals rise over run, or change in *y* over change in *x*, you can interpret the value of 0.113 in the following way: As *x* increases on average by 1 pound, *y* increases by 0.113 pounds.

## Making point estimates by using the regression line

When you have a line that estimates $y$ given $x$, you can use it to give a one-number estimate for the (average) value of $y$ for a given value of $x$. This is called making a *point estimate.* The basic idea is to take a reasonable value of $x$, plug it into the equation of the regression line, and see what you get for the value of $y$.

In the textbook-weight example, the best-fitting line (or model) is the line $y = 3.69 + 0.113x$. For an average student who weighs 60 pounds, for example, a one-number point estimate of the average textbook weight is $3.69 + (0.113 * 60) = 10.47$ pounds (those poor little kids!). If the average student weighs 100 pounds, the estimated average textbook weight is $3.69 + (0.113 * 100) = 14.99$, or nearly 15 pounds, plus or minus something. (You find out what that something is in the following section.)

# No Conclusion Left Behind: Tests and Confidence Intervals for Regression

After you have the slope of the best-fitting regression line for your data (see the previous sections), you need to step back and take into account the fact that sample results will vary. You shouldn't just say, "Okay, the slope of this line is 2. I'm done!" It won't be exactly 2 the next time. This variability is why statistics professors harp on adding a margin of error to your sample results; you want to be sure to cover yourself by adding that plus or minus.

In hypothesis testing, you don't just compare your sample mean to the population mean and say, "Yep, they're different alright!" You have to standardize your sample result using the standard error so that you can put your results in the proper perspective (see Chapter 3 for a review of confidence intervals and hypothesis tests).

The same idea applies here with regression. The data were used to figure out the best-fitting line, and you know it fits well for that data. That's not to say that the best-fitting line will work perfectly well for a new data set taken from the same population. So, in regression, all your results should involve the standard error with them in order to allow for the fact that sample results vary. That goes for estimating and testing for the slope and $y$-intercept and for any predictions that you make.

Many times in Stats I courses the concept of margin of error is skipped over after the best-fitting regression line is found, But these are very important ideas and should always be included. (Okay, enough of the soap box for now. Let's get out there and do it!)

## Scrutinizing the slope

Recall the *slope* of the regression line is the amount by which you expect the *y* variable to change on average as the *x* variable increases by 1 unit — the old rise-over-run idea (see the section "The slope of the regression line" earlier in this chapter). Now, how do you deal with knowing the best-fitting line will change with a new data set? You just apply the basic ideas of confidence intervals and hypothesis tests (see Chapter 3).

### A confidence interval for slope

A *confidence interval* in general has this form: your statistic plus or minus a margin of error. The margin of error includes a certain number of standard deviations (or standard errors) from your statistic. How many standard errors you add and subtract depends on what confidence level, $1 - \alpha$, you want. The size of the standard error depends on the sample size and other factors.

The equation of the best-fitting simple linear regression line, $y = a + bx$, includes a slope *(b)* and a *y*-intercept *(a)*. Because these were found using the data, they're only estimates of what's going on in the population, and therefore they need to be accompanied by a margin of error.

The formula for a $1 - \alpha$ level confidence interval for the slope of a regression line is $b \pm t^*_{n-2} * SE_b$, where the standard error is denoted

$$SE_b = \frac{s}{\sqrt{\sum_i \left( x_i - \bar{x} \right)^2}}, \text{ where } s = \sqrt{\frac{1}{n-2} \sum_i \left( y_i - \hat{y}_i \right)^2}. \text{ The value of } t^* \text{ comes}$$

from the *t*-distribution with $n - 2$ degrees of freedom and area to its right equal to $\alpha + 2$. (See Chapter 3 regarding the concept of $\alpha$.)

In case you wonder why you see $n - 2$ degrees of freedom here, as opposed to $n - 1$ degrees of freedom used in *t*-tests for the population mean in Stats I, here's the scoop. From Stats I you know that a *parameter* is a number that describes the population; it's usually known, and it can change from scenario to scenario. For each parameter in a model, you lose 1 degree of freedom. The regression line contains two parameters — the slope and the *y*-intercept — and you lose 1 degree of freedom for each one. With the *t*-test from Stats I you only have one parameter, the population mean, to worry about, hence you use $n - 1$ degrees of freedom.

**REMEMBER**

You can find the value of $t^*$ in any $t$-distribution table (check your textbook for one). For example, suppose you want to find a 95 percent confidence interval based on sample size $n = 10$. The value of $t^*$ is found in Table A-1 in the appendix in the row marked $10 - 2 = 8$ degrees of freedom, and the column marked 0.025 (because $\alpha \div 2 = 0.05 \div 2 = 0.025$). This value of $t^*$ is 2.306. (*Statistics For Dummies* can tell you a lot more about the $t$-distribution and the $t$-table.)

To put together a 95 percent confidence interval for the slope using computer output, you pull off the pieces that you need. For the textbook-weight example, in Figure 4-2 you see that the slope is equal to 0.11337. (Recall that slope is the coefficient of the $x$ variable in the equation, which is why you see the abbreviation Coef in the output.)

Because the slope changes from sample to sample, it's a random variable with its own distribution, its own mean, and its own standard error. (Recall from Stats I the standard error of a statistic is likened to the standard deviation of a random variable.) If you look just to the right of the slope in Figure 4-2, you see SE Coef; this stands for the standard error of the slope (which is 0.01456 in this case.)

Now all you need is the value of $t^*$ from the $t$-table (Table A-1 in the appendix). Because $n = 12$, you look in the row where degrees of freedom is $12 - 2 = 10$. You want a 95 percent confidence interval, so you look in the column for $(1 - 0.95) \div 2 = 0.25$. The $t^*$ value you get is 2.228.

Putting these pieces together, a 95 percent confidence interval for the slope of the best-fitting regression line for the textbook-weight example is $0.11337 \pm 2.228 * 0.01456$ which goes from 0.0809 to 0.1458. The units are in pounds (textbook) per pounds (child weight). Note this interval is large due to the small sample size, which increases the standard error.

### A hypothesis test for slope

You may be interested in conducting a hypothesis test for the slope of a regression line as another way to assess how well the line fits. If the slope is zero or close to it, the regression line is basically flat, signifying that no matter the value of $x$, you'll always estimate $y$ by using its mean. This means that $x$ and $y$ aren't related at all, so a specific value of $x$ doesn't help you predict a specific value for $y$. You can also test to see if the slope is some value other than zero, but that's atypical. So for all intents and purposes, I use the hypotheses Ho: $\beta = 0$ versus Ha: $\beta \neq 0$, where $\beta$ is the slope of the true model.

To conduct a hypothesis test for the slope of a simple linear regression line, you follow the basic steps of any hypothesis test. You take the statistic *(b)* from your data, subtract the value in Ho (in this case it's zero), and standardize it by dividing by the standard error (see Chapter 3 for more on this process).

Using the formula for standard error for $b$, the test statistic for the hypothesis test of whether or not the slope equals zero is $\frac{b-0}{SE_b}$, where $SE_b = \frac{s}{\sqrt{\sum_i (x_i - \bar{x})^2}}$, and $s = \sqrt{\frac{1}{n-2} \sum_i (y_i - \hat{y}_i)^2}$. On the Minitab output from Figure 4-2, the test statistic is located right next to the SE Coef column; it's cleverly marked T. In this case T = 7.78. Compare this value to $t^* = 2.228$ from the $t$-table. Because $T > t^*$, you have strong evidence to reject Ho and conclude that the slope of the regression line for the textbook-weight data is not zero. (In fact, it has to be greater than that, according to your data.)

You can also just find the exact $p$-value on the output, right next to the T column, in the column is marked P. The $p$-value for the test for slope in this case is 0.000, which means it's less than 0.001. You conclude that the slope of this line is not zero, so textbook weight is significantly related to student weight. (See Chapter 3 to brush up on $p$-values.)

To test to see whether the slope is some value other than zero, just plug that value in for $b_0$ in the formula for the test statistic. Also, you may conduct one-sided hypothesis tests to see whether the slope is strictly greater than zero or strictly less than zero. In those cases, you find the same test statistic but compare it to the value $t^*$ where the area to the right (or left, respectively) is $\alpha$.

## Inspecting the y-intercept

The $y$-intercept is the place where the regression line $y = a + bx$ crosses the $y$-axis and is denoted by $a$ (see the earlier section "The $y$-intercept of the regression line"). Sometimes the $y$-intercept can be interpreted in a meaningful way, and sometimes not. This differs from slope, which is always interpretable. In fact, between the two elements of slope and intercept, the slope is the star of the show, with the $y$-intercept serving as the somewhat less famous but still noticeable sidekick.

There are times when the $y$-intercept makes no sense. For example, suppose you use rain to predict bushels per acre of corn; if you have zero rain, you have zero corn, but if the regression line crosses the $y$-axis somewhere else besides zero (and it most likely will), the $y$-intercept will make no sense.

Another situation is where no data were collected near the value of $x = 0$; interpreting the $y$-intercept at that point is not appropriate. For example, using a student's score on midterm 1 to predict her score on midterm 2, unless the student didn't take the exam at all (in which case it doesn't count), she'll get at least some points.

Many times, however, the $y$-intercept is of interest to you and has a value that you can interpret, such as when you're talking about predicting coffee sales using temperature for football games. Some games get cold enough to have zero and subzero temperatures (like Packers games for example — Go Pack Go!).

Suppose I collect data on ten of my students who recorded their study time (in minutes) for a 10-point quiz, along with their quiz scores. The data have a strong linear relationship by all the methods used in this chapter (for example, refer to the earlier section "Exploring Relationships with Scatterplots and Correlations"). I went ahead and conducted a regression analysis, and the results are shown in Figure 4-3.

Because there are students who (heaven forbid!) didn't study at all for the quiz, the $y$-intercept of 3.29 points (where study time $x = 0$) can be interpreted safely. Its value is shown in the Coef column in the row marked Constant (see the section "The $y$-intercept of the regression line" for more information). The next step is to give a confidence interval for the $y$-intercept of the regression line, where you can take conclusions beyond just this sample of ten students.

The formula for a $1 - \alpha$ level confidence interval for the $y$-intercept $(a)$ of a simple linear regression line is $a \pm t^*_{n-2}SE_a$. The standard error, $SE_a$, is equal to

$$SE_a = s\sqrt{\frac{1}{n} + \frac{\bar{x}^2}{\sum_i (x_i - \bar{x})^2}}, \text{ where } s = \sqrt{\frac{1}{n-2}\sum_i (y_i - \hat{y}_i)^2}, \text{ where again the value}$$

of $t^*$ comes from the $t$-distribution with $n - 2$ degrees of freedom whose area to the right is equal to $\alpha + 2$. Using the output from Figure 4-3 and the $t$-table, I'm 95 percent confident that the quiz score $(y)$ for someone with a study time of $x = 0$ minutes is $3.29 \pm 2.306 * 0.4864$, which is anywhere from 2.17 to 4.41, on average. Note that 2.306 comes from the $t$-table with $10 - 2 = 8$ degrees of freedom and 0.4864 is the SE for the $y$-intercept from Figure 4-3. (So studying for zero minutes for my quiz is not something to aspire to.)

By the way, to find out how much time studying affected the quiz score for these students, you can get an estimate of the slope on the output from Figure 4-3 that the coefficient for slope is 0.1793, which says each minute of studying is related to an increase in score of 0.1793 of a point, plus or minus the margin of error, of course. Or, 10 more minutes relates to 1.793 more points. On a 10-point quiz, it all adds up!

```
The regression equation is
quiz score = 3.29 + 0.179 minutes studying

Predictor              Coef    SE Coef        T        P

Constant             3.2931    0.4864      6.77    0.000

Minutes studying     0.17931   0.02103     8.53    0.000

S = 0.877153         R-Sq = 90.1%          R-Sq (adj) = 88.8%
```

**Figure 4-3:** Regression analysis for study time and quiz score data.

Testing a hypothesis about the $y$-intercept isn't really something you'll find yourself doing much because most of the time you don't have a preconceived notion about what the $y$-intercept would be (nor do you really care ahead of time). The confidence interval is much more useful. However, if you do need to conduct a hypothesis test for the $y$-intercept, you take your $y$-intercept, subtract the value in Ho, and divide by the standard error, found on the computer output in the row for Constant and the column for SE Coef. (The default value is to test to see whether the $y$-intercept is zero.) The test is in the T column of the output, and its $p$-value is shown in the P column. In the study time and quiz score example, the $p$-value is 0.000, so the $y$-intercept is significantly different from zero. All this means is that the line crosses the $y$-axis somewhere else.

## Building confidence intervals for the average response

When you have the slope and $y$-intercept for the best-fitting regression line, you put them together to get the line $y = a + bx$. The value of $y$ here really represents the average value of $y$ for a particular value of $x$. For example, in the textbook-weight data, Figure 4-2 shows the regression line $y = 3.69 + 0.11337x$ where $x$ = average student weight and $y$ = average textbook weight. If you put in 100 pounds for $x$, you get $y = 3.69 + 0.1137 * 100 = 15.02$ pounds of textbook weight for the group averaging 100 pounds. This number, 15.02, is an estimate of the average weight of textbooks for children of this weight.

But you can't stop there. Because you're getting an estimate of the average textbook weight using $y$, you also need a margin of error for $y$ to go with it, to create a confidence interval for the average $y$ at a given $x$ that generalizes to the population.

Take your estimate, $y$, which you get by plugging your given $x$ value into the regression line, and then add and subtract the margin of error for $y$. The formula for a $1 - \alpha$ confidence interval for the mean of $y$ for a given value of $x$ (call it $x^*$) is equal to $y \pm t^*_{n-2} SE_{\hat{\mu}}$, where $y$ is the value of the equation of the line when you plug in $x^*$ for $x$. The standard error for $y$ is equal to

$$SE_y = s\sqrt{1 + \frac{1}{n} + \frac{\left(x^* - \bar{x}\right)^2}{\sum_i \left(x_i - \bar{x}\right)^2}}, \text{ where } s = \sqrt{\frac{1}{n-2}\sum_i \left(y_i - \hat{y}_i\right)^2}.$$ Luckily Minitab does

these calculations for you and reports a confidence interval for the mean of $y$ for a given $x^*$.

To find a confidence interval for the mean value of $y$ using Minitab, you ask for a regression analysis (see instructions in the earlier section "Finding the best-fitting line to model your data") and click on Options. You see a box called Prediction Intervals for New Observations; enter the value of $x^*$ that you want, and just below that, put in your confidence level (the default is 95 percent). Even though this box is labeled Prediction Intervals, it finds both a confidence interval and a prediction interval for you as well. (Prediction intervals are different from confidence intervals, and I discuss them in the next section.) On the computer output, the confidence interval is labeled 95% CI.

Returning to the textbook-weight example, the computer output for finding a 95 percent confidence interval for the average textbook weight for 100-pound children is shown in Figure 4-4. The result is (14.015, 16.048) pounds. I'm 95 percent confident that the average textbook weight for the group of children averaging 100 pounds is between 14.015 and 16.048 pounds. (Get out those rolling backpacks, kids!)

You should only make predictions for the average value of $y$ for $x$ values that are within the range of where the data was collected. Failure to do so will result in the statistical no-no called *extrapolation* (see the later section "Knowing the Limitations of Your Regression Analysis").

## Making the band with prediction intervals

Suppose instead of the mean value of $y$, you want to take a guess at what $y$ would be for some future value of $x$. Because you're looking into the future, you have to make a prediction, and to do that, you need a range of likely values of $y$ for a given $x^*$. This is what statisticians call a *prediction interval*.

The formula for a $1 - \alpha$ level prediction interval for $y$ at a given value $x^*$ is

$$SE_y = s\sqrt{1 + \frac{1}{n} + \frac{\left(x^* - \bar{x}\right)^2}{\sum_i \left(x_i - \bar{x}\right)^2}}, \text{ where } s = \sqrt{\frac{1}{n-2}\sum_i \left(y_i - \hat{y}_i\right)^2}.$$ Again, Minitab easily

makes these calculations for you.

To find a 1 – α level prediction interval for the value of *y* for a given *x*\* using Minitab, you ask for a regression analysis (see instructions in the earlier section "Finding the best-fitting line to model your data") and click Options. In the box Prediction Intervals for New Observations, enter the value of *x* that you want, and just below that, put in your confidence level (the default is 95 percent). On the computer output, the prediction interval is labeled 95% PI, and it appears right next to the confidence interval for the mean of *y* for that same *x*\*.

## Predicting textbook weight using student weight

For the textbook-weight data, suppose you've already made your regression line and now a new student comes on the scene. You want to predict this student's textbook weight. This means you want a prediction interval rather than a confidence interval, because you want to predict the textbook weight for one person, not the average weight for a group.

Suppose this new student weighs 100 pounds. To find the prediction interval for the textbook weight for this student, you use *x*\* = 100 pounds and let Minitab do its thing.

The computer output in Figure 4-4 shows the 95 percent prediction interval for textbook weight for a single 100-pound child is (11.509, 18.533) pounds. Note this is wider than the confidence interval of (14.015, 16.048) for the mean textbook weight for 100-pound children found in the earlier section "Building confidence intervals for the average response." This difference is due to the increased variability in looking at one child and predicting one textbook weight.

**Figure 4-4:** Prediction interval of textbook weight for a 100-pound child.

```
Predicted Values for New Observations

New

Obs    Fit    SE Fit    95% CI           95% PI
1      15.031  0.456    (14.015, 16.048) (11.509, 18.553)
```

## Comparing prediction and confidence intervals

Note that the formulas for prediction intervals and confidence intervals are very similar. In fact, the prediction interval formula is exactly the same as the confidence interval formula except it adds a 1 under the square root. Because of this difference in the formulas, the margin of error for a prediction interval is larger than for a confidence interval.

This difference also makes sense from a statistical point. A prediction interval has more variability than a confidence interval because it's harder to make a prediction about $y$ for a single value of $x^*$ than it is to estimate the average value of $y$ for a given $x^*$. (For example, individual test scores vary more than average test scores do.) A prediction interval will be wider than a confidence interval; it will have a larger margin of error.

A similarity between prediction intervals and confidence intervals is that their margin of error formulas both contain $x^*$, which means the margin of error in either case depends on which value of $x^*$ you use. It turns out in both cases that if you use the mean value of $x$ as your $x^*$, the margin of error for each interval is at its smallest because there's more data around the mean of $x$ than at any other value. As you move away from the mean of $x$, the margin of error increases for each interval.

# Checking the Model's Fit (The Data, Not the Clothes!)

After you've established a relationship between $x$ and $y$ and have come up with an equation of a line that represents that relationship, you may think your job is done. (Many researchers erringly stop here, so I'm depending on you to break the cycle!) The most-important job remains to be completed: checking to be sure that the conditions of the model are truly met and that the model fits well in more specific ways than the scatterplot and correlation measure (which I cover in the earlier section "Exploring Relationships with Scatterplots and Correlations").

This section presents methods for defining and assessing the fit of a simple linear regression model.

## Defining the conditions

Two major conditions must be met before you apply a simple linear regression model to a data set:

- ✔ The $y$'s must have an approximately normal distribution for each value of $x$.

- ✔ The $y$'s must have a constant amount of spread (standard deviation) for each value of $x$.

### Normal y's for every x

For any value of $x$, the population of possible $y$-values must have a normal distribution. The mean of this distribution is the value for $y$ that's on the best-fitting line for that $x$-value. That is, some of your data fall above the best-fitting line, some data fall below the best fitting line, and a few may actually land right on the line.

If the regression model is fitting well, the data values should be scattered around the best-fitting line in such a way that about 68 percent of the values lie within one standard deviation of the line, about 95 percent of the values lie within two standard deviations of the line, and about 99.7 percent of the values lie within three standard deviations of the line. This specification, as you may recall from your Stats I course, is called the *68-95-99.7 rule,* and it applies to all bell-shaped data (for which the normal distribution applies).

You can see in Figure 4-5 how for each $x$-value, the $y$-values you may observe tend to be located near the best-fitting line in greater numbers, and as you move away from the line, you see fewer and fewer $y$-values, both above and below the line. More than that, they're scattered around the line in a way that reflects a bell-shaped curve, the normal distribution. This indicates a good fit.

Why does this condition makes sense? The data you collect on $y$ for any particular $x$-value vary from individual to individual; for example, not all students' textbooks weigh the same, even for students who weigh the exact same amount. But those values aren't allowed to vary any way they want to. To fit the conditions of a linear regression model, for each given value of $x$, the data should be scattered around the line according to a normal distribution. Most of the points should be close to the line, and as you get farther from the line, you can expect fewer data points to occur. So condition number one is that the data have a normal distribution for each value of $x$.

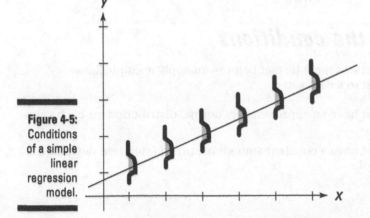

**Figure 4-5:**
Conditions of a simple linear regression model.

### Same spread for every x

In order to use the simple linear regression model, as you move from left to right on the x-axis, the spread in the y-values around the line should be the same, no matter which value of x you're looking at. This requirement is called the *homoscedasticity condition*. (How they came up with that mouthful of a word just for describing the fact that the standard deviations stay the same across the x-values, I'll never know.) This condition ensures that the best-fitting line works well for all relevant values of x, not just in certain areas.

You can see in Figure 4-5 that no matter what the value of x is, the spread in the y-values stays the same throughout. If the spread got bigger and bigger as x got larger and larger, for example, the line would lose its ability to fit well for those large values of x.

## Finding and exploring the residuals

To check to see whether the y-values come from a normal distribution, you need to measure how far off your predictions were from the actual data that came in. These differences are called *errors*, or *residuals*. To evaluate whether a model fits well, you need to check those errors and see how they stack up.

In a model-fitting context, the word *error* doesn't mean "mistake." It just means a difference between the data and the prediction based on the model. The word I like best to describe this difference is *residual*, however. It sounds more upbeat.

The following sections focus on finding a way to measure these residuals that the model makes. You also explore the residuals to identify particular problems that occurred in the process of trying to fit a straight line to the data. In other words, you can discover that looking at residuals helps you assess the fit of the model and diagnose problems that caused a bad fit, if that was the case.

### Finding the residuals

A *residual* is the difference between the observed value $\hat{y}$ of y (from the best-fitting line) and the predicted value of y, also known as y (from the data set). Its notation is $(y - \hat{y})$. Specifically, for any data point, you take its observed y-value (from the data) and subtract its expected y-value (from the line). If the residual is large, the line doesn't fit well in that spot. If the residual is small, the line fits well in that spot.

For example, suppose you have a point in your data set (2, 4) and the equation of the best-fitting line is $y = 2x + 1$. The expected value of y in this case is $(2 * 2) + 1 = 5$. The observed value of y from the data set is 4. Taking the observed value minus the estimated value, you get $4 - 5 = -1$. The residual for that particular data point (2, 4) is –1. If you observe a y-value of 6 and use the same straight line to estimate y, then the residual is $6 - 5 = +1$.

In general, a positive residual means you underestimated *y* at that point; the line is below the data. A negative residual means you overestimated *y* at that point; the line is above the data.

### Standardizing the residuals

Residuals in their raw form are in the same units as the original data, making them hard to judge out of context. To make interpreting the residuals easier, statisticians typically *standardize* them — that is, subtract the mean of the residuals (zero) and divide by the standard deviation of all the residuals. The residuals are a data set just like any other data set, so you can find their mean and standard deviation like you always do. Standardizing just means converting to a *Z*-score so that you see where it falls on the standard normal distribution. (See your Stats I text or *Statistics For Dummies* for info on *Z*-scores.)

### Making residual plots

You can plot the residuals on a graph called a *residual plot.* (If you've standardized the residuals, you call it a *standardized residual plot.*) Figure 4-6 shows the Minitab output for a variety of standardized residual plots, all getting at the same idea: checking to be sure the conditions of the simple linear regression model are met.

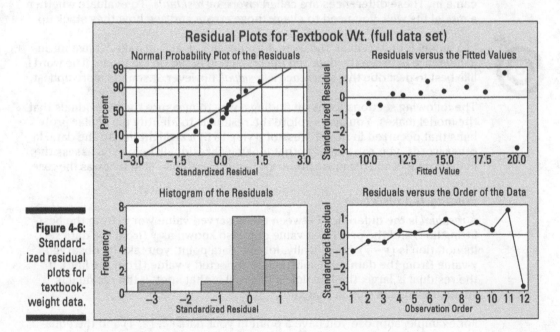

**Figure 4-6:** Standardized residual plots for textbook-weight data.

### Checking normality

If the condition of normality is met, you can see on the residual plot lots of (standardized) residuals close to zero; as you move farther away from zero, you can see fewer residuals. *Note:* You shouldn't expect to see a standardized residual at or beyond +3 or –3. If this occurs, you can consider that point an outlier, which warrants further investigation. (For more on outliers, see the section "Scoping for outliers" later in this chapter.)

The residuals should also occur at random — some above the line, and some below the line. If a pattern occurs in the residuals, the line may not be fitting right.

The plots in Figure 4-6 seem to have an issue with the very last observation, the one for 12th graders. In this observation, the average student weight (142) seemed to follow the pattern of increasing with each grade level, but the textbook weight (16.06) was less than for 11th graders (20.79) and is the first point to break the pattern.

You can also see in the plot in the upper-right corner of Figure 4-6 that the very last data value has a standardized residual that sticks out from the others and has a value of –3 (something that should be a very rare occurrence). So the value you expected for *y* based on your line was off by a factor of 3 standard deviations. And because this residual is negative, what you observed for *y* was much lower than you may have expected it to be using the regression line.

The other residuals seem to fall in line with a normal distribution, as you can see in the upper-right plot of Figure 4-6. The residuals concentrate around zero, with fewer appearing as you move farther away from zero. You can also see this pattern in the upper-left plot of Figure 4-6, which shows how close to normal the residuals are. The line in this graph represents the equal-to-normal line. If the residuals follow close to the line, then normality is okay. If not, you have problems (in a statistical sense, of course). You can see the residual with the highest magnitude is –3, and that number falls outside the line quite a bit.

The lower-left plot in Figure 4-6 makes a histogram of the standardized residuals, and you can see it doesn't look much like a bell-shaped distribution. It doesn't even look *symmetric* (the same on each side when you cut it down the middle). The problem again seems to be the residual of –3, which skews the histogram to the left.

The lower-right plot of Figure 4-6 plots the residuals in the order presented in the data set in Table 4-1. Because the data was ordered already, the lower-right residual plot looks like the upper-right residual plot in Figure 4-6, except the dots are connected. This lower-right residual plot makes the residual of –3 stand out even more.

### Checking the spread of the y's for each x

The graph in the upper-right corner of Figure 4-6 also addresses the homoscedasticity condition. If the condition is met, then the residuals for every *x*-value have about the same spread. If you cut a vertical line down through each *x*-value, the residuals have about the same spread (standard deviation) each time, except for the last *x*-value, which again represents grade 12. That means the condition of equal spread in the *y*-values is met for the textbook-weight example.

If you look at only one residual plot, choose the one in the upper-right corner of Figure 4-6, the plot of the fitted values (the values of $y$ on the line) versus the standardized residuals. Most problems with model fit will show up on that plot because a residual is defined as the difference between the observed value of $y$ and the fitted value of $y$. In a perfect world, all the fitted values have no residual at all; a large residual (such as the one where the estimated textbook weight is 20 pounds for students averaging 142 pounds; see Figure 4-1) is indicated by a point far off from zero. This graph also shows you deviations from the overall pattern of the line; for example, if large residuals are on the extremes of this graph (very low or very high fitted values), the line isn't fitting in those areas. On balance, you can say this line fits well at least for grades 1 through 11.

# Using $r^2$ to measure model fit

One important way to assess how well the model fits is to use a statistic called the *coefficient of determination*, or $r^2$. This statistic takes the value of the correlation, $r$, and squares it to give you a percentage. You interpret $r^2$ as the percentage of variability in the $y$ variable that's explained by, or due to, its relationship with the $x$ variable.

The $y$-values of the data you collect have a great deal of variability in and of themselves. You look for another variable $(x)$ that helps you explain that variability in the $y$-values. After you put that $x$ variable into the model and find that it's highly correlated with $y$, you want to find out how well this model did at explaining why the values of $y$ are different.

Note that you have to interpret $r^2$ using different standards than those for interpreting $r$. Because squaring a number between $-1$ and $+1$ results in a smaller number (except for $+1$, $-1$, and 0, which stay the same or switch signs), an $r^2$ of 0.49 isn't too bad, because it's the square of $r = 0.7$, which is a fairly strong correlation.

The following are some general guidelines for interpreting the value of $r^2$:

- ✓ If the model containing $x$ explains a lot of the variability in the $y$-values, then $r^2$ is high (in the 80 to 90 percent range is considered to be extremely high). Values like 0.70 are still considered fairly high. A high percentage of variability means that the line fits well because there's not much left to explain about the value of $y$ other than using $x$ and its relationship to $y$. So a larger value of $r^2$ is a good thing.

- ✓ If the model containing $x$ doesn't help much in explaining the difference in the $y$-values, then the value of $r^2$ is small (closer to zero; between, say, 0.00 and 0.30 roughly). The model, in this case, wouldn't fit well. You need another variable to explain $y$ other than the one you already tried.

- ✓ Values of $r^2$ that fall in the middle (between, say, 0.30 and 0.70) mean that $x$ does help somewhat in explaining $y$, but it doesn't do the job well enough on its own. In this case, statisticians would try to add one or more variables to the model to help explain $y$ more fully as a group (read more about this in Chapter 5).

For the textbook-weight example, the value of $r$ (the correlation coefficient) is 0.93. Squaring this result, you get $r^2 = 0.8649$. That number means approximately 86 percent of the variability you find in average textbook weights for all students ($y$-values) is explained by the average student weight ($x$-values). This percentage tells you that the model of using year in school to estimate backpack weight is a good bet.

In the case of simple linear regression, you have only one $x$ variable, but in Chapter 5, you can see models that contain more than one $x$ variable. In that situation, you use $r^2$ to help sort out the contribution that those $x$ variables as a group bring to the model.

## Scoping for outliers

Sometimes life isn't perfect (oh really?), and you may find a residual in your otherwise tidy data set that totally sticks out. It's called an *outlier,* and it has a standardized value at or beyond +3 or –3. It threatens to blow the conditions of your regression model and send you crying to your professor.

Before you panic, the best thing to do is to examine that outlier more closely. First, can you find an error in that data value? Did someone report her age as 642, for instance? (After all, mistakes do happen.) If you do find a certifiable error in your data set, you remove that data point (or fix it if possible) and analyze the data without it. However, if you can't explain away the problem by finding a mistake, you must think of another approach.

If you can't find a mistake that caused the outlier, you don't necessarily have to trash your model; after all, it's only one data point. Analyze the data with that data point, and analyze the data again without it. Then report and compare both analyses. This comparison gives you a sense of how influential that one data point is, and it may lead other researchers to conduct more research to zoom in on the issue you brought to the surface.

In Figure 4-1, you can see the scatterplot of the full data set for the textbook-weight example. Figure 4-7 shows the scatterplot for the data set minus the outlier. The scatterplot fits the data better without the outlier. The correlation increases to 0.993, and the value of $r^2$ increases to 0.986. The equation for the regression line for this data set is $y = 1.78 + 0.139x$.

The slope of the regression line doesn't change much by removing the outlier (compare it to Figure 4-2, where the slope is 0.113). However, the $y$-intercept changes: It's now 1.78 without the outlier compared to 3.69 with the outlier. The slopes of the lines are about the same, but the lines cross the $y$-axis in different places. It appears that the outlier (the last point in the data set) has quite an effect on the best-fitting line.

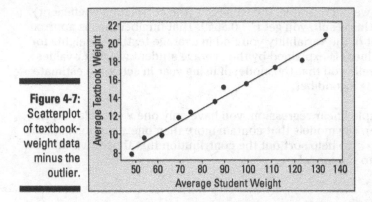

**Figure 4-7:**
Scatterplot of textbook-weight data minus the outlier.

Figure 4-8 shows the residual plots for the regression line for the data set without the outlier. Each of these plots shows a much better fit of the data to the model compared to Figure 4-6. This result tells you that the data for grade 12 is influential in this data set and that the outlier needs to be noted and perhaps explored further. Do students peak when they're juniors in high school? Or do they just decide when they're seniors that it isn't cool to carry books around? (A statistician's job isn't to wonder why, but to do and analyze.)

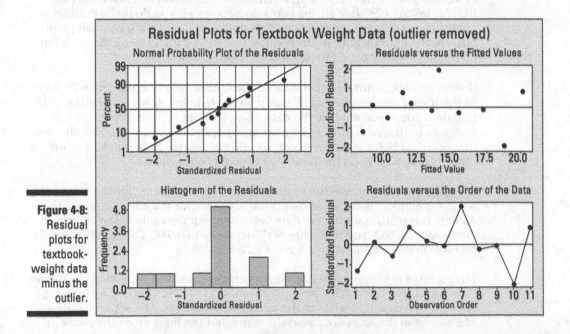

**Figure 4-8:**
Residual plots for textbook-weight data minus the outlier.

# Knowing the Limitations of Your Regression Analysis

The bottom line of any data analysis is to make the correct conclusions given your results. When you're working with a simple linear regression model, there's the potential to make three major errors. This section shows you those errors and tells you how to avoid them.

## Avoiding slipping into cause-and-effect mode

In a simple linear regression, you investigate whether $x$ is related to $y$, and if you get a strong correlation and a scatterplot that shows a linear trend, then you find the best-fitting line and use it to estimate the value of $y$ for reasonable values of $x$.

There's a fine line, however (no pun intended), that you don't want to cross with your interpretation of regression results. Be careful to not automatically interpret slope in a cause-and-effect mode when you're using the regression line to estimate the value of $y$ using $x$. Doing so can result in a leap of faith that can send you into the frying pan. Unless you have used a controlled experiment to get the data, you can only assume that the variables are correlated; you can't really give a stone-cold guarantee about why they're related.

In the textbook-weight example, you estimate the average weight of the students' textbooks by using the students' average weight, but that doesn't mean increasing a particular child's weight causes his textbook weight to increase. For example, because of the strong positive correlation, you do know that students with lower weights are associated with lower total textbook weights, and students with higher weights tend to have higher textbook weights. But you can't take one particular third-grade student, increase his weight, and presto — suddenly his textbooks weigh more.

The variable underlying the relationship between a child's weight and the weight of his backpack is the grade level of the student from an academic standpoint; as grade level increases, so might the size and number of his books, as well as the homework coming home. Student grade level drives both student weight and textbook weight. In this situation, student grade level is what statisticians call a *lurking variable;* it's a variable that wasn't included in the model but is related to both the outcome and the response. A lurking variable confuses the issue of what's causing what to happen.

If the collected data was the result of a well-designed experiment that controls for possible confounding variables, you can establish a cause-and-effect relationship between $x$ and $y$ if they're strongly correlated. Otherwise, you can't establish such a relationship. (See your Stats I text or *Statistics For Dummies* for info regarding experiments.)

## Extrapolation: The ultimate no-no

Plugging values of $x$ into the model that fall outside of the reasonable boundaries of $x$ is called *extrapolation*. And one of my colleagues sums up this idea very well: "Friends don't let friends extrapolate."

When you determine a best-fitting line for your data, you come up with an equation that allows you to plug in a value for $x$ and get a predicted value for $y$. In algebra, if you find the equation of a line and graph it, the line typically has an arrow on each end indicating it goes on forever in either direction. But that doesn't work for statistical problems (because statistics represents the *real* world). When you're dealing with real-world units like height, weight, IQ, GPA, house prices, and the weight of your statistics textbook, only certain numbers make sense.

So the first point is, don't plug in values for $x$ that don't make any sense. For example, if you're estimating the price of a house $(y)$ by using its square footage $(x)$, you wouldn't think of plugging in a value of $x$ like 10 square feet or 100 square feet, because houses simply aren't that small.

You also wouldn't think about plugging in values like 1,000,000 square feet for $x$ (unless your "house" is the Ohio State football stadium or something). It wouldn't make sense. Likewise, if you're estimating tomorrow's temperature using today's temperature, negative numbers for $x$ could possibly make sense, but if you're estimating the amount of precipitation tomorrow given the amount of precipitation today, negative numbers for $x$ (or $y$ for that matter) don't make sense.

Choose only reasonable values of $x$ for which you try to make estimates about $y$ — that is, look at the values of $x$ for which your data was collected, and stay within those bounds when making predictions. In the textbook-weight example, the smallest average student weight is 48.5 pounds, and the largest average student weight is 142 pounds. Choosing student weights between 48.5 and 142 to plug in for $x$ in the equation is okay, but choosing values less than 48.5 or more than 142 isn't a good idea. You can't guarantee that the same linear relationship (or any linear relationship for that matter) continues outside the given boundaries.

Think about it: If the relationship you found actually continued for any value of $x$, no matter how large, then a 250-pound lineman from OSU would have to carry $3.69 + 0.113 * 250 = 31.94$ pounds of books around in his backpack. Of course this would be easy for him, but what about the rest of us?

## Sometimes you need more than one variable

A simple linear regression model is just what it says it is: simple. I don't mean easy to work with, necessarily, but simple in the uncluttered sense. The model tries to estimate the value of $y$ by only using one variable, $x$. However, the number of real-world situations that can be explained by using a simple, one-variable linear regression is small. Often one variable just can't do all the predicting.

If one variable alone doesn't result in a model that fits well enough, you can try to add more variables. It may take many variables to make a good estimate for $y$, and you have to be careful in how you choose them. In the case of stock market prices, for example, they're still looking for that ultimate prediction model.

As another example, health insurance companies try to estimate how long you'll live by asking you a series of questions (each of which represents a variable in the regression model). You can't find one single variable that estimates how long you'll live; you must consider many factors: your health, your weight, whether or not you smoke, genetic factors, how much exercise you do each week, and the list goes on and on and on.

The point is that regression models don't always use just one variable, $x$, to estimate $y$. Some models use two, three, or even more variables to estimate $y$. Those models aren't called simple linear regression models; they're called *multiple linear regression models* because of their employment of multiple variables to make an estimate. (You explore multiple linear regression models in Chapter 5.)

# Multiple Regression with Two X Variables

• • • • • • • • • • • • • • • • • • • • • • • • • • • • • • • • • • • • • • • • • • • • •

## In This Chapter

▶ Getting the basic ideas behind a multiple regression model

▶ Finding, interpreting, and testing coefficients

▶ Checking model fit

• • • • • • • • • • • • • • • • • • • • • • • • • • • • • • • • • • • • • • • • • • • • • •

The idea of regression is to build a model that estimates or predicts one quantitative variable *(y)* by using at least one other quantitative variable *(x)*. Simple linear regression uses exactly one *x* variable to estimate the *y* variable. (See Chapter 4 for all the information you need on simple linear regression.) *Multiple linear regression,* on the other hand, uses more than one *x* variable to estimate the value of *y*.

In this chapter, you see how multiple regression works and how to apply it to build a model for *y*. You see all the steps necessary for the process, including determining which *x* variables to include, estimating their contributions to the model, finding the best model, using the model for estimating *y*, and assessing the fit of the model. It may seem like a mountain of information, but you won't regress on the topic of regression if you take this chapter one step at a time.

## *Getting to Know the Multiple Regression Model*

Before you jump right into using the multiple regression model, get a feel for what it's all about. In this section, you see the usefulness of multiple regression as well as the basic elements of the multiple regression model. Some of the ideas are just an extension of the simple linear regression model (see Chapter 4). Some of the concepts are a little more complex, as you may guess because the model is more complex. But the concepts and the results should make intuitive sense, which is always good news.

## Discovering the uses of multiple regression

One situation in which multiple regression is useful is when the *y* variable is hard to track down — that is, its value can't be measured straight up, and you need more than one other piece of information to help get a handle on what its value will be. For example, you may want to estimate the price of gold today. It would be hard to imagine being able to do that with only one other variable. You may base your estimate on recent gold prices, the price of other commodities on the market that move with or against gold, and a host of other possible economic conditions associated with the price of gold.

Another case for using multiple regression is when you want to figure out what factors play a role in determining the value of *y*. For example, you want to find out what information is important to real estate agents in setting a price for a house going on the market.

## Looking at the general form of the multiple regression model

The general idea of simple linear regression is to fit the best straight line through that data that you possibly can and use that line to make estimates for *y* based on certain *x*-values. The equation of the best-fitting line in simple linear regression is $y = b_0 + b_1x_1$, where $b_0$ is the *y*-intercept and $b_1$ is the slope. (The equation also has the form $y = a + bx$; see Chapter 4.)

In the multiple regression setting, you have more than one *x* variable that's related to *y*. Call these *x* variables $x_1, x_2, \ldots x_k$. In the most basic multiple regression model, you use some or all of these *x* variables to estimate *y* where each *x* variable is taken to the first power. This process is called finding the best-fitting linear function for the data. This linear function looks like the following: $y = b_0 + b_1x_1 + b_2x_2 + \ldots + b_kx_k$, and you can call it the *multiple (linear) regression model*. You use this model to make estimates about *y* based on given values of the *x* variables.

A *linear* function is an equation whose *x* terms are taken to the first power only. For example $y = 2x_1 + 3x_2 + 24x_3$ is a linear equation using three *x* variables. If any of the *x* terms are squared, the function is a *quadratic* one; if an *x* term is taken to the third power, the function is a *cubic* function, and so on. In this chapter, I consider only linear functions.

## Stepping through the analysis

Your job in conducting a multiple regression analysis is to do the following (the computer can help you do steps three through six):

1. **Come up with a list of possible $x$ variables that may be helpful in estimating $y$.**

2. **Collect data on the $y$ variable and your $x$ variables from step one.**

3. **Check the relationships between each $x$ variable and $y$ (using scatterplots and correlations), and use the results to eliminate those $x$ variables that aren't strongly related to $y$.**

4. **Look at possible relationships between the $x$ variables to make sure you aren't being redundant (in statistical terms, you're trying to avoid the problem of multicolinearity).**

   If two $x$ variables relate to $y$ the same way, you don't need both in the model.

5. **Use those $x$ variables (from step four) in a multiple regression analysis to find the best-fitting model for your data.**

6. **Use the best-fitting model (from step five) to predict $y$ for given $x$-values by plugging those $x$-values into the model.**

I outline each of these steps in the sections to follow.

# Looking at x's and y's

The first step of a multiple regression analysis comes way before the number crunching on the computer; it occurs even before the data is collected. Step one is where you sit down and think about what variables may be useful in predicting your response variable $y$. This step will likely take more time than any other step, except maybe the data-collection process. Deciding which $x$ variables may be candidates for consideration in your model is a deal-breaking step, because you can't go back and collect more data after the analysis is over.

Always check to be sure that your response variable, $y$, and at least one of the $x$ variables are quantitative. For example, if $y$ isn't quantitative but at least one $x$ is, a logistic regression model may be in order (see Chapter 8).

Suppose you're in the marketing department for a major national company that sells plasma TVs. You want to sell as many TVs as you can, so you want to figure out which factors play a role in plasma TV sales. In talking with your advertising people and remembering what you learned in those business classes in college, you know that one powerful way to get sales is through advertising. You think of the types of advertising that may be related to sales of plasma TVs and your team comes up with two ideas:

- ✔ **TV ads:** Of course, how better to sell a TV than through a TV ad?
- ✔ **Newspaper sales:** Hit 'em on Sunday when they're reading the paper before watching the game through squinty eyes that are missing all the good plays and the terrible calls the referees are making.

By coming up with a list of possible $x$ variables to predict $y$, you have just completed step one of a multiple regression analysis, according to the list in the previous section. Note that all three variables I use in the TV example are quantitative (the TV ad and newspaper sales variables and the TV sales response variable), which means you can go ahead and think about a multiple regression model by using the two types of ads to predict TV sales.

# Collecting the Data

Step two in the multiple regression analysis process is to collect the data for your $x$ and $y$ variables. To do this, make sure that for each individual in the data set, you collect all the data for that individual at the same time (including the $y$-value and all $x$-values) and keep the data all together for each individual, preserving any relationships that may exist between the variables. You must then enter the data into a table format by using Minitab or any other software package (each column represents a variable and each row represents all the data from a single individual) to get a glimpse of the data and to organize it for later analyses.

To continue with the TV sales example from the preceding section, suppose that you start thinking about all the reams of data you have available to you regarding the plasma TV industry. You remember working with the advertising department before to do a media blitz by using, among other things, TV and newspaper ads. So you have data on these variables from a variety of store locations. You take a sample of 22 store locations in different parts of the country and put together the data on how much money was spent on each type of advertising, along with the plasma TV sales for that location. You can see the data in Table 5-1.

| Table 5-1 | Advertising Dollars and Sales of Plasma TVs | | |
|-----------|------------------|------------------|------------------|
| Location | Sales (In Millions of Dollars) | TV Ads (In Thousands of Dollars) | Newspaper Ads (In Thousands of Dollars) |
| 1 | 9.73 | 0 | 20 |
| 2 | 11.19 | 0 | 20 |
| 3 | 8.75 | 5 | 5 |
| 4 | 6.25 | 5 | 5 |
| 5 | 9.10 | 10 | 10 |
| 6 | 9.71 | 10 | 10 |
| 7 | 9.31 | 15 | 15 |
| 8 | 11.77 | 15 | 15 |
| 9 | 8.82 | 20 | 5 |
| 10 | 9.82 | 20 | 5 |
| 11 | 16.28 | 25 | 25 |
| 12 | 15.77 | 25 | 25 |
| 13 | 10.44 | 30 | 0 |
| 14 | 9.14 | 30 | 0 |
| 15 | 13.29 | 35 | 5 |
| 16 | 13.30 | 35 | 5 |
| 17 | 14.05 | 40 | 10 |
| 18 | 14.36 | 40 | 10 |
| 19 | 15.21 | 45 | 15 |
| 20 | 17.41 | 45 | 15 |
| 21 | 18.66 | 50 | 20 |
| 22 | 17.17 | 50 | 20 |

In reviewing this data, the question is whether the amount of money spent on these two forms of advertising can do a good job of estimating sales (in other words, are the ads worth the money?). And if so, do you need to include spending for both types of ads to estimate sales, or is one of them enough? Looking at the numbers in Table 5-1, you can see that higher sales may be related at least to higher amounts spent on TV advertising; the situation with newspaper advertising may not be so clear. So will the final multiple regression model contain both *x* variables or only one? In the following sections, you can find out.

# Pinpointing Possible Relationships

The third step in doing a multiple regression analysis (see the list in the "Stepping through the analysis" section) is to find out which (if any) of your possible *x* variables are actually related to *y*. If an *x* variable has no relationship with *y*, including it in the model is pointless. Data analysts use a combination of scatterplots and correlations to examine relationships between pairs of variables (as you can see in Chapter 4). Although you can view these two techniques under the heading of looking for relationships, I walk you through each one separately in the following sections to discuss their nuances.

## Making scatterplots

You make scatterplots in multiple linear regression to get a handle on whether your possible *x* variables are even related to the *y* variable you're studying. To investigate these possible relationships, you make one scatterplot of each *x* variable with the response variable *y*. If you have *k* different *x* variables being considered for the final model, you make *k* different scatterplots.

To make a scatterplot in Minitab, enter your data in columns, where each column represents a variable and each row represents all the data from one individual. Go to Graph>Scatterplots>Simple. Select your *y* variable on the left-hand side, and click Select. That variable appears in the *y*-variable box on the right-hand side. Then select your *x* variable on the left-hand side, and click Select. That variable appears in the *x*-variable box on the right-hand side. Click OK.

Scatterplots of TV ad spending versus TV sales and newspaper ad spending versus TV sales are shown in Figure 5-1.

**Figure 5-1:** Scatterplots of TV and newspaper ad spending versus plasma TV sales.

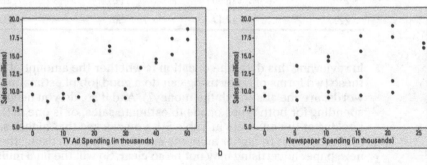

You can see from Figure 5-1a that TV spending does appear to have a fairly strong linear relationship with sales. This observation provides evidence that TV ad spending may be useful in estimating plasma TV sales. Figure 5-1b shows a linear relationship between newspaper ad spending and sales, but the relationship isn't as strong as the one between TV ads and sales. However, it still may be somewhat helpful in estimating sales.

# Correlations: Examining the bond

The second portion of step three involves calculating and examining the correlations between the $x$ variables and the $y$ variable. (Of course, if a scatterplot of an $x$ variable and the $y$ variable fails to come up with a pattern, then you drop that $x$ variable altogether and don't proceed to find the correlation.)

Whenever you employ scatterplots to explore possible linear relationships, correlations are typically not far behind. The *correlation coefficient* is a number that measures the strength and direction of the linear relationship between two variables, $x$ and $y$. (See Chapter 4 for the lowdown on correlation.)

This step involves two parts:

✔ Finding and interpreting the correlations

✔ Testing the correlations to see which ones are statistically significant (thereby determining which $x$ variables are significantly related to $y$)

### Finding and interpreting correlations

You can calculate a set of all possible correlations between all pairs of variables — which is called a *correlation matrix* — in Minitab. You can see the correlation matrix output for the TV data from Table 5-1 in Figure 5-2. Note the correlations between the $y$ variable (sales) and each $x$ variable, as well as the correlation between TV ads and newspaper ads.

**Figure 5-2:**
Correlation values and *p*-values for the TV sales example.

```
Correlations: Sales, TV, Newspaper

              Sales   TV
TV            0.791
              0.000

Newspaper     0.594 0.058
              0.004 0.799
```

Minitab can find a correlation matrix between any pairs of variables in the model, including the *y* variable and all the *x* variables as well. To calculate a correlation matrix for a group of variables in Minitab, first enter your data in columns (one for each variable). Then go to Stat>Basic Statistics>Descriptive Statistics>Correlation. Highlight the variables from the left-hand side for which you want correlations, and click Select.

To find the values of the correlation matrix from the computer output, intersect the row and column variables for which you want to find the correlation, and the top number in that intersection is the correlation of those two variables. For example, the correlation between TV ads and TV sales is 0.791, because it intersects the TV row with the Sales column in the correlation matrix in Figure 5-2.

### Testing correlations for significance

By the rule-of-thumb approach from Stats I (also reviewed in Chapter 4), a correlation that's close to 1 or –1 (starting around ± 0.75) is strong; a correlation close to 0 is very weak/nonexistent; and around ± 0.6 to 0.7, the relationships become moderately strong. The correlation between TV ads and TV sales of 0.791 indicates a fairly strong positive linear relationship between these two variables, based on the rule-of-thumb. The correlation between newspaper ads and TV sales seen in Figure 5-2 is 0.594, which is moderate by my rule-of-thumb.

Many times in statistics a rule-of-thumb approach to interpreting a correlation coefficient is sufficient. However, you're in the big leagues now, so you need a more precise tool for determining whether or not a correlation coefficient is large enough to be statistically significant. That's the real test of any statistic: not that the relationship is fairly strong or moderately strong in the sample, but whether or not the relationship can be generalized to the population.

Now, that phrase *statistically significant* should ring a bell. It's your old friend the hypothesis test calling to you (see Chapter 3 for a brush-up on hypothesis testing). Just like a hypothesis test for the mean of a population or the difference in the means of two populations, you also have a test for the correlation between two variables within a population.

The null hypothesis to test a correlation is Ho: $\rho = 0$ (no relationship) versus Ha: $\rho \neq 0$ (a relationship exists). The letter $\rho$ is the Greek version of *r* and represents the true correlation of *x* and *y* in the entire population; *r* is the correlation coefficient of the sample.

✔ **If you can't reject Ho based on your data,** you can't conclude that the correlation between *x* and *y* differs from zero, indicating you don't have evidence that the two variables are related and *x* shouldn't be in the multiple regression model.

> ✔ **If you can reject Ho based on your data,** you conclude that the correlation isn't equal to zero, so the variables are related. More than that, their relationship is deemed to be statistically significant — that is, the relationship would occur very rarely in your sample just by chance.

Any statistical software package can calculate a hypothesis test of a correlation for you. The actual formulas used in that process are beyond the scope of this book. However the interpretation is the same as for any test: If the *p*-value is smaller than your predetermined value of α (typically 0.05), reject Ho and conclude *x* and *y* are related. Otherwise you can't reject Ho, and you conclude you don't have enough evidence to indicate that the variables are related.

In Minitab, you can conduct a hypothesis test for a correlation by clicking on Stat>Basic Statistics>Correlation, and checking the Display *p*-values box. Choose the variables you want to find correlations for, and click Select. You'll get output in the form of a little table that shows the correlations between the variables for each pair with the respective *p*-values under each one. You can see the correlation output for the ads and sales example in Figure 5-2.

Looking at Figure 5-2, the correlation of 0.791 between TV ads and sales has a *p*-value of 0.000, which means it's actually less than 0.001. That's a highly significant result, much less than 0.05 (your predetermined α level). So TV ad spending is strongly related to sales. The correlation between newspaper ad spending and sales was 0.594, which is also found to be statistically significant with a *p*-value of 0.004.

# Checking for Multicolinearity

You have one more very important step to complete in the relationship-exploration process before going on to using the multiple regression model. You need to complete step four: looking at the relationship between the *x* variables themselves and checking for redundancy. Failure to do so can lead to problems during the model-fitting process.

*Multicolinearity* is a term you use if two *x* variables are highly correlated. Not only is it redundant to include both related variables in the multiple regression model, but it's also problematic. The bottom line is this: If two *x* variables are significantly correlated, only include one of them in the regression model, not both. If you include both, the computer won't know what numbers to give as coefficients for each of the two variables because they share their contribution to determining the value of *y*. Multicolinearity can really mess up the model-fitting process and give answers that are inconsistent and often not repeatable in subsequent studies.

To head off the problem of multicolinearity, along with the correlations you examine regarding each $x$ variable and the response variable $y$, also find the correlations between all pairs of $x$ variables. If two $x$ variables are highly correlated, don't leave them both in the model, or multicolinearity will result. To see the correlations between all the $x$ variables, have Minitab calculate a correlation matrix of all the variables (see the section "Finding and interpreting correlations"). You can ignore the correlations between the $y$ variable and the $x$ variables and only choose the correlations between the $x$ variables shown in the correlation matrix. Find those correlations by intersecting the rows and columns of the $x$ variables for which you want correlations.

**REMEMBER**

If two $x$ variables $x_1$ and $x_2$ are strongly correlated (that is, their correlation is beyond +0.7 or –0.7), then one of them would do just about as good a job of estimating $y$ as the other, so you don't need to include them both in the model. If $x_1$ and $x_2$ aren't strongly correlated, then both of them working together would do a better job of estimating sales than either variable alone.

For the ad-spending example, you have to examine the correlation between the two $x$ variables, TV ad spending and newspaper ad spending, to be sure no multicolinearity is present. The correlation between these two variables (as you can see in Figure 5-2) is only 0.058. You don't even need a hypothesis test to tell you whether or not these two variables are related; they're clearly not.

The $p$-value for the correlation between the spending for the two ad types is 0.799 (see Figure 5-2), which is much, much larger than 0.05 ever thought of being and therefore isn't statistically significant. The large $p$-value for the correlation between spending for the two ad types confirms your thoughts that both variables together may be helpful in estimating $y$ because each makes its own contribution. It also tells you that keeping them both in the model won't create any multicolinearity problems. (This completes step four of the multiple regression analysis, as listed in the "Stepping through the analysis" section.)

# Finding the Best-Fitting Model for Two x Variables

After you have a group of $x$ variables that are all related to $y$ and not related to each other (refer to previous sections), you're ready to perform step five of the multiple regression analysis (as listed in the "Stepping through the analysis" section). You're ready to find the best-fitting model for the data.

In the multiple regression model with two $x$ variables, you have the general equation $y = b_0 + b_1x_1 + b_2x_2$, and you already know which $x$ variables to include in the model (by doing step four in the previous section); the task now is to figure out which coefficients (numbers) to put in for $b_0$, $b_1$, and $b_2$, so you can use the resulting equation to estimate *y*. This specific model is the *best-fitting multiple linear regression model.* This section tells you how to get, interpret, and test those coefficients in order to complete step five in the multiple regression analysis.

Finding the best-fitting linear equation is like finding the best-fitting line in simple linear regression, except that you're not finding a line. When you have two *x* variables in multiple regression, for example, you're estimating a best-fitting plane for the data.

## *Getting the multiple regression coefficients*

In the simple linear regression model, you have the straight line $y = b_0 + b_1 x$; the coefficient of *x* is the slope, and it represents the change in *y* per unit change in *x*. In a multiple linear regression model, the coefficients $b_1$, $b_2$, and so on quantify in a similar matter the sole contribution that each corresponding *x* variable ($x_1$, $x_2$) makes in predicting *y*. The coefficient $b_0$ indicates the amount by which to adjust all these values in order to provide a final fit to the data (like the *y*-intercept does in simple linear regression).

Computer software does all the nitty-gritty work for you to find the proper coefficients ($b_0$, $b_1$, and so on) that fit the data best. The coefficients that Minitab settles on to create the best-fitting model are the ones that, as a group, minimize the sum of the squared residuals (sort of like the variance in the data around the selected model). The equations for finding these coefficients by hand are too unwieldy to include in this book; a computer can do all the work for you. The results appear in the regression output in Minitab. You can find the multiple regression coefficients ($b_0$, $b_1$, $b_2$, . . . , $b_k$) on the computer output under the column labeled COEF.

To run a multiple regression analysis in Minitab, click on Stat>Regression> Regression. Then choose the response variable *(y)* and click on Select. Then choose your predictor variables (*x* variables), and click Select. Click on OK, and the computer will carry out the analysis.

For the plasma TV sales example from the previous sections, Figure 5-3 shows the multiple regression coefficients in the COEF column for the multiple regression model. The first coefficient (5.257) is just the constant term (or $b_0$ term) in the model and isn't affiliated with any *x* variable. This constant just sort of goes along for the ride in the analysis; it's the number that you tack on the end to make the numbers work out right. The second coefficient in the COEF column is 0.162; this value is the coefficient of the $x_1$ (TV ad amount) term, also known as $b_1$. The third coefficient in the COEF column is 0.249, which is the value for $b_2$ in the multiple regression model and is the coefficient that goes with $x_2$ (newspaper ad amount).

**Figure 5-3:**
Regression
output for
the ads
and plasma
TV sales
example.

```
The regression equation is
Sales = 5.267 + 0.162 TV ads + 0.249 Newsp ads

Predictor       Coef  SE Coef      T      P
Constant      5.2574   0.4984  10.55  0.000
TV ads       0.16211  0.01319  12.29  0.000
Newsp ads    0.24887  0.02792   8.91  0.000

S = 0.976613   R-Sq = 92.8%   R-Sq(adj) = 92.0%
```

Putting these coefficients into the multiple regression equation, you see the regression equation is Sales = 5.267 + 0.162 (TV ads) + 0.249 (Newspaper ads), where sales are in millions of dollars and ad spending is in thousands of dollars.

So you have your coefficients (no sweat, right?), but where do you go from here? What does it all mean? The next section guides you through interpretation.

## *Interpreting the coefficients*

In simple linear regression (covered in Chapter 4), the coefficients represent the slope and $y$-intercept of the best-fitting line and are straightforward to interpret. The slope in particular represents the change in $y$ due to a one-unit increase in $x$ because you can write any slope as a number over one (and *slope* is rise over run).

In the multiple regression model, the interpretation's a little more complicated. Due to all the mathematical underpinnings of the model and how it's finalized (believe me, you don't want to go there unless you're looking for a PhD in statistics), the coefficients have a different meaning.

The coefficient of an $x$ variable in a multiple regression model is the amount by which $y$ changes if that $x$ variable increases by one unit and the values of all other $x$ variables in the model *don't change*. So basically, you're looking at the marginal contribution of each $x$ variable when you hold the other variables in the model constant.

In the ads and sales regression analysis (see Figure 5-3), the coefficient of $x_1$ (TV ad spending) equals 0.16211. So $y$ (plasma TV sales) increases by 0.16211 million dollars when TV ad spending increases by 1.0 thousand dollars and spending on newspaper ads doesn't change. (Note that keeping more digits after the decimal point reduces rounding error when in units of millions.)

You can more easily interpret the number "0.16211 million dollars" by converting it to a dollar amount without the decimal point: $0.16211 million is equal to $162,110. (To get this value, I just multiplied $0.16211 by 1,000,000.) So plasma TV sales increase by $162,110 for each $1,000 increase in TV ad spending and newspaper ad spending remains the same. Similarly, the coefficient of $x_2$ (newspaper ad spending) equals 0.24887. So plasma TV sales increase by 0.24887 million dollars (or $248,870) when newspaper ad spending increases by $1,000 and TV ad spending remains the same.

Don't forget the units of each variable in a multiple regression analysis. This mistake is one of the most common in Stats II. If you were to forget about units in the ads and sales example, you would think that sales increased by 0.24887 dollars with $1 in newspaper ad spending!

Knowing the multiple regression coefficients ($b_1$ and $b_2$, in this case) and their interpretation, you can now answer the original question: Is the money spent on TV or newspaper ads worth it? The answer is a resounding *yes!* Not only that, but you also can say how much you expect sales to increase per $1,000 you spend on TV or newspaper advertising. Note that this conclusion assumes the model fits the data well. You have some evidence of that through the scatterplots and correlation tests, but more checking needs to be done before you can run to your manager and tell her the good news. The next section tells you what to do next.

## Testing the coefficients

To officially determine whether you have the right $x$ variables in your multiple regression model, do a formal hypothesis test to make sure the coefficients aren't equal to zero. Note that if the coefficient of an $x$ variable is zero, when you put that coefficient into the model, you get zero times that $x$ variable, which equals zero. This result is essentially saying that if an $x$ variable's coefficient is equal to zero, you don't need that $x$ variable in the model.

With any regression analysis, the computer automatically performs all the necessary hypothesis tests for the regression coefficients. Along with the regression coefficients you can find on the computer output, you see the test statistics and $p$-values for a test of each of those coefficients in the same row for each coefficient. Each one is testing Ho: Coefficient = 0 versus Ha: Coefficient ≠ 0.

The general format for finding a test statistic in most any situation is to take the statistic (in this case, the coefficient), subtract the value in Ho (zero), and divide by the standard error of that statistic (for this example, the standard error of the coefficient). (For more info on the general format of hypothesis tests, see Chapter 3.)

To test a regression coefficient, the test statistic (using the labels from Figure 5-3) is (Coef − 0)/SE Coef. In noncomputer language, that means you take the coefficient, subtract zero, and divide by the standard error (SE) of the coefficient. The standard error of a coefficient here is a measure of how much the coefficient is expected to vary when you take a new sample. (Refer to Chapter 3 for more on standard error.)

The test statistic has a $t$-distribution with $n - k - 1$ degrees of freedom, where $n$ equals the sample size and $k$ is the number of predictors ($x$ variables) in the model. This number of degrees of freedom works for any coefficient in the model (except you don't bother with a test for the constant, because it has no $x$ variable associated with it).

The test statistic for testing each coefficient is listed in the column marked T (because it has a $t$-distribution) on the Minitab output. You compare the value of the test statistic to the $t$-distribution with $n - k - 1$ degrees of freedom (using Table A-1 in the appendix) and come up with your $p$-value. If the $p$-value is less than your predetermined $\alpha$ (usually 0.05), then you reject Ho and conclude that the coefficient of that $x$ variable isn't zero and that variable makes a significant contribution toward estimating $y$ (given the other variables are also included in the model). If the $p$-value is larger than 0.05, you can't reject Ho, so that $x$ variable makes no significant contribution toward estimating $y$ (when the other variables are included in the model).

In the case of the ads and plasma TV sales example, Figure 5-3 shows that the coefficient for the TV ads is 0.1621 (the second number in column two). The standard error is listed as being 0.0132 (the second number in column three). To find the test statistic for TV ads, take 0.1621 minus zero and divide by the standard error, 0.0132. You get a value of $t = 12.29$, which is the second number in column four. Comparing this value of $t$ to a $t$-distribution with $n - k - 1 = 22 - 2 - 1 = 19$ degrees of freedom (Table A-1 in the appendix), you see the value of $t$ is way off the scale. That means the $p$-value is smaller than can be measured on the $t$-table. Minitab lists the $p$-value in column five of Figure 5-3 as 0.000 (meaning it's less than 0.001). This result leads you to conclude that the coefficient for TV ads is statistically significant, and TV ads should be included in the model for predicting TV sales.

The newspaper ads coefficient is also significant with a $p$-value of 0.000 by the same reasoning; you find these results by looking across the newspaper ads row of Figure 5-3. Based on your coefficient tests and the lack of multicolinearity between TV and newspaper ads (see the earlier section "Checking for Multiconlinearity"), you should include both the TV ads variable and the newspaper ads variable in the model for estimating TV sales.

# Predicting y by Using the x Variables

When you have your multiple regression model, you're finally ready to complete step six of the multiple regression analysis: to predict the value of $y$ given a set of values for the $x$ variables. To make this prediction, you take those $x$ values for which you want to predict $y$, plug them into the multiple regression model, and simplify.

In the ads and plasma TV sales example (see analysis from Figure 5-3), the best-fitting model is $y = 5.26 + 0.162x_1 + 0.249x_2$. In the context of the problem, the model is Sales $= 5.26 + 0.162$ TV ad spending $(x_1) + 0.249$ newspaper ad spending $(x_2)$.

Remember that the units for plasma TV sales is in millions of dollars and the units for ad spending for both TV and newspaper ads is in the thousands of dollars. That is, \$20,000 spent on TV ads means $x_1 = 20$ in the model. Similarly, \$10,000 spent on newspaper ads means $x_2 = 10$ in the model. Forgetting the units involved can lead to serious miscalculations.

Suppose you want to estimate plasma TV sales if you spend \$20,000 on TV ads and \$10,000 on newspaper ads. Plug $x_1 = 20$ and $x_2 = 10$ into the multiple regression model, and you get $y = 5.26 + 0.162(20) + 0.249(10) = 10.99$. In other words, if you spend \$20,000 on TV advertising and \$10,000 in newspaper advertising, you estimate that sales will be \$10.99 million.

This estimate at least makes sense in terms of the data from the 22 store locations shown in Table 5-1. Location 10 spent \$20,000 on TV ads and \$5,000 on newspaper ads (short of what you had) and got sales of \$9.82 million. Location 11 spent a little more on TV ads and a lot more on newspaper ads than what you had and got sales of \$16.28 million. Your estimates of sales for Store Locations 10 and 11 are $5.26 + 0.162 * 20 + 0.249 * 5 = \$9.745$ million, and $5.26 + 0.162 * 25 + 0.249 * 25 = \$15.535$ million, respectively. These estimates turned out to be pretty close to the actual sales at those two locations (\$9.82 million and \$16.28 million, respectively, as shown in Table 5-1), giving at least some confidence that your estimates will be close for the other store locations not chosen for the study.

Be careful to put in only values for the $x$ variables that fall in the range of where the data lies. In other words, Table 5-1 shows data for TV ad spending between \$0 and \$50,000; newspaper ad spending goes from \$0 to \$25,000. It wouldn't be appropriate to try to estimate sales for spending amounts of \$75,000 for TV ads and \$50,000 for newspaper ads, respectively, because the regression model you came up with only fits the data that you collected. You have no way of knowing whether that same relationship continues outside that area. This no-no of estimating $y$ for values of the $x$ variables outside their range is called *extrapolation*. As one of my colleagues says, "Friends don't let friends extrapolate."

# Checking the Fit of the Multiple Regression Model

Before you run to your boss in triumph saying you've slam-dunked the question of how to estimate plasma TV sales, you first have to make sure all your i's are dotted and all your t's are crossed, as you do with any other statistical procedure. In this case, you have to check the conditions of the multiple regression model. These conditions mainly focus on the *residuals* (the difference between the estimated values for y and the observed values of y from your data). If the model is close to the actual data you collected, you can feel somewhat confident that if you were to collect more data, it would fall in line with the model as well, and your predictions should be good.

In this section, you see what the conditions are for multiple regression and specific techniques statisticians use to check each of those conditions. The main character in all this condition-checking is the residual.

## Noting the conditions

The conditions for multiple regression concentrate on the error terms, or residuals. The residuals are the amount that's left over after the model has been fit. They represent the difference between the actual value of y observed in the data set and the estimated value of y based on the model. Following are the conditions for the residuals of the multiple regression model; note that all conditions need to be met in order to give the go-ahead for a multiple regression model:

- They have a normal distribution with a mean of zero.
- They have the same variance for each fitted (predicted) value of y.
- They're independent (meaning they don't affect each other).

## Plotting a plan to check the conditions

It may sound like you have a ton of things to check here and there, but luckily, Minitab gives you all the info you need to know in a series of four graphs, all presented at one time. These plots are called the *residual plots*, and they graph the residuals so that you can check to see whether the conditions from the previous section are met.

You can get the set of residual plots in two flavors:

✔ **Regular residuals:** The regular residual plots (the vanilla-flavored ones) show you exactly what the residuals are for each value of *y*. Their units depend on the variables in the model; use them *only* if you want to mainly look for patterns in the data. Figure 5-4 shows the plots of the regular residuals for the TV sales example. These residuals are in units of millions of dollars.

✔ **Standardized residuals:** The standardized residual plots (the strawberry-flavored kind) take each residual and convert it to a *Z*-score by subtracting the mean and dividing by the standard deviation of all the residuals. Figure 5-5 shows the plots of the standardized residuals for the TV sales example. Use these plots if you want to not only look for patterns in the data but also assess the standardized values of the residuals in terms of values on a *Z*-distribution to check for outliers. (Most statisticians use standardized residual plots.)

Note that the plots in Figure 5-5 look almost exactly the same as those in Figure 5-4. It's not surprising that the shapes of all graphs are the same for both types of residuals. Note however that the values of the regular residuals in Figure 5-4 are in millions of dollars and the standardized residuals in Figure 5-5 are from the standard normal distribution, which has no units.

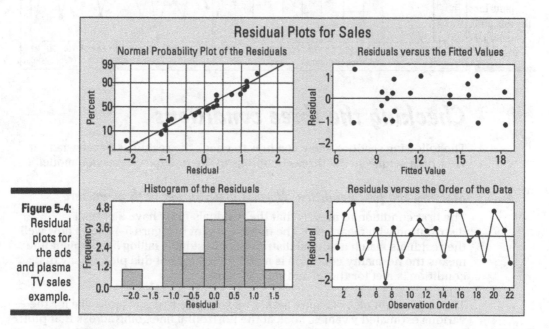

**Figure 5-4:** Residual plots for the ads and plasma TV sales example.

To make residual plots in Minitab, go to Stat>Regression>Regression. Select your response *(y)* variable and your predictor *(x)* variables. Click on Graphs, and choose either Regular or Standardized for the residuals, depending on which one you want. Then click on Four-in-one, which indicates you want to get all four residual plots shown in Figure 5-4 (using regular residuals) and Figure 5-5 (using standardized residuals).

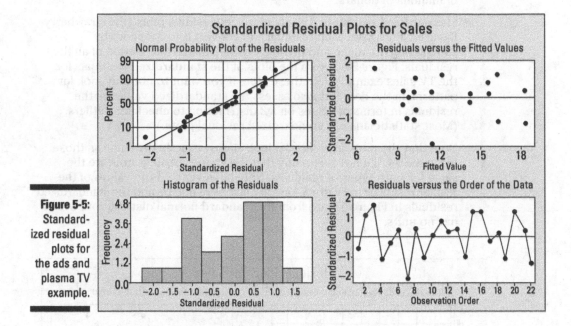

**Figure 5-5:**
Standard-
ized residual
plots for
the ads and
plasma TV
example.

# Checking the three conditions

The following sections show you how to check the residuals to see whether your data set meets the three conditions of the multiple regression model.

### Meeting the first condition: Normal distribution with mean zero

The first condition to meet is that the residuals must have a normal distribution with mean zero. The upper-left plot of Figure 5-4 shows how well the residuals match a normal distribution. Residuals falling in a straight line means the normality condition is met. By the looks of this plot, I'd say that condition is met for the ad and sales example.

The upper-right plot of Figure 5-4 shows what the residuals look like for the various estimated *y* values. Look at the horizontal line going across that plot: It's at zero as a marker. The residuals should average out to be at that line (zero). This Residuals versus Fitted Values plot checks the mean-of-zero condition and holds for the ads and sales example looking at Figure 5-4.

As an alternative check for normality apart from using the regular residuals, you can look at the standardized residuals plot (see Figure 5-5) and check out the upper-right plot. It shows how the residuals are distributed across the various estimated (fitted) values of $y$. Standardized residuals are supposed to follow a standard normal distribution — that is, they should have mean of zero and standard deviation of one. So when you look at the standardized residuals, they should be centered around zero in a way that has no predictable pattern, with the same amount of variability around the horizontal line that crosses at zero as you move from left to right.

In looking at the upper-right plot of Figure 5-5, you should also find that most (95 percent) of the standardized residuals fall within two standard deviations of the mean, which in this case is –2 to +2 (via the 68-95-99.7 Rule — remember that from Stats I?). You should see more residuals hovering around zero (where the middle lump would be on a standard normal distribution), and you should have fewer and fewer of the residuals as you go away from zero. The upper-right plot in Figure 5-5 confirms a normal distribution for the ads and sales example on all the counts mentioned here.

The lower-left plots of Figures 5-4 and 5-5 show histograms of the regular and standardized residuals, respectively. These histograms should reflect a normal distribution; the shape of the histograms should be approximately symmetric and look like a bell-shaped curve. If the data set is small (as is the case here with only 22 observations), the histogram may not be as close to normal as you would like; in that case, consider it part of the body of evidence that all four residual plots show you. The histograms shown in the lower-left plots of Figure 5-4 and 5-5 aren't terribly normal looking; however, because you can't see any glaring problems with the upper-right plots, don't be worried.

### Satisfying the second condition: Variance

The second condition in checking the multiple regression model is that the residuals have the same variance for each fitted (predicted) value of $y$. Look again at the upper-right plot of Figure 5-4 (or Figure 5-5). You shouldn't see any change in the amount of spread (variability) in the residuals around that horizontal line as you move from left to right. Looking at the upper-right graph of Figure 5-4, there's no reason to say condition number two hasn't been met.

One particular problem that raises a red flag with the second condition is if the residuals fan out, or increase in spread, as you move from left to right on the upper-right plot. This fanning out means that the variability increases more and more for higher and higher predicted values of $y$, so the condition of equal variability around the fitted line isn't met, and the regression model wouldn't fit well in that case.

### Checking the third condition

The third condition is that the residuals are independent; in other words, they don't affect each other. Looking at the lower-right plot on either Figure 5-4 or 5-5, you can see the residuals plotted by *observation number,* which is the order in which the data came in the sample. If you see a pattern, you have trouble; for example, if you were to connect the dots, so to speak, you might see a pattern of a straight line, a curve, or any kind of predictable up or down trend. You can see no patterns in the lower-right plots, so the independence condition is met for the ads and plasma TV sales example.

If the data must be collected over time, such as stock prices over a ten-year period, the independence condition may be a big problem because the data from the previous time period may be related to the data from the next time period. This kind of data requires time series analysis and is beyond the scope of this book.

# Chapter 6

# How Can I Miss You If You Won't Leave? Regression Model Selection

---

## In This Chapter

▶ Evaluating different methods for choosing a multiple regression model

▶ Understanding how forward selection and backward selection work

▶ Using the best subsets methods to find a good model

---

**S**uppose you're trying to estimate some quantitative variable, *y*, and you have many *x* variables available at your disposal. You have so many variables related to *y*, in fact, that you feel like I do in my job every day — overwhelmed with opportunity. Where do you go? What do you do? Never fear, this chapter is for you.

In this chapter, you uncover criteria for determining when a model fits well. I discuss different model selection procedures and all the details of the most statistician-approved method for selecting the best model. Plus, you get to find out what factors come into play when a punter kicks a football. (You can think about that while you're reading.)

Note that the term *best* has many connotations here. You can't find one end-all-be-all model that everyone comes up with in the end. That's to say that each data analyst can come up with a different model, and each model still could do a good job of predicting *y*.

# *Getting a Kick out of Estimating Punt Distance*

Before you jump into a model selection procedure to predict *y* by using a set of *x* variables, you have to do some legwork. The variable of interest is *y*, and that's a given. But where do the *x* variables come from? How do you choose which ones to investigate as being possible candidates for predicting *y*? And how do those possible *x* variables interact with each other toward making that prediction?

You must answer all these questions before using any model selection procedure. However, this part is the most challenging and the most fun; a computer can't think up *x* variables for you!

Suppose you're at a football game and the opposing team has to punt the ball. You see the punter line up and get ready to kick the ball, and some questions come to you: "Gee, I wonder how far this punt will go? I wonder what factors influence the distance of a punt? Can I use those factors in a multiple regression model to try to estimate punt distance? Hmm, I think I'll consult my *Statistics II For Dummies* book on this and analyze some data during halftime. . . ."

Well, maybe that's pushing it, but it's still an interesting line of questioning for football players, golfers, soccer players, and even baseball players. Everyone's looking for more distance and a way to get it.

In the following sections, you can see how to identify and assess different *x* variables in terms of their potential contribution to predicting *y*.

## *Brainstorming variables and collecting data*

Starting with a blank slate and trying to think of a set of *x* variables that may be related to *y* may sound like a daunting task, but in reality, it's probably not as bad as you think. Most researchers who are interested in predicting some variable *y* in the first place have some ideas about which variables may be related to it. After you come up with a set of logical possibilities for *x*, you collect data on those variables, as well as on *y*, to see what their actual relationship with *y* may be.

The Virginia Polytechnic Institute did a study to try to estimate the distance of a punt in football (something Ohio State fans aren't familiar with). Possible variables they thought may be related to the distance of a punt included the following:

✔ Hang time (time in the air, in seconds)

✔ Right leg strength (measured in pounds of force)

✔ Left leg strength (in pounds of force)

✔ Right leg flexibility (in degrees)

✔ Left leg flexibility (in degrees)

✔ Overall leg strength (in pounds)

The data collected on a sample of 13 punts (by right-footed punters) is shown in Table 6-1.

| Table 6-1 | | Data Collected for Punt Distance Study | | | | |
|---|---|---|---|---|---|---|
| Distance (In Feet) | Hang Time | Right Leg Strength | Left Leg Strength | Right Leg Flexibility | Left Leg Flexibility | Overall Leg Strength |
| 162.50 | 4.75 | 170 | 170 | 106 | 106 | 240.57 |
| 144.00 | 4.07 | 140 | 130 | 92 | 93 | 195.49 |
| 147.50 | 4.04 | 180 | 170 | 93 | 78 | 152.99 |
| 163.50 | 4.18 | 160 | 160 | 103 | 93 | 197.09 |
| 192.00 | 4.35 | 170 | 150 | 104 | 93 | 266.56 |
| 171.75 | 4.16 | 150 | 150 | 101 | 87 | 260.56 |
| 162.00 | 4.43 | 170 | 180 | 108 | 106 | 219.25 |
| 104.93 | 3.20 | 110 | 110 | 86 | 92 | 132.68 |
| 105.67 | 3.02 | 120 | 110 | 90 | 86 | 130.24 |
| 117.59 | 3.64 | 130 | 120 | 85 | 80 | 205.88 |
| 140.25 | 3.68 | 120 | 140 | 89 | 83 | 153.92 |
| 150.17 | 3.60 | 140 | 130 | 92 | 94 | 154.64 |
| 165.17 | 3.85 | 160 | 150 | 95 | 95 | 240.57 |

Other variables you may think of that are related to punt distance may include the direction and speed of the wind at the time of the punt, the angle at which the ball was snapped, the average distance of punts made in the past by a particular punter, whether the game is at home or away in a hostile environment, and so on. However, these researchers seem to have enough information on their hands to build a model to estimate punt distance.

For the sake of simplicity, you can assume the kicker is right-footed, which isn't always the case, but it represents the overwhelming majority of kickers.

Looking just at this raw data set in Table 6-1, you can't figure out which variables, if any, are related to distance of the punt or how those variables may be related to punt distance. You need more analyses to get a handle on this.

## Examining scatterplots and correlations

After you've identified a set of possible *x* variables, the next step is to find out which of these variables are highly related to *y* in order to start trimming down the set of possible candidates for the final model. In the punt distance example, the goal is to see which of the six variables in Table 6-1 are strongly related to punt distance. The two ways to look at these relationships are

- **Scatterplot:** A graphical technique
- **Correlation:** A one-number measure of the linear relationship between two variables

### Seeing relationships through scatterplots

To begin examining the relationships between the *x* variables and *y*, you use a series of scatterplots. Figure 6-1 shows all the scatterplots — not only of each *x* variable with *y* but also of each *x* variable with the other *x* variables. The scatterplots are in the form of a *matrix,* which is a table made of rows and columns. For example, the first scatterplot in row two of Figure 6-1 looks at the variables of distance (which appears in column one) and hang time (which appears in row two). This scatterplot shows a possible positive (uphill) linear relationship between distance and hang time.

Note that Figure 6-1 is essentially a symmetric matrix across the diagonal line. The scatterplot for distance and hang time is the same as the scatterplot for hang time and distance; the *x* and *y* axes are just switched. The essential relationship shows up either way. So you only have to look at all the scatterplots below the diagonal (where the variable names appear) or above the diagonal. You don't need to examine both.

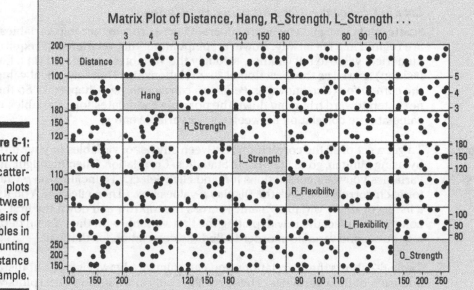

Matrix Plot of Distance, Hang, R_Strength, L_Strength . . .

**Figure 6-1:**
A matrix of all scatterplots between pairs of variables in the punting distance example.

To get a matrix of all scatterplots between a set of variables in Minitab, go to Graph>Matrix Plot and choose Matrix of Plots>Simple. Highlight all the variables in the left-hand box for which you want scatterplots by clicking on them; click Select, and then click OK. You'll see the matrix of scatterplots with a format similar to Figure 6-1.

Looking across row one of Figure 6-1, you can see that all the variables seem to have a positive linear relationship with punt distance except left leg flexibility. Perhaps the reason left leg flexibility isn't much related to punt distance is because the left foot is planted into the ground when the kick is made — for a right-footed kicker, the left leg doesn't have to be nearly as flexible as the right leg, which does the kicking. So it doesn't appear that left leg flexibility contributes a great deal to the estimation of punt distance on its own.

You can also see in Figure 6-1 that the scatterplots showing relationships between pairs of *x* variables are to the right of column one and below row one. (Remember, you need to look on only the bottom part of the matrix or the top part of the matrix to see the relevant scatterplots.) It appears that hang time is somewhat related to each of the other variables (except left leg flexibility, which doesn't contribute to estimating *y*). So hang time could possibly be the most important single variable in estimating the distance of a punt.

### *Looking for connections by using correlations*

Scatterplots can give you some general ideas as to whether two variables are related in a linear way. However, pinpointing that relationship requires a numerical value to tell you how strongly the variables are related (in a linear fashion) as well as the direction of that relationship. That numerical value is the *correlation* (also known as *Pearson's correlation;* see Chapter 4). So the next step toward trimming down the possible candidates for *x* variables is to calculate the correlation between each *x* variable and *y*.

To get a set of all the correlations between any set of variables in your model by using Minitab, go to Stat>Basic Statistics>Correlation. Then highlight all the variables you want correlations for, and click Select. (To include the *p*-values for each correlation, click the Display *p*-values box.) Then click OK. You'll see a listing of all the variables' names across the top row and down the first column. Intersect the row depicting the first variable with the column depicting the second variable in order to find the correlation for that pair.

Table 6-2 shows the correlations you can calculate between *y* = punt distance and each of the *x* variables. These results confirm what the scatterplots were telling you. Distance seems to be related to all the variables except left leg flexibility because that's the only variable that didn't have a statistically significant correlation with distance using the α level 0.05. (For more on the test for correlation, see Chapter 5.)

| Table 6-2 | Correlations between Distance of a Punt and Other Variables | |
|---|---|---|
| *x* Variable | Correlation with Punt Distance | *p*-value |
| Hang time | 0.819 | 0.001* |
| Right leg strength | 0.791 | 0.001* |
| Left leg strength | 0.744 | 0.004* |
| Right leg flexibility | 0.806 | 0.001* |
| Left leg flexibility | 0.408 | 0.167 |
| Overall leg strength | 0.796 | 0.001* |

\* Statistically significant at level α = 0.05

If you take a look at Figure 6-1, you can see that hang time is related to other *x* variables such as right foot and left foot strength, right leg flexibility, and so on. This is where things start to get sticky. You have hang time related to distance, and lots of other variables related to hang time. Although hang time is clearly the most related to distance, the final multiple regression model may not include hang time.

Here's one possible scenario: You find a combination of other $x$ variables that can do a good job estimating $y$ together. And all those other variables are strongly related to hang time. This result may mean that in the end you don't need to include hang time in the model. Strange things happen when you have many different $x$ variables to choose from.

After you narrow down the set of possible $x$ variables for inclusion in the model to predict punt distance, the next step is to put those variables through a selection procedure to trim down the list to a set of essential variables for predicting $y$.

# Just Like Buying Shoes: The Model Looks Nice, But Does It Fit?

When you get into model selection procedures, you find that many different methods exist for selecting the best model, according to a wide range of criteria. Each one can result in models that differ from each other, but that's something I love about statistics: Sometimes there's no one single best answer.

The three model selection procedures covered in this section are

- Best subsets procedure
- Forward selection
- Backward selection

Of all the model selection procedures out there, the one that gets the most votes with statisticians is the *best subsets procedure,* which examines every single possible model and determines which one fits best, using certain criteria.

In this section, you see different methods statisticians use to assess and compare the fit of different models. You see how the best subsets procedure works for model selection in a step-by-step manner. Then I show you how to take all the information given to you and wade through it to make your way to the answer — the best-fitting model based on a subset of the available $x$ variables. Finally, you see how this procedure is applied to find a model to predict punt distance.

## Assessing the fit of multiple regression models

For any model selection procedure, assessing the fit of each model being considered is built into the process. In other words, as you go through all the possible models, you're always keeping an eye on how well each model fits. So before you get into a discussion of how to do the best subsets procedure, you need criteria to assess how well a particular model fits a data set.

Although there are tons of different statistics for assessing the fit of regression models, I discuss the most popular ones: $R^2$ (simple linear regression only), $R^2$ adjusted, and Mallow's C-p. All three appear on the bottom line of the Minitab output when you do any sort of model selection procedure. Here's a breakdown of the assessment techniques:

✔ **$R^2$:** $R^2$ is the percentage of the variability in the $y$ values that's explained by the model. It falls between 0 and 100 percent (0 and 1.0). In simple linear regression (see Chapter 4), a high value of $R^2$ means the line fits well, and a low value of $R^2$ means the line doesn't fit well.

When you have multiple regression, however, there's a bit of a catch here. As you add more and more variables (no matter how significant), the value of $R^2$ increases or stays the same — it never goes down. This can result in an inflated measure of how well the model fits. Of course, statisticians have a fix for the problem, which leads me to the next item on this list.

✔ **$R^2$ adjusted:** $R^2$ adjusted takes the value of $R^2$ and adjusts it downward according to the number of variables in the model. The higher the number of variables in the model, the lower the value of $R^2$ adjusted will be, compared to the original $R^2$.

A high value of $R^2$ adjusted means the model you have is fitting the data very well (the closer to 1, the better). I typically find a value of 0.70 to be considered okay for $R^2$ adjusted, and the higher the better.

Always use $R^2$ adjusted rather than the regular $R^2$ to assess the fit of a multiple regression model. With every addition of a new variable into a multiple regression model, the value of $R^2$ stays the same or increases. It will never go down because a new variable will either help explain some of the variability in the $y$'s (thereby increasing $R^2$ by definition), or it will do nothing (leaving $R^2$ exactly where it was before). So theoretically, you could just keep adding more and more variables into the model just for the sake of getting a larger value of $R^2$.

$R^2$ adjusted is important because it keeps you from adding more and more variables by taking into account how many variables there already are in the model. The value of $R^2$ adjusted can actually decrease if the added value of the additional variable is outweighed by the number of variables in the model. This gives you an idea of how much or how little added value you get from a bigger model (bigger isn't always better).

✔ **Mallow's C-p:** Mallow's C-p takes the amount of error left unexplained by a model of $p$ with the $x$ variables, divides that number by the average amount of error left over from the full model (with all the $x$ variables), and adjusts that result for the number of observations $(n)$ and the number of $x$ variables used $(p)$. In general, the smaller Mallow's C-p is, the better, because when it comes to the amount of error in your model, less is more. A C-p value close to $p$ (the number of $x$ variables in the model) reflects a model that fits well.

# Model selection procedures

The process of finding the "best" model is not cut and dry. (Heck, even the definition of "best" here isn't cut and dry.) Many different procedures exist for going through different models in a systematic way, evaluating each one, and stopping at the right model. Three of the more common model selection procedures are forward selection, backward selection, and the best subsets model. In this section you get a very brief overview of the forward and backward selection procedures, and then you get into the details of the best subsets model, which is the one statisticians use most.

### Going with the forward selection procedure

The *forward selection procedure* starts with a model with no variables in it and adds variables one at a time according to the amount of contribution they can make to the model.

Start with an entry level value of $\alpha$. Then run hypothesis tests (see Chapter 3 for instructions) for each $x$ variable to see how it's related to $y$. The $x$ variable with the smallest $p$-value wins and is added to the model, as long as its $p$-value is smaller than the entry level. You keep doing this with the remaining variables until the one with the smallest $p$-value doesn't make the entry level. Then you stop.

The drawback of the forward selection procedure is that it starts with nothing and adds variables one at a time as you go along; after a variable is added, it's never removed. The best model might not even get tested.

### Opting for the backward selection procedure

The *backward selection procedure* does the opposite of the forward selection method. It starts with a model with all the $x$ variables in it and removes variables one at a time. Those that make the least amount of contribution to the model are removed first. You choose a removal level to begin; then you test all the $x$ variables and find the one with the largest $p$-value. If the $p$-value of this $x$ variable is higher than the removal level, that variable is taken out of the model.

You continue removing variables from the model until the one with the largest $p$-value doesn't exceed the removal level. Then you stop.

The drawback of the backward selection procedure is that it starts with everything and removes variables one at a time as you go along; after a variable is removed, it never comes back. Again, the best model might not even be tested.

### Using the best subsets procedure

The best subsets procedure has fewer steps than the forward or backward selection model because the computer formulates and analyzes all possible models in a single step. In this section, you see how to get the results and then use them to come up with a best multiple regression model for predicting $y$.

Here are the steps for conducting the best subsets model selection procedure to select a multiple regression model; note that Minitab does all the work for you to crunch the numbers:

1. **Conduct the best subsets procedure in Minitab, using all possible subsets of the $x$ variables being considered for inclusion in the final model.**

   To carry out the best subsets selection procedure in Minitab, go to Stat>Regression>Best Subsets. Highlight the response variable $(y)$, and click Select. Highlight all the predictor $(x)$ variables, click Select, and then click OK.

   The output contains a listing of all models that contain one $x$ variable, all models that contain two $x$ variables, all models that contain three $x$ variables, and so on, all the way up to the full model (containing all the $x$ variables). Each model is presented in one row of the output.

2. **Choose the best of all the models shown in the best subsets Minitab output by finding the model with the largest value of $R^2$ adjusted and the smallest value of Mallow's C-p; if two competing models are about equal, choose the model with the fewer number of variables.**

   If the model fits well, $R^2$ adjusted is high. So you also want to look for the smallest possible model that has a high value of $R^2$ adjusted and a small value of Mallow's C-p compared to its competitors. And if it comes down to two similar models, always make your final model as easy to interpret as possible by selecting the model with fewer variables.

## The secret to a punter's success: An example

Returning to the punt distance example from earlier in this chapter, suppose that you analyzed the punt distance data by using the best subsets model selection procedure. Your results are shown in Figure 6-2. This section follows Minitab's footsteps in getting these results and provides you with a guide for interpreting the results.

Assuming that you already used Minitab to carry out the best subsets selection procedure on the punt distance data, you can now analyze the output from Figure 6-2. Each variable shows up as a column on the right side of the output. Each row represents the results from a model containing the number of variables shown in column one. The X's at the end of each row tell you which variables were included in that model. The number of variables in the model starts at 1 and increases to 6 because six $x$ variables are available in the data set.

The models with the same number of variables are ordered by their values of $R^2$ adjusted and Mallows C-p, from best to worst. The top-two models (for each number of variables) are included in the computer output.

For example, rows one and two of Figure 6-2 (both marked 1 in the Vars column) show the top-two models containing one $x$ variable; rows three and four show the top two models containing two $x$ variables; and so on. Finally, the last row shows the results of the full model containing all six variables. (Only one model contains all six variables, so you don't have a second-best model in this case.)

Looking at the first two rows of Figure 6-2, the top one-variable model is the one including hang time only. The second-best one-variable model includes only right foot flexibility. The right foot flexibility model has a lower value of $R^2$ and a higher Mallow's C-p than the hang time model, which is why it's the second best.

Row three shows that the best two-variable model for estimating punt distance is the model containing right leg strength and overall leg strength. The best three-variable model is in row five; it shows that the best three-variable model includes right foot strength, right foot flexibility, and overall leg strength. The best four-variable model is found in row seven and includes right foot strength, right and left foot flexibility, and overall foot strength. The best five-variable model is found in row nine and includes every variable except left foot strength. The only six-variable model with all variables included is listed in the last row.

Among the best one-variable, two-variable, three-variable, four-variable, and five-variable models, which one should you choose for your final multiple regression model? Which model is the best of the best? With all these results, it would be easy to have a major freakout over which one to pick, but never fear — Mallow's is here (along with his friendly sidekick, the $R^2$ adjusted).

Looking at Figure 6-2 column three, you see that as the number of variables in the model increases, $R^2$ adjusted peaks out and then drops way off. That's because $R^2$ adjusted takes into account the number of variables in the model and reduces $R^2$ accordingly. You can see that $R^2$ adjusted peaks out at a level of 74.1 percent for two models. The corresponding models are the top two-variable model (right leg strength and overall leg strength) and the best three-variable model (right foot strength, right foot flexibility, and overall leg strength).

Now look at Mallow's C-p for these two models. Notice that Mallow's C-p is zero for the best two-variable model and 1.3 for the best three-variable model. Both values are small compared to others in Figure 6-2, but because Mallow's C-p is smaller for the two-variable model, and because it has one less variable in it, you should choose the two-variable model (right leg strength and overall leg strength) as the final model, using the best subsets procedure.

```
Best Subsets Regression: Distance versus Hang, R_Strength ...

Response is Distance
                                                    R L
                                                    F F
                                              R L  1 1 0
                                                   e e
                                              S S  x x S
                                              t t  i i t
                                              r r  b b r
                                              e e  i i e
                                          H n n  l l n
                                          a g g  i i g
                                 Mallows  n t t  t t t
    Vars  R-Sq  R-Sq(adj)    C-p      S  g h h  y y h
      1   67.1    64.1      1.7   15.570  X
      1   65.0    61.8      2.3   16.043       X
      2   78.5    74.1     -0.0   13.206  X        X
      2   78.2    73.8      0.1   13.294     X     X
      3   80.6    74.1      1.3   13.214  X  X     X
      3   79.5    72.7      1.6   13.581  X X      X
      4   81.4    72.1      3.0   13.724  X   X X X
      4   80.7    72.0      3.3   13.977  X X X   X
      5   81.5    68.2      5.0   14.643  X X   X X X
      5   81.4    68.2      5.0   14.650  X   X X X X
      6   81.5    62.9      7.0   15.812  X X X X X X
```

**Figure 6-2:** Best subsets procedure results for the punt distance example.

# Chapter 7

# Getting Ahead of the Learning Curve with Nonlinear Regression

· · · · · · · · · · · · · · · · · · · · · · · · · · · · · · · · · · · · · · · · · · · · ·

*In This Chapter*

▶ Getting a feel for nonlinear regression

▶ Making use of scatterplots

▶ Fitting a polynomial to your data set

▶ Exploring exponential models to fit your data

· · · · · · · · · · · · · · · · · · · · · · · · · · · · · · · · · · · · · · · · · · · · ·

*I*n Stats I, you concentrate on the *simple linear regression model,* where you look for one quantitative variable, *x*, that you can use to make a good estimate of another quantitative variable, *y*, using a straight line. The examples you look at in Stats I fall right in line with this kind of model, such as using height to estimate weight or using study time to estimate exam score. (For more information and examples for using simple linear regression models, see Chapter 4.)

But not all situations fall into the straight line category. Take gas mileage and speed, for example. At low speeds, gas mileage is lower, and at high speeds, gas mileage is lower; but at medium speeds, gas mileage is higher. This low-high-low relationship between speed and gas mileage represents a curved relationship. Relationships that don't resemble straight lines are called *nonlinear relationships* (clever, huh?). Looked at simply, nonlinear regression takes the stage when you want to predict some quantitative variable *(y)* by using another quantitative variable *(x)* but the pattern you see in the data collected resembles a curve, not a straight line.

In this chapter, you see how to make your way around the curved road of data that leads to nonlinear regression models. The good news is twofold: You can use many of the same techniques you use for regular regression, and in the end, Minitab does the analysis for you.

# Anticipating Nonlinear Regression

Nonlinear regression comes into play in situations where you have graphed your data on a *scatterplot* (a two-dimensional graph showing the *x* variable on the *x*-axis and the *y* variable on the *y*-axis; see the next section "Starting Out with Scatterplots"), and you see a pattern emerging that looks like some type of curve. Examples of data that follow a curve include changes in population size over time, demand for a product as a function of supply, or the length of time that a battery lasts. When a data set follows a curved pattern, the time has come to move away from the linear regression models (covered in Chapters 4 and 5) and move on to a nonlinear regression model.

Suppose a manager is considering the purchase of new office management software but is hesitating. She wants to know how long it typically takes someone to get up to speed using the software.

What's the statistical question here? She wants a model that shows what the learning curve looks like (on average). (A *learning curve* shows the decrease in time to do a task with more and more practice.) In this scenario, you have two variables: time to complete the task and trial number (for example, the first try is designated by 1, the second try by 2, and so on). Both variables are *quantitative* (numerical) and you want to find a connection between two quantitative variables. At this point, you can start thinking regression.

A *regression model* produces a function (be it a line or otherwise) that describes a pattern or relationship. The relationship here is task time versus number of times the task is practiced. But what type of regression model do you use? After all, you can see four types in this book: simple linear regression, multiple regression, nonlinear regression, and logistic regression. You need more clues.

The word "curve" in learning curve is a clue that the relationship being modeled here may not be linear. That word signals that you're talking about a nonlinear regression model. If you think about what a possible learning curve may look like, you can imagine task time on the *y*-axis and the number of the trial on the *x*-axis.

You may guess that the *y*-values will be high at first, because the first couple of times you try a new task, it takes longer to perform. Then, as the task is repeated, the task time decreases, but at some point more practice doesn't reduce task time much. So the relationship may be represented by some sort of curve, like the one I simulate in Figure 7-1 (which can be fit by using an exponential function).

This example illustrates the basics of nonlinear regression; the rest of the chapter shows you how the model breaks down.

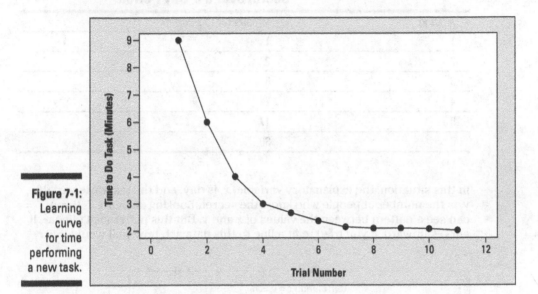

**Figure 7-1:**
Learning
curve
for time
performing
a new task.

# Starting Out with Scatterplots

As with any type of data analysis, before you dive in and select a model that you think fits the data (or that's supposed to fit the data), you have to step back and take a look at the data to see whether any patterns emerge. To do this, look at a scatterplot of the data, and see whether or not you can draw a smooth curve through the data and find that most of the points follow along that curve.

Suppose you're interested in modeling how quickly a rumor spreads. One person knows a secret and tells it to another person, and now two know the secret; each of them tells a person, and now four know the secret; some of those people may pass it on, and so it goes on down the line. Pretty soon, a large number of people know the secret, which is a secret no longer.

To collect your data, you count the number of people who know a secret by tracking who tells whom over a six-day period. The data are shown in Table 7-1. Note that the spread of the secret catches fire on day 5 — this is how an exponential model works. You can see a scatterplot of the data in Figure 7-2.

| Table 7-1 | Number of People Knowing a Secret over a 6-Day Period |
|-----------|------------------------------------------------------|
| *x (Day)* | *y (Number of People)* |
| 1 | 1 |
| 2 | 2 |
| 3 | 5 |
| 4 | 7 |
| 5 | 17 |
| 6 | 30 |

In this situation, the explanatory variable, $x$, is day, and the response variable, $y$, is the number of people who know the secret. Looking at Figure 7-2, you can see a pattern between the values of $x$ and $y$. But this pattern isn't linear. It curves upward. If you tried to fit a line to this data set, how well would it fit?

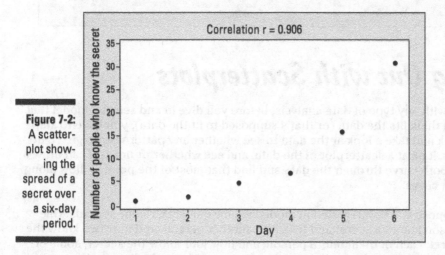

**Figure 7-2:**
A scatter-plot showing the spread of a secret over a six-day period.

To figure this out, look at the correlation coefficient between $x$ and $y$, which is found on Figure 7-2 to be 0.906 (see Chapter 4 for more on correlation). You can interpret this correlation as a strong, positive (uphill) linear relationship between $x$ and $y$. However, in this case, the correlation is misleading because the scatterplot appears to be curved.

If the correlation looks good (close to +1 or –1), don't stop there. As with any regression analysis, it's very important to take into account both the scatterplot and the correlation when making a decision about how well the model being considered would fit the data. The contradiction in this example between the scatterplot and the correlation is a red flag that a straight-line model isn't the best idea.

The correlation coefficient measures only the strength and direction of the linear relationship between $x$ and $y$ (see Chapter 4). However, you may run into situations (like the one shown in Figure 7-2) where a correlation is strong, yet the scatterplot shows a curve would fit better. Don't rely solely on either the scatterplot or the correlation coefficient alone to make your decision about whether to go ahead and fit a straight line to your data.

The bottom line here is that fitting a line to data that appear to have a curved pattern isn't the way to go. Instead, explore models that have curved patterns themselves.

The following sections address two major types of nonlinear (or curved) models that are used to model curved data: polynomials (that are not straight lines — that is, curves like quadratics or cubics), and exponential models (that start out small and quickly increase, or the other way around). Because the pattern of the data in Figure 7-2 starts low and bends upward, the correct model to fit this data is an exponential regression model. (This model is also appropriate for data that start out high and bend down low.)

# Handling Curves in the Road with Polynomials

One major family of nonlinear models is the *polynomial* family. You use these models when a polynomial function (beyond a straight line) best describes the curve in the data. (For example, the data may follow the shape of a parabola, which is a second-degree polynomial.) You typically use polynomial models when the data follow a pattern of curves going up and down a certain number of times.

For example, suppose a doctor examines the occurrence of heart problems in patients as it relates to their blood pressure. She finds that patients with very low or very high blood pressure had a higher occurrence of problems, while patients whose blood pressure fell in the middle, constituting the normal range, had fewer problems. This pattern of data has a U-shape, and a parabola would fit this data well.

In this section, you see what a polynomial regression model is, how you can search for a good-fitting polynomial for your data, and how you can assess polynomial models.

## Bringing back polynomials

You may recall from algebra that a *polynomial* is a sum of $x$ terms raised to a variety of powers, and each $x$ is preceded by a constant called the *coefficient* of that term. For example, the model $y = 2x + 3x^2 + 6x^3$ is a polynomial. The general form for a polynomial regression model is $y = \beta_0 + \beta_1 x^1 + \beta_2 x^2 + \beta_3 x^3 + \ldots + \beta_k x^k + \varepsilon$. Here, $k$ represents the total number of terms in the model. The $\varepsilon$ represents the error that occurs simply due to chance. (Not a bad kind of error, just random fluctuations from a perfect model.)

Here are a few of the more common polynomials you run across when analyzing data and fitting models. Remember, the simplest model that fits is the one you use (don't try to be a hero in statistics — save that for Batman and Robin). The models I discuss in this book are some of your old favorites from algebra: second-, third-, and fourth-degree polynomials.

✔ **Second-degree (or quadratic) polynomial:** This model is called a *second-degree (or quadratic) polynomial,* because the largest exponent is 2. An example model is $y = 2x + 3x^2$. A second-degree polynomial forms a parabola shape — either an upside-down or right-side up bowl; it changes direction one time (see Figure 7-3).

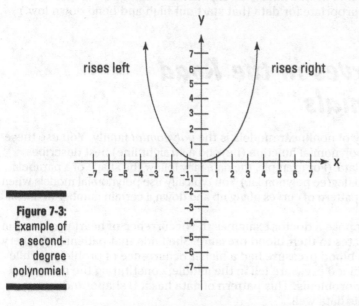

**Figure 7-3:** Example of a second-degree polynomial.

✔ **Third-degree polynomial:** This model has 3 as the highest power of $x$. It typically has a sideways S-shape, changing directions two times (see Figure 7-4).

✔ **Fourth-degree polynomial:** Fourth-degree polynomials involve $x^4$. They typically change directions in curvature three times to look like the letter W or the letter M, depending on whether they're upside down or right-side up (see Figure 7-5).

In general, if the largest exponent on the polynomial is $n$, the number of curve changes in the graph is typically $n - 1$. For more information on graphs of polynomials, refer to your algebra textbook or *Algebra For Dummies* by Mary Jane Sterling (Wiley).

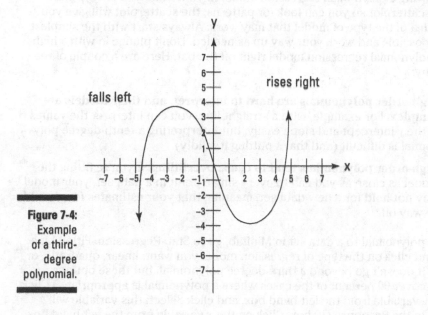

**Figure 7-4:** Example of a third-degree polynomial.

The nonlinear models in this chapter involve only one explanatory variable, *x*. You can include more explanatory variables in a nonlinear regression, raising each separate variable to a power, but those models are beyond the scope of this book. I give you information on basic multiple regression models in Chapter 5.

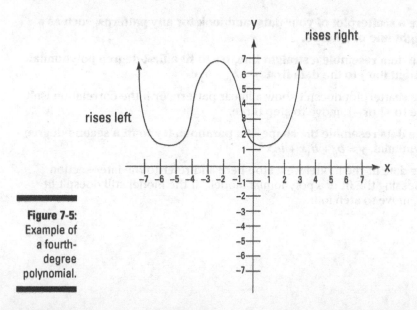

**Figure 7-5:** Example of a fourth-degree polynomial.

# Searching for the best polynomial model

When fitting a polynomial regression model to your data, you always start with a scatterplot so you can look for patterns; the scatterplot will give you some idea of the type of model that may work. Always start with the simplest model possible and work your way up as needed. Don't plunge in with a high-order polynomial regression model right off the bat. Here are a couple of reasons why:

- **High-order polynomials are hard to interpret, and their models are complex.** For example, with a straight line, you can interpret the values of the y-intercept and slope easily, but interpreting a tenth-degree polynomial is difficult (and that's putting it mildly).

- **High-order polynomials tend to cause overfitting.** If you're fitting the model as close as you can to every single point in a data set, your model may not hold for a new data set, meaning that your estimates for y could be way off.

To fit a polynomial to a data set in Minitab, go to Stat>Regression>Fitted Line Plot> and click on the type of regression model you want: linear, quadratic, or cubic. (It doesn't go beyond a third-degree polynomial, but these options should cover 90 percent of the cases where a polynomial is appropriate.) Click on the y variable from the left-hand box, and click Select; this variable will appear in the Response *(y)* box. Click on the x variable from the left-hand box, and click Select; it will appear in the Predictor *(x)* box. Click OK.

Following are steps that you can use to see if a polynomial fits your data. (Statistical software can jump in and fit the models for you after you tell it which ones to fit.)

1. **Make a scatterplot of your data, and look for any patterns, such as a straight line or a curve.**

2. **If the data resemble a straight line, try to fit a first-degree polynomial (straight line) to the data first: $y = b_0 + b_1 x$.**

    If the scatterplot doesn't show a linear pattern, or if the correlation isn't close to +1 or –1, move to step three.

3. **If the data resemble the shape of a parabola, try to fit a second-degree polynomial: $y = b_0 + b_1 x + b_2 x^2$.**

    If the data fit the model well, stop here and refer to the later section "Assessing the fit of a polynomial model." If the model still doesn't fit well, move to step four.

4. **If you see curvature that's more complex than a parabola, try to fit a third-degree polynomial:** $y = b_0 + b_1x + b_2x^2 + b_3x^3$.

   If the data fit the model well, stop here and refer to the later section "Assessing the fit of a polynomial model." If the model still doesn't fit well, move to step five.

5. **Continue trying to fit higher-order polynomials until you find one that fits or until the order of the polynomial (largest exponent) simply gets too large to find a reliable pattern.**

   How large is too large? Typically, if you can't fit the data by the time the degree of the polynomial reaches three, then perhaps a different type of model would work better. Or you may determine that you observe too much scatter and haphazard behavior in the data to try to fit any model.

Minitab can do each of these steps for you up to degree two (that's step three); from there, you need a more sophisticated statistical software program, such as SAS or SPSS. However, most of the models you need to fit go up to the second-degree polynomials.

## Using a second-degree polynomial to pass the quiz

The first step in fitting a polynomial model is to graph the data in a scatterplot and see whether the data fall into a particular pattern. Many different types of polynomials exist to fit data that have a curved type of pattern. One of the most common patterns found in curved data is the quadratic pattern, or second-degree polynomial, which goes up and comes back down, or goes down and comes back up, as the $x$ values move from left to right (see Figure 7-3). The second-degree polynomial is the simplest and most commonly used polynomial beyond the straight line, so it deserves special consideration. (After you master the basic ideas based on second-degree polynomials, you can apply them to polynomials with higher powers.)

Suppose 20 students take a statistics quiz. You record the quiz scores (which have a maximum score of ten) and the number of hours students reported studying for the quiz. You can see the results in Figure 7-6.

Looking at Figure 7-6, it appears that three camps of students are in this class. Camp 1, on the left end of the $x$-axis, understands the stuff (as reflected in their higher scores) but didn't have to study hardly at all (see that their study time on the $x$-axis is low). Camp 3 also did very well on the quiz (as indicated by their high quiz scores) but had to study a great deal to get those grades (as seen on the far-right end of the $x$-axis). The students in the middle, Camp 2, didn't seem to fare well.

All in all, based on the scatterplot, it does appear that study time may explain quiz scores on some level in a way indicative of a second-degree polynomial. So a quadratic regression model may fit this data.

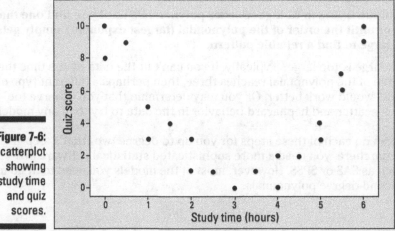

**Figure 7-6:** Scatterplot showing study time and quiz scores.

Suppose a data analyst (not you!) doesn't know about polynomial regression and just tries to fit a straight line to the quiz-score data. In Figure 7-7, you can see the data and the straight line that he tried to fit to the data. The correlation as shown in the figure is –0.033, which is basically zero. This correlation means that no linear relationship lies between $x$ and $y$. It doesn't mean that no relationship is present at all, just that it's not a linear relationship (see Chapter 4 for more on linear relationships). So trying to fit a straight line here was indeed a bad idea.

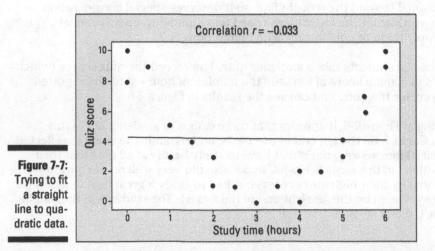

**Figure 7-7:** Trying to fit a straight line to quadratic data.

After you know that a quadratic polynomial seems to be a good fit for the data, the next challenge is finding the equation for that particular parabola that fits the data from among all the possible parabolas out there.

Remember from algebra that the general equation of a parabola is $y = ax^2 + bx + c$. Now you have to find the values of $a$, $b$, and $c$ that create the best-fitting parabola to the data (just like you find the $a$ and the $b$ that create the best-fitting line to data in a linear regression model). That's the object of any regression analysis.

Suppose that you fit a quadratic regression model to the quiz-score data by using Minitab (see the Minitab output in Figure 7-8 and the instructions for using Minitab to fit this model in the previous section). On the top line of the output, you can see that the equation of the best-fitting parabola is quiz score = $9.82 - 6.15 *$ (study time) + $1.00 *$ (study time)$^2$. (Note that $y$ is quiz score and $x$ is study time in this example because you're using study time to predict quiz score.)

**Figure 7-8:**
Minitab
output for
fitting a
parabola
to the
quiz-score
data.

```
Polynomial Regression Analysis: Quiz Score versus Study Time

The regression equation is
Quiz score = 9.823 - 6.149 study time + 1.003 study time**2

S = 1.04825    R-Sq = 91.7%   R-Sq(adj) = 90.7%
```

The scatterplot of the quiz-score data and the parabola that was fit to the data via the regression model is shown in Figure 7-9. From algebra, you may remember that a positive coefficient on the quadratic term (here $a = 1.00$) means the bowl is right-side up, which you can see is the case here.

Looking at Figure 7-9, it appears that the quadratic model fits this data pretty well, because the data fall close to the curve that Minitab found. However, data analysts can't live by scatterplots alone, so the next section helps you figure out how to assess the fit of a polynomial model in more detail.

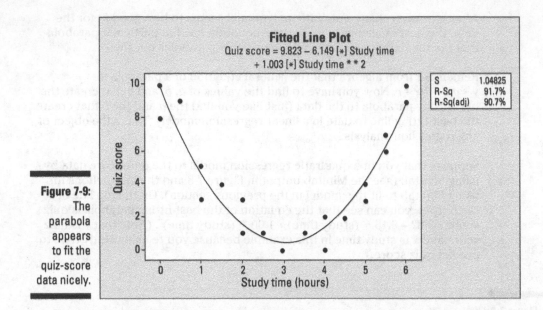

**Figure 7-9:**
The
parabola
appears
to fit the
quiz-score
data nicely.

## Assessing the fit of a polynomial model

You make a scatterplot of your data, and you see a curved pattern. So you use polynomial regression to fit a model to the data; the model appears to fit well because the points follow closely to the curve Minitab found, but don't stop there. To make sure your results can be generalized to the population from which your data was taken, you need to do a little more checking beyond just the graph to make sure your model fits well.

To assess the fit of any model beyond the usual suspect, a scatterplot of the data, you look at two additional items, typically in this order: the value of $R^2$ adjusted and the residual plots.

All three assessments must agree before you can conclude that the model fits. If the three assessments don't agree, you'll likely have to use a different model to fit the data besides a polynomial model, or you'll have to change the units of the data to help a polynomial model fit better. However, the latter fix is outside the scope of Stats II, and you probably won't encounter that situation.

In the following sections, you take a deeper look at the value of $R^2$ adjusted and the residual plots and figure out how you can use them to assess your model's fit. (You can find more info on the scatterplot in the section "Starting out with Scatterplots" earlier in this chapter.)

### Examining $R^2$ and $R^2$ adjusted

Finding $R^2$, the *coefficient of determination* (see Chapter 5 for full details), is like the day of reckoning for any model. You can find $R^2$ on your regression output, listed as "R-Sq" right under the portion of the output where the coefficients of the variables appear. Figure 7-8 shows the Minitab output for the quiz-score data example; the value of $R^2$ in this case is 91.7 percent.

The value of $R^2$ tells you what percentage of the variation in the $y$-values the model can explain. To interpret this percentage, note $R^2$ is the square of $r$, the correlation coefficient (see Chapter 5). Because values of $r$ beyond ± 0.80 are considered to be good, $R^2$ values above 0.64 are considered pretty good also, especially for models with only one $x$ variable.

You can consider values of $R^2$ over 80 percent good, and values under 60 percent aren't good. Those in between I'd consider so-so; they could be better. (This assessment is just my rule of thumb; opinions may vary a bit from one statistician to another.)

However, you can find such a thing in statistics as too many variables spoiling the pot. Every time you add another $x$ variable to a regression model, the value of $R^2$ automatically goes up, whether the variable really helps or not (this is just a mathematical fact). Right beside $R^2$ on the computer output from any regression analysis is the value of $R^2$ adjusted, which adjusts the value of $R^2$ down a notch for each variable (and each power of each variable) entered into the model. You can't just throw a ton of variables into a model whose tiny increments all add up to an acceptable $R^2$ value without taking a hit for throwing everything in the model but the kitchen sink.

To be on the safe side, you should always use $R^2$ adjusted to assess the fit of your model, rather than $R^2$, especially if you have more than one $x$ variable in your model (or more than one power of an $x$ variable). The values of $R^2$ and $R^2$ adjusted are close if you have only a couple of different variables (or powers) in the model, but as the number of variables (or powers) increases, so does the gap between $R^2$ and $R^2$ adjusted. In that case, $R^2$ adjusted is the most fair and consistent coefficient to use to examine model fit.

In the quiz-score example (analysis shown in Figure 7-8), the value of $R^2$ adjusted is 90.7 percent, which is still a very high value, meaning that the quadratic model fits this data very well. (See Chapter 6 for more on $R^2$ and $R^2$ adjusted.)

### Checking the residuals

You've looked at the scatterplot of your data and the value of $R^2$ is high. What's next? Now you examine how well the model fits each individual point in the data to make sure you can't find any spots where the model is way off or places where you missed another underlying pattern in the data.

A *residual* is the amount of error, or leftover, that occurs when you fit a model to a data set. The residuals are the distances between the predicted values in the model and the observed values of the data themselves. For each observed *y*-value in the data set, you also have a predicted value from the model, typically called *y-hat,* denoted $\hat{y}$. The residual is the difference between the values of *y* and *y*-hat. Each *y*-value in the data set has a residual; you examine all the residuals together as a group, looking for patterns or unusually high values (indicating a big difference between the observed *y* and the predicted *y* at that point; see Chapter 4 for the full info on residuals and their plots).

In order for the model to fit well, the residuals need to meet two conditions:

✔ **The residuals are independent.** The independence of residuals means that you don't see any pattern as you plot the residuals. The residuals don't affect each other and should be random.

✔ **The residuals have a normal distribution centered at zero, and the standardized residuals follow suit.** Having a normal distribution with mean zero means that most of the residuals should be centered around zero, with fewer of them occurring the farther from zero you get. You should observe about as many residuals above the zero line as below it. If the residuals are standardized, this means that as a group their standard deviation is 1; you should expect about 95 percent of them to lie between –2 and +2, following the 68-95-99.7 Rule (see your Stats I text).

You determine whether or not these two conditions are met for the residuals by using a series of four graphs called *residual plots.* Most statisticians prefer to standardize the residuals (meaning they convert them to *Z*-scores by subtracting their mean and dividing by their standard deviation) before looking at them, because then they can compare the residuals with values on a *Z*-distribution. If you take this step also, you can ask Minitab to give you a series of four standardized residual plots with which to check the conditions. (See Chapter 4 for full details on standardized residuals and residual plots.)

Figure 7-10 shows the standardized residual plots for the quadratic model, using the quiz-score data from the previous sections.

✔ The upper-left plot shows that the standardized residuals resemble a normal distribution because your data and the normal distribution match up pretty well, point for point.

✔ The upper-right plot shows that most of the standardized residuals fall between –2 and +2 (see Chapter 4 for more on standardized residuals).

✔ The lower-left plot shows that the residuals bear some resemblance to a normal distribution.

✔ The lower-right plot demonstrates how the residuals have no pattern. They appear to occur at random.

When taken together, all these plots suggest that the conditions on the residual are met to apply the selected quadratic regression model.

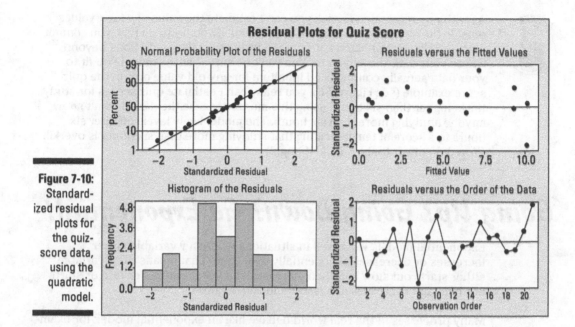

**Figure 7-10:**
Standard-
ized residual
plots for
the quiz-
score data,
using the
quadratic
model.

## Making predictions

After you've found the model that fits well, you can use that model to make predictions for *y* given *x*. Simply plug in the desired *x*-value, and out comes your predicted value for *y*. (Make sure any values you plug in for *x* occur within the range of where data were collected; if not, you can't guarantee the model holds.)

Returning to the quiz-score data from previous sections, can you use study time to predict quiz score by using a quadratic regression model? By looking at the scatterplot and the value of $R^2$ adjusted (review Figures 7-8 and 7-9, respectively), you can see that the quadratic regression model appears to fit the data well. (Isn't it nice when you find something that fits?) The residual plots in Figure 7-10 indicate that the conditions seem to be met to fit this model; you can find no major patterns in the residuals, they appear to center at 1, and most of them stay within the normal boundaries of standardized residuals of –2 and +2.

Considering all this evidence together, study time does appear to have a quadratic relationship with quiz score in this case. You can now use the model to make estimates of quiz score given study time. For example, because the model (shown in Figure 7-8) is $y = 9.82 – 6.15x + 1.00x^2$, if your study time is 5.5 hours, then your estimated quiz score is $9.82 – 6.15 * 5.5 + 1.00 * 5.5^2 = 9.82 – 33.83 + 30.25 = 6.25$. This value makes sense according to what you see on the graph in Figure 7-6 if you look at the place where *x* = 5.5; the *y*-values are in the vicinity of 6 to 7.

As with any regression model, you can't estimate the value of *y* for *x*-values outside the range of where data was collected. If you try to do this, you commit a no-no called *extrapolation*. It refers to trying to make predictions beyond where your data allows you to. You can't be sure that the model you fit to your data actually continues ad infinitum for any old value of *x*. In the quiz-score example (see Figure 7-6), you really can't estimate quiz scores for study times higher than six hours using this model because the data doesn't show anyone studying more than six hours. The model likely levels off after six hours to a score of ten, indicating that studying more than six hours is overkill. (You didn't hear that from me though!)

# Going Up? Going Down? Go Exponential!

Exponential models work well in situations where a *y* variable either increases or decreases exponentially over time. That means the *y* variable either starts out slow and then increases at a faster and faster rate or starts out high and decreases at a faster and faster rate.

Many processes in the real world behave like an exponential model: for example, the change in population size over time, average household incomes over time, the length of time a product lasts, or the level of patience one has as the number of statistics homework problems goes up.

In this section, you familiarize yourself with the exponential regression model and see how to use it to fit data that either rise or fall at an exponential rate. You also discover how to build and assess exponential regression models in order to make accurate predictions for a response variable *y*, using an explanatory variable *x*.

## Recollecting exponential models

Exponential models have the form $y = \alpha\beta^x$. These models involve a constant, $\beta$, raised to higher and higher powers of *x* multiplied by a constant, $\alpha$. The constant $\beta$ represents the amount of curvature in the model. The constant $\alpha$ is a multiplier in front of the model that shows where the model crosses the *y*-axis (because when $x = 0$, $y = \alpha * 1$).

An exponential model generally looks like the upper part of a hyperbola (remember those from advanced algebra?). A *hyperbola* is a curve that crosses the *y*-axis at a point and curves downward toward zero or starts at some point and curves upward to infinity (see Figure 7-11 for examples). If $\beta$ is greater than 1 in an exponential model, the graph curves upward toward infinity. If $\beta$ is less than 1, the graph curves downward toward zero. All exponential models stay above the *x*-axis.

For example, the model $y = 1 * 3^x$ is an exponential model. Here, suppose you make $\alpha = 1$, indicating that the model crosses the $y$-axis at 1 (because plugging $x = 0$ into the equation gives you 1). You set the value of $\beta$ equal to 3, indicating that you want a bit of curvature to this model. The $y$-values curve upward quickly from the point (0, 1). For example, when $x = 1$, you get $1 * 3^1 = 3$; for $x = 2$, you get $1 * 3^2 = 9$; for $x = 3$, you get $1 * 3^3 = 27$; and so on. Figure 7-11a shows a graph of this model. Notice the huge scale needed on the $y$-axis when $x$ is only 10.

Now suppose you let $\alpha = 1$ and $\beta = 0.5$. These values give you the model $y = 1 * 0.5^x$. This model takes 0.5 (a fraction between 0 and 1) to higher and higher powers starting at $1 * 0.5^0 = 1$, which makes the $y$-values smaller and smaller, never reaching zero but always getting closer. (For example, 0.5 to the second power is 0.25, which is less than 0.50, and 0.50 to the tenth power is 0.00098.) Figure 7-11b shows a graph of this model.

**Figure 7-11:**
The exponential regression model for different values of $\beta$.

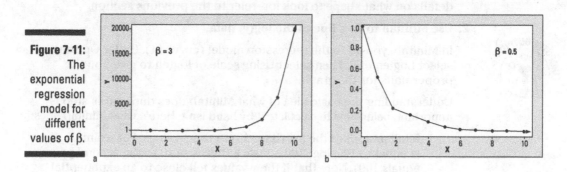

# Searching for the best exponential model

Finding the best-fitting exponential model requires a bit of a twist compared to finding the best-fitting line by using simple linear regression (see Chapter 4). Because fitting a straight-line model is much easier than fitting an exponential model directly from data, you transform the data into something for which a line fits. Then you fit a straight-line model to that transformed data. Finally you undo the transformation, getting you back to an exponential model.

For the transformation, you use *logarithms* because they're the inverse of exponentials. But before you start sweating, don't worry; these math gymnastics aren't something you do by hand — the computer does most of the grunt work for you.

The exponential model looks like this (if you're using base 10): $y = 10^{b_0 + b_1 x}$; note the equation of the line is in the exponent. Follow these steps for fitting an exponential model to your data and using it to make predictions:

The math magic used in these steps is courtesy of the definition of logarithm, which says $\log_b(a) = y \Leftrightarrow b^y = a$. Suppose you have the equation $\log_{10} y = 2 + 3x$. If you take ten to the power of each side, you get $10^{\log_{10}(y)} = 10^{2+3x}$. By the definition of logarithm, the tens cancel out on the left side and you get $y = 10^{2+3x}$. This model is exponential because $x$ is in the exponent. You can take step two up another notch to include the general form of the straight line model $y = b_0 + b_1 x$. Using the definition of logarithm on this line, you get $\log_{10}(y) = b_0 + b_1 x \Leftrightarrow 10^{b_0 + b_1 x} = y$.

1. **Make a scatterplot of the data and see whether the data appears to have a curved pattern that resembles an exponential curve.**

   If the data follows an exponential curve, proceed to step two; otherwise, consider alternative models (such as multiple regression in Chapter 5).

   Chapter 4 tells you how to make a scatterplot in Minitab. For more details on what shape to look for, refer to the previous section.

2. **Use Minitab to fit a line to the log($y$) data.**

   In Minitab, you go to the regression model (curve fit). Under Options, select Logten of $y$. Then select Using scale of logten to give you the proper units for the graph.

   Understanding the basic idea of what Minitab does during this step is important; being able to calculate it by hand isn't. Here's what Minitab does:

   1. Minitab applies the log (base 10) to the $y$-values. For example, if $y$ is equal to 100, $\log_{10} 100$ equals 2 (because 10 to the second power equals 100). Note that if the $y$-values fell close to an exponential model before, the log($y$) values will fall close to a straight-line model. This phenomenon occurs because the logarithm is the inverse of the exponential function, so they basically cancel each other out and leave you with a straight line.

   2. Minitab fits a straight line to the log($y$) values by using simple linear regression (see Chapter 4). The equation of the best-fitting straight line for the log($y$) data is log($y$) = $b_0 + b_1 x$. Minitab passes this model on to you in its output, and you take it from there.

3. **Transform the model back to an exponential model by starting with the straight-line model, log($y$) = $b_0 + b_1 x$, that was fit to the $\log_{10}(y)$ data and then applying ten to the power of the left side of the equation and ten to the power of the right side.**

   By the definition of logarithm, you get $y$ on the left side of the model and ten to the power of $b_0 + b_1 x$ on the right side. The resulting exponential model for $y$ is $y = 10^{b_0 + b_1 x}$.

4. **Use the exponential model in step three to make predictions for $y$ (your original variable) by plugging your desired value of $x$ into the model.**

   Only plug in values for $x$ that are in the range of where the data are located.

5. **Assess the fit of the model by looking at the scatterplot of the log(*y*) data, checking out the value of $R^2$ (adjusted) for the straight-line model for log(*y*), and checking the residual plots for the log(*y*) data.**

   The techniques and criteria you use to do this are the same as those I discuss in the previous section "Assessing the fit of a polynomial model."

If these steps seem dubious to you, stick with me. The example in the next section lets you see each step firsthand, which helps a great deal. In the end, actually finding predictions by using an exponential model is a lot easier to do than it is to explain.

## Spreading secrets at an exponential rate

Often, the best way to figure something out is to see it in action. Using the secret-spreading example from Figure 7-2, you can work through the series of steps from the preceding section to find the best-fitting exponential model and use it to make predictions.

### Step one: Check the scatterplot

Your goal in step one is to make a scatterplot of the secret-spreading data and determine whether the data resemble the curved function of an exponential model. Figure 7-2 shows the data for the spread of a number of people knowing the secret, as a function of the number of days. You can see that the number of people who know the secret starts out small, but then as more and more people tell more and more people, the number grows quickly until the secret isn't a secret anymore. This is a good situation for an exponential model, due to the amount of upward curvature in this graph.

### Step two: Let Minitab do its thing to log (y)

In step two, you let Minitab find the best-fitting line to the log(*y*) data (see the section "Searching for the best exponential model" to find out how to do this in Minitab). The output for the analysis of the secret-spreading data is in Figure 7-12; you can see that the best-fitting line is log(*y*) = –0.19 + 0.28 * *x*, where *y* is the number of people knowing the secret and *x* is the number of days.

**Figure 7-12:** Minitab fits a line to the log(*y*) for the secret-spreading data.

```
Regression Analysis: Day versus Number

The regression equation is
logten (number) = -0.1883 + 0.2805 day

S = 0.157335    R-Sq = 93.3%    R-Sq(adj) = 91.6%
```

### Step three: Go exponential

After you have your Minitab output, you're ready for step three. You transform the model $\log(y) = -0.19 + 0.28 * x$ into a model for $y$ by taking 10 to the power of the left-hand side and 10 to the power of the right-hand side. Transforming the $\log(y)$ equation for the secret-spreading data, you get $y = 10^{-0.19 + 0.28x}$.

### Step four: Make predictions

By using the exponential model from step three, you can move on to step four: Make predictions for appropriate values of $x$ (within the range of where data was collected). Continuing to use the secret-spreading data, suppose you want to estimate the number of people knowing the secret on day five (see Figure 7-2). Just plug $x = 5$ into the exponential model to get $y = 10^{-0.19 + 0.28 * 5} = 10^{1.21} = 16.22$. Looking back at Figure 7-2, you can see that this estimate falls right in line with the data on the graph.

### Step five: Assess the fit of your exponential model

Now that you've found the best-fitting exponential model, you have the worst behind you. You've arrived at step five and are ready to further assess the model fit (beyond the scatterplot of the original data) to make sure no major problems arise.

In general, to assess the fit of an exponential model, you're really looking at the straight-line fit of $\log(y)$. Just use these three items (in any order) in the same way as described in the earlier section "Assessing the fit of a polynomial model":

✔ **Check the scatterplot of the $\log(y)$ data to see how well it resembles a straight line.** You assess the fit of the $\log(y)$ for the secret-spreading data first through the scatterplot shown in Figure 7-13. The scatterplot shows that the model appears to fit the data well, because the points are scattered in a tight pattern around a straight line.

During this process the data were transformed also. You started with $x$ and $y$ data, and now you have $x$ and $\log(y)$ for your data. You see $x$, $y$, and $\log(y)$ for the secret-spreading data in Table 7-2.

✔ **Examine the value of $R^2$ adjusted for the model of the best-fitting line for $\log(y)$, done by Minitab.** The value of $R^2$ adjusted for this model is found in Figure 7-13 to be 91.6 percent. This value also indicates a good fit because it's very close to 100 percent. Therefore, 91.6 percent of the variation in the number of people knowing the secret is explained by how many days it has been since the secret-spreading started. (Makes sense.)

✔ **Look at the residual plots from the fit of a line to the $\log(y)$ data.** The residual plots from this analysis (see Figure 7-14) show no major departures from the conditions that the errors are independent and have a normal distribution. Note that the histogram in the lower-left corner doesn't look all that bell-shaped, but you don't have a lot of data in this example, and the rest of the residual plots seem okay. So, you have little cause to really worry.

| Table 7-2 | Log($y$) Values for the Secret-Spreading Data | |
| --- | --- | --- |
| x (Day) | y (Number of People) | log(y) |
| 1 | 1 | 0.00 |
| 2 | 2 | 0.30 |
| 3 | 5 | 0.70 |
| 4 | 7 | 0.85 |
| 5 | 17 | 1.23 |
| 6 | 30 | 1.48 |

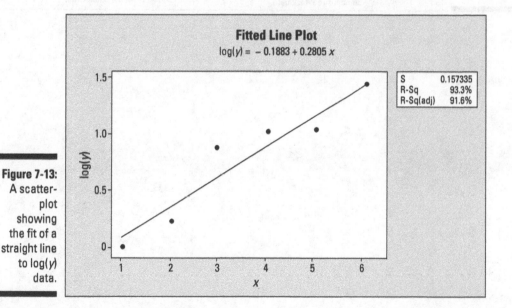

**Figure 7-13:**
A scatter-plot showing the fit of a straight line to log($y$) data.

All in all, it appears that the secret's out on the secret-spreading data, now that you have an exponential model that explains how it happens.

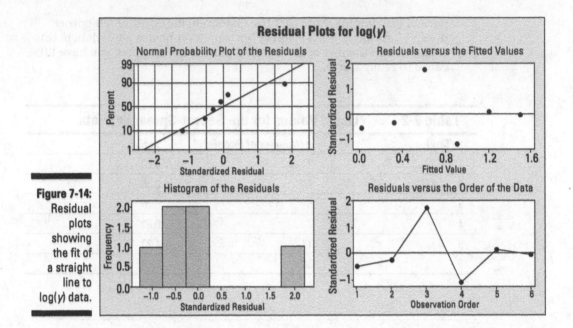

**Figure 7-14:**
Residual plots showing the fit of a straight line to log(*y*) data.

# Chapter 8

# Yes, No, Maybe So: Making Predictions by Using Logistic Regression

*E*veryone (even yours truly) tries to make predictions about whether or not a certain event is going to happen. For example, what's the chance it's going to rain this weekend? What are your team's chances of winning the next game? What's the chance that I'll have complications during this surgery? These predictions are often based on *probability*, the long-term percentage of time an event is expected to happen.

In the end, you want to estimate *p*, the probability of an event occurring. In this chapter, you see how to build and test models for *p* based on a set of explanatory *(x)* variables. This technique is called *logistic regression*, and in this chapter, I explain how to put it to use.

# Understanding a Logistic Regression Model

In a logistic regression, you're estimating the probability that an event occurs for a randomly selected individual versus the probability that the event doesn't occur. In essence, you're looking at yes or no data: yes it occurred (probability = $p$); or no, it didn't occur (probability = $1 - p$). Yes or no data that come from a random sample have a binomial distribution with probability of success (the event occurring) equal to $p$.

In the binomial problems you saw in Stats I, you had a sample of size $n$ trials, you had yes or no data, and you had a probability of success on each trial, denoted by $p$. In your Stats I course, for any binomial problem the value of $p$ was somehow given to be a certain value, like a fair coin has probability $p$ = 0.50 for coming up heads. But in Stats II, you operate under the much more realistic scenario that it's not. In fact, because $p$ isn't known, your job is to estimate what it is and use a model to do that.

To estimate $p$, the chance of an event occurring, you need data that come in the form of yes or no, indicating whether or not the event occurred for each individual in the data set.

Because yes or no data don't have a normal distribution, which is a condition needed for other types of regression, you need a new type of regression model to do this job; that model is *logistic regression*.

## How is logistic regression different from other regressions?

You use logistic regression when you use a quantitative variable to predict or guess the outcome of some categorical variable with only two outcomes (for example, using barometric pressure to predict whether or not it will rain).

A logistic regression model ultimately gives you an estimate for $p$, the probability that a particular outcome will occur in a yes or no situation (for example, the chance that it will rain versus not). The estimate is based on information from one or more explanatory variables; you can call them $x_1, x_2, x_3, \ldots x_k$. (For example, $x_1$ = humidity, $x_2$ = barometric pressure, $x_3$ = cloud cover, . . . . and $x_k$ = wind speed.)

Because you're trying to use one variable $(x)$ to make a prediction for another variable $(y)$, you may think about using regression — and you would be right. However, you have many types of regression to choose from, and you need to determine what kind is most appropriate here. You need the type of regression that uses a quantitative variable *(x)* to predict the outcome of some categorical variable *(y)* that has only two outcomes (yes or no).

So being the good Stats II student that you are, you go to your trusty list of statistical techniques, and you look under regression — and immediately see more than one type.

- ✔ You see simple linear regression. No, you use that when you have one quantitative variable predicting another (see Chapter 4).

- ✔ Multiple regression? No, that method just expands simple linear regression to add more *x* variables (see Chapter 5).

- ✔ Nonlinear regression? Well no, that still works with two quantitative variables; it's just that the data form a curve, not a line.

But then you come across logistic regression, and . . . eureka! You see that logistic regression handles situations where the *x* variable is numerical and the *y* variable is categorical with two possible categories. Just what you're looking for!

Logistic regression, in essence, estimates the probability of *y* being in one category or the other, based on the value of some quantitative variable, *x*. For example, suppose you want to predict someone's height based on gender. Because gender is a categorical variable, you use logistic regression to make these predictions. Suppose a 1 indicates a male. People who receive a probability of more than 0.5 of being male (based on their heights) are predicted to be male, and people who receive a probability of less than 0.5 of being male (based on their heights) are predicted to be female.

In this chapter, I present only the case where you use one explanatory variable to predict the outcome. You can extend the ideas in exactly the same way as you can extend the simple linear regression model to a multiple regression model.

## Using an S-curve to estimate probabilities

In a simple linear regression model, the general form of a straight line is $y = \beta_0 + \beta_1 x$ and *y* is a quantitative variable. In the logistic regression model, the *y* variable is categorical, not quantitative. What you're estimating, however, is not which category the individual lies in, but rather what the probability is that the individual lies in a certain category. So, the model for logistic regression is based on estimating this probability, called *p*.

If you were to estimate $p$ using a simple linear regression model, you may think you should try to fit a straight line, $p = \beta_0 + \beta_1 x$. However, it doesn't make sense to use a straight line to estimate the probability of an event occurring based on another variable, due to the following reasons:

- **The estimated values of $p$ can never be outside of [0, 1], which goes against the idea of a straight line (a straight line continues on in both directions).**

- **It doesn't make sense to force the values of $p$ to increase in a linear way based on $x$.** For example, an event may occur very frequently with a range of large values of $x$ and very frequently with a range of small values of $x$, with very little chance of the event happening in an area in between. This type of model would have a U shape rather than a straight-line shape.

To come up with a more appropriate model for $p$, statisticians created a new function of $p$ whose graph is called an S-curve. The *S-curve* is a function that involves $p$, but it also involves $e$ (the natural logarithm) as well as a ratio of two functions.

The values of the S-curve always fit between 0 and 1, which allows the probability, $p$, to change from low to high or high to low, according to a curve that's shaped like an S. The general form of the logistic regression model based on an S-curve is $p = \dfrac{e^{\beta_0 + \beta_1 x}}{1 + e^{\beta_0 + \beta_1 x}}$.

# Interpreting the coefficients of the logistic regression model

The sign on the parameter $\beta_1$ tells you the direction of the S-curve. If $\beta_1$ is positive, the S-curve goes from low to high (see Figure 8-1a); if $\beta_1$ is negative, the S-curve goes from high to low (Figure 8-1b).

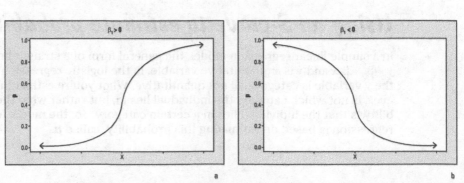

**Figure 8-1:**
Two basic types of S-curves.

The magnitude of $\beta_1$ (indicated by its absolute value) tells you how much curvature is in the model. High values indicate a steep curvature, and low values indicate gradual curvature. The parameter $\beta_0$ just shifts the S-curve to the proper location to fit your data. It shows you the cutoff point where $x$-values change from high to low probability and vice versa.

## The logistic regression model in action

Often, the best way to figure something out is to see it in action. In this section, I give you an example of a situation where you can use a logistic regression model to estimate a probability. (I expand on this example later in this chapter; for now, I'm just setting up a scenario for logistic regression.)

Suppose movie marketers want to estimate the chance that someone will enjoy a certain family movie, and you believe age may have something to do with it. Translating this research question into $x$'s and $y$'s, the response variable $(y)$ is whether or not a person will enjoy the movie, and the explanatory variable $(x)$ is the person's age. You want to estimate $p$, the chance of someone enjoying the movie.

You collect data on a random sample of 40 people, shown in Table 8-1. Based on your data, it appears that younger people enjoyed the movie more than older people and that at a certain age, the trend switches from liking the movie to disliking it. Armed with this data, you can build a logistic regression model to estimate $p$.

| Table 8-1 | Movie Enjoyment (Yes or No Data) Based on Age | |
|---|---|---|
| *Age* | *Enjoyed the Movie* | *Total Number Sampled* |
| 10 | 3 | 3 |
| 15 | 4 | 4 |
| 16 | 3 | 3 |
| 18 | 2 | 3 |
| 20 | 2 | 3 |
| 25 | 2 | 4 |
| 30 | 2 | 4 |
| 35 | 1 | 5 |
| 40 | 1 | 6 |
| 45 | 0 | 3 |
| 50 | 0 | 2 |

# Carrying Out a Logistic Regression Analysis

The basic idea of any model-fitting process is to look at all possible models you can have under the general format and find the one that fits your data best.

The general form of the best-fitting logistic regression model is $\hat{p} = \dfrac{e^{b_0+b_1x}}{1+e^{b_0+b_1x}}$, where $\hat{p}$ is the estimate of $p$, $b_0$ is the estimate of $\beta_0$, and $b_1$ is the estimate of $\beta_1$ (from the previous section "Using an S-curve to estimate probabilities"). The only values you have a choice about to form your particular model are the values of $b_0$ and $b_1$. These values are the ones you're trying to estimate through the logistic regression analysis.

To find the best-fitting logistic regression model for your data, complete the following steps:

1. **Run a logistic regression analysis on the data you collected (see the next section).**

2. **Find the coefficients of constant and $x$, where $x$ is the name of your explanatory variable.**

   These coefficients are $b_0$ and $b_1$, the estimates of $\beta_0$ and $\beta_1$ in the logistic regression model.

3. **Plug the coefficients from step one into the logistic regression model:**
   $$\hat{p} = \frac{e^{b_0+b_1x}}{1+e^{b_0+b_1x}}.$$

   This equation is your best-fitting logistic regression model for the data. Its graph is an S-curve (for more on the S-curve, see the section "Using an S-curve to estimate probabilities" earlier in this chapter).

In the sections that follow, you see how to ask Minitab to do the above steps for you. You also see how to interpret the resulting computer output, find the equation of the best-fitting logistic regression model, and use that model to make predictions (being ever mindful that all conditions are met).

## Running the analysis in Minitab

Here's how to perform a logistic regression using Minitab (other statistical software packages are similar):

1. **Input your data in the spreadsheet as a table that lists each value of the** $x$ **variable in column one, the number of yeses for that value of** $x$ **in column two, and the total number of trials at that** $x$**-value in column three.**

   These last two columns represent the outcome of the response variable $y$. (For an example of how to enter your data, see Table 8-1 based on the movie and age data.)

2. **Go to Stat>Regression>Binary Logistic Regression.**

3. **Beside the Success option, select your variable name from column two, and beside Trial, select your variable name for column three.**

4. **Under Model, select your variable name from column one, because that's the column containing the explanatory** $(x)$ **variable in your model.**

5. **Click OK, and you get your logistic regression output.**

When you fit a logistic regression model to your data, the computer output is composed of two major portions:

✔ **The model-building portion:** In this part of the output, you can find the coefficients $b_0$ and $b_1$. (I describe coefficients in the next section.)

✔ **The model-fitting portion:** You can see the results of a Chi-square goodness-of-fit test (see Chapter 15) as well as the percentage of concordant and discordant pairs in this section of the output. (A *concordant pair* means the predicted outcome from the model matches the observed outcome from the data. A *discordant pair* is one that doesn't match.)

In the case of the movie and age data, the model-building part of the Minitab output is shown in Figure 8-2. The model-fitting part of the Minitab output from the logistic regression analysis is in Figure 8-4.

In the following sections, you see how to use this output to build the best-fitting logistic regression model for your data and to check the model's fit.

**Figure 8-2:**
The model-building part of the movie and age data's logistic regression output.

```
Logistic Regression Table
                                                    Odds      95% CI
Predictor      Coef     SE Coef      Z       P     Ratio   Lower   Upper
Constant     4.86539   1.43434     3.39    0.001
Age         -0.175745  0.0499620  -3.52    0.000    0.84    0.76    0.93
```

## *Finding the coefficients and making the model*

After you have Minitab run a logistic regression analysis on your data, you can find the coefficients $b_0$ and $b_1$ and put them together to form the best-fitting logistic regression model for your data.

Figure 8-2 shows part of the Minitab output for the movie enjoyment and age data. (I discuss the remaining output in the section "Checking the fit of the model.") The first column of numbers is labeled Coef, which stands for the coefficients in the model. The first coefficient, $b_0$, is labeled Constant. The second coefficient is in the row labeled by your explanatory variable, $x$. (In the movie and age data, the explanatory variable is age. This age coefficient represents the value of $b_1$ in the model.)

According to the Minitab output in Figure 8-2, the value of $b_0$ is 4.87 and the value of $b_1$ is –0.18. After you've determined the coefficients $b_0$ and $b_1$ from the Minitab output to find the best-fitting S-curve for your data, you put these two coefficients into the general logistic regression model: $\hat{p} = \dfrac{e^{b_0 + b_1 x}}{1 + e^{b_0 + b_1 x}}$.

For the movie and age data, you get $\hat{p} = \dfrac{e^{4.87 - 0.18x}}{1 + e^{4.87 - 0.18x}}$, which is the best-fitting logistic regression model for this data set.

The graph of the best-fitting logistic regression model for the movie and age data is shown in Figure 8-3. Note that the graph is a downward-sloping S-curve because higher probabilities of liking the movie are affiliated with lower ages and lower probabilities are affiliated with higher ages.

The movie marketers now have the answer to their question. This movie has a higher chance of being well liked by kids (and the younger, the better) and a lower chance of being well liked by adults (and the older they are, the lower the chance of liking the movie).

The point where the probability changes from high to low (that is, at the $\hat{p} = 0.50$ mark) is between ages 25 and 30. That means that the tide of probability of liking the movie appears to turn from higher to lower in that age range. Using calculus terms, this point is called the *saddle point* of the S-curve, which is the point where the graph changes from concave up to concave down, or vice versa.

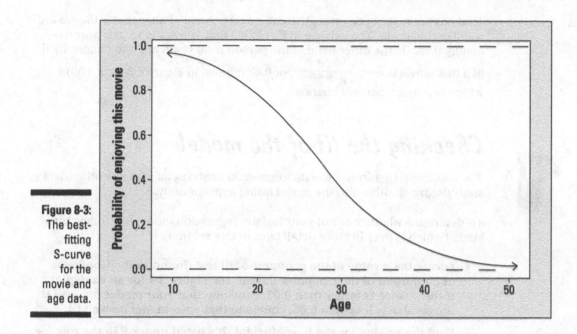

**Figure 8-3:**
The best-fitting
S-curve
for the
movie and
age data.

*(Graph: y-axis labeled "Probability of enjoying this movie" from 0.0 to 1.0; x-axis labeled "Age" from 10 to 50.)*

# Estimating p

You've determined the best-fitting logistic regression model for your data, obtained the values of $b_0$ and $b_1$ from the logistic regression analysis, and know the precise S-curve that fits your data best (check out the previous sections). You're now ready to estimate $p$ and make predictions about the probability that the event of interest will happen, given the value of the explanatory variable $x$.

To estimate $p$ for a particular value of $x$, plug that value of $x$ into your equation (the best-fitting logistic regression model) and simplify it by using your algebra skills. The number you get is the estimated chance of the event occurring for that value of $x$, and it should be a number between 0 and 1, being a probability and all.

Continuing with the movie and age example from the preceding sections, suppose you want to predict whether a 15-year-old would enjoy the movie. To estimate $p$, plug 15 in for $x$ in the logistic regression model $\hat{p} = \dfrac{e^{b_0 + b_1 x}}{1 + e^{b_0 + b_1 x}}$ to get

$$\hat{p} = \frac{e^{4.87 - 0.18 \cdot 15}}{1 + e^{4.87 - 0.18 \cdot 15}} = \frac{e^{2.17}}{1 + e^{2.17}} = \frac{8.76}{9.76} = 0.90.$$

That answer means you estimate there's a 90 percent chance that a 15-year-old will like the movie. You can see in Figure 8-3 that when $x$ is 15, $p$ is approximately 0.90. On the other hand, if the person is 50 years old, the chance he'll like this movie is $\hat{p} = \dfrac{e^{4.87-0.18*50}}{1+e^{4.87-0.18*50}}$, or 0.02 (shown in Figure 8-3 for $x$ = 50), which is only a 2 percent chance.

## Checking the fit of the model

The results you get from a logistic regression analysis, as with any other data analysis, are all subject to the model fitting appropriately.

To determine whether or not your logistic regression model fits, follow these steps (which I cover in more detail later in this section):

1. **Locate the $p$-value of the goodness-of-fit test (found in the Goodness-of-Fit portion of the computer output; see Figure 8-4 for an example). If the $p$-value is larger than 0.05, conclude that your model fits, and if the $p$-value is less than 0.05, conclude that your model doesn't fit.**

2. **Find the $p$-value for the $b_1$ coefficient (it's listed under $P$ in the row for your column one (explanatory) variable in the model-building portion of the output; see Figure 8-2 for an example). If the $p$-value is less than 0.05, the $x$ variable is statistically significant in the model, so it should be included. If the $p$-value is greater than or equal to 0.05, the $x$ variable isn't statistically significant and shouldn't be included in the model.**

3. **Look later in the output at the percentage of concordant pairs. This percentage reflects the proportion of time that the data and the model actually agree with each other. The higher the percentage, the better the model fits.**

The conclusion in step one based on the $p$-value may seem backward to you, but here's what's happening: Chi-square goodness-of-fit tests measure the overall difference between what you expect to see via your model and what you actually observe in your data. (Chapter 15 gives you the lowdown on Chi-square tests.) The null hypothesis (Ho) for this test says you have a difference of zero between what you observed and what you expected from the model; that is, your model fits. The alternative hypothesis, denoted Ha, says that the model doesn't fit. If you get a small $p$-value (under 0.05), reject Ho and conclude the model doesn't fit. If you get a larger $p$-value (above 0.05), you can stay with your model.

Failure to reject Ho here (having a large $p$-value) only means that you can't say your model doesn't fit the population from which the sample came. It doesn't necessarily mean the model fits perfectly. Your data could be unrepresentative of the population just by chance.

# Fitting the movie model

You're ready to check out the fit of the movie data to make sure you still have a job when the box office totals come in.

### Step one: p-value for Chi-squared

Using Figure 8-4 to complete the first step of checking the model's fit, you can see many different goodness-of-fit tests. The particulars of each of these tests are beyond the scope of this book; however, in this case (as with most cases), each test has only slightly different numerical results and the same conclusions.

All the *p*-values in column four of Figure 8-4 are over 0.80, which is much higher than the 0.05 you need to reject the model. After looking at the *p*-values, the model using age to predict movie likeability appears to fit this data.

**Figure 8-4:**
The model-fitting part of the movie and age data's logistic regression output.

```
Goodness-of-Fit Test
Method             Chi-Square   DF       P
Pearson            2.83474       9     0.970
Deviance           3.63590       9     0.934
Hosmer-Lemeshow    2.75232       6     0.839

Measures of Association:
 (Between the Response Variable and Predicted Probabilities)
Pairs         Number   Percent   Summary Measures
Concordant      349     87.3     Somers' D               0.80
Discordant       30      7.5     Goodman-Kruskal Gamma   0.84
Ties             21      5.3     Kendall's Tau-a         0.41
Total           400    100.0
```

### Step two: p-value for the x variable

For step two, you look at the significance of the *x* variable age. Back in Figure 8-2, you can see the constant for age, –0.18, and farther along in its row, you can see that the *Z*-value is –3.52; this *Z*-value is the test statistic for testing Ho: $\beta_1 = 0$ versus Ha: $\beta_1 \neq 0$. The *p*-value is listed as 0.000, which means it's smaller than 0.001 (a highly significant number). So you know that the coefficient in front of *x*, also known as $\beta_1$, is statistically significant (not equal to zero), and you should include *x* (age) in the model.

### Step three: Concordant pairs

To complete step three of the fit-checking process, look at the percentage of concordant pairs reported in Figure 8-4. This value shows the percentage of times the data actually agreed with the model (87.3). To determine concordance, the computer makes predictions as to whether the event should have occurred for each individual based on the model and compares those results to what actually happened.

The logistic regression model is for $p$, the probability of the event occurring, so if $p$ is estimated to be $> 0.50$ for some value of $x$, the computer predicts that the event will occur (versus not occurring). If the estimated value of $p$ is $< 0.50$ for a particular $x$-value, the computer predicts that it won't occur.

For the movie and age data, the percentage of concordant pairs (that is, the percentage of times the model made the right decision in predicting what would happen) is 87.3 percent, which is quite high.

The percentage of concordant pairs was obtained by taking the number of concordant pairs and dividing by the total number of pairs. I'd start getting excited if the percentage of concordant pairs got over 75 percent; the higher, the better.

Figure 8-5 shows the logistic regression model for the movie and age data, with the actual values of the observed data added as circles. The S-curve shows the probability of liking the movie for each age level, and the computer will predict "1" = they will like the movie, if $\hat{p} > 0.50$. Circles indicate whether the people of those age levels actually liked the movie ($y = 1$) or not ($y = 0$).

Much of the time, the model made the right decision; probabilities above 0.50 are associated with more circles at the value of 1, and probabilities below 0.50 are associated with more circles at the value of zero. It's the outcomes that have $p$ near 0.50 that are hard to predict because the results can go either way.

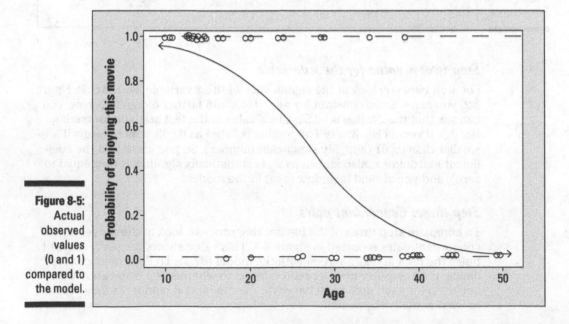

**Figure 8-5:**
Actual observed values (0 and 1) compared to the model.

## Which method to use to compare? Sorting out similar situations

Data come in a variety of forms, and each form has its own analysis to use to make comparisons. It can get difficult to decide which type of analysis to use when.

It may help to sort out some situations that sound similar but have subtle differences that lead to very different analyses. You can use the following list to compare these subtle, but important, differences:

✔ **If you want to compare three or more groups of numerical variables,** use ANOVA (see Chapter 10). For only two groups, use a *t*-test (see Chapters 3 and 9).

✔ **If you want to estimate one numerical variable based on another,** use simple linear regression (see Chapter 4).

✔ **If you want to estimate one numerical variable using many other numerical variables,** use multiple regression (see Chapter 5).

✔ **If you want to estimate a categorical variable with two categories by using a numerical variable,** use logistic regression, which is the focus of this chapter, of course.

✔ **If you want to compare two categorical variables to each other,** head straight for a Chi-square test (see Chapter 14).

All this evidence helps confirm that your model fits your data well. You can go ahead and make predictions based on this model for the next individual that comes up, whose outcome you don't know (see the section "Estimating *p*" earlier in this chapter).

# Part III
# Analyzing Variance with ANOVA

The 5th Wave                    By Rich Tennant

"This is my old Statistics II professor
his wife Doris, and their two children,
Wilcox and Kruskal."

## In this part . . .

Y ou get all the nuts and bolts you need to understand one-way and two-way analyses of variance (also known as ANOVA), which compare the means of several populations at one time based on one or two different characteristics. You see how to read and understand ANOVA tables and computer output, and you get to go behind the scenes to understand the big ideas behind the formulas used in ANOVA. Finally, you see lots of different multiple comparison procedures to zoom in on which means are different. (Don't sweat it. I always present formulas only on a need-to-know basis.)

# Chapter 9

# Testing Lots of Means? Come On Over to ANOVA!

*O*ne of the most commonly used statistical techniques at the Stats II level is *analysis of variance* (affectionately known as ANOVA). Because the name has the word *variance* in it, you may think that this technique has something to do with variance — and you would be right. Analysis of variance is all about examining the amount of variability in a *y* (response) variable and trying to understand where that variability is coming from.

One way you can use ANOVA is to compare several populations regarding some quantitative variable, *y*. The populations you want to compare constitute different groups (denoted by an *x* variable), such as political affiliations, age groups, or different brands of a product. ANOVA is also particularly suitable for situations involving an experiment in which you apply certain treatments *(x)* to subjects and measure a response *(y)*.

In this chapter, you start with the *t*-test for two population means, the precursor to ANOVA. Then you move on to the basic concepts of ANOVA to compare more than two means: sums of squares, the *F*-test, and the ANOVA table. You apply these basics to the *one-factor* or *one-way ANOVA*, where you compare the responses based only on one treatment variable. (In Chapter 11, you can see the basics applied to a two-way ANOVA, which has two treatment variables.)

# Comparing Two Means with a t-Test

The *two-sample t-test* is designed to test to see whether two population means are different. The conditions for the two-sample t-test are the following:

✔ The two populations are independent. In other words, their outcomes don't affect each other.

✔ The response variable *(y)* is a quantitative variable, meaning its values have numerical meaning and represent quantities of some kind.

✔ The y-values for each population have a normal distribution. However, their means may be different; that's what the t-test determines.

✔ The variances of the two normal distributions are equal.

For large sample sizes when you know the variances, you use a Z-test for the two population means. However, a t-test allows you to test two population means when the variances are unknown or the sample sizes are small. This occurs quite often in situations where an experiment is performed and the number of subjects is limited. (See your Stats I text or *Statistics For Dummies* (Wiley) for info on the Z-test.)

Although you've seen t-tests before in your Stats I class, it may be good to review the main ideas. The t-test tests the hypotheses Ho: $\mu_1 = \mu_2$ versus Ha: $\mu_1$ is ≤, ≥, or ≠ $\mu_2$, where the situation dictates which of these hypotheses you use. (*Note:* With ANOVA, you extend this idea to $k$ different means from $k$ different populations, and the only version of Ha of interest is ≠.)

To conduct the two-sample t-test, you collect two data sets from the two populations, using two independent samples. To form the test statistic (the t-statistic), you subtract the two sample means and divide by the *standard error* (a combination of the two standard deviations from the two samples and their sample sizes). You compare the t-statistic to the t-distribution with $n_1 + n_2 - 2$ degrees of freedom and find the p-value. If the p-value is less than the predetermined $\alpha$ level, say 0.05, you have enough evidence to say the population means are different. (For information on hypothesis tests, see Chapter 3.)

For example, suppose you're at a watermelon seed-spitting contest where contestants each put watermelon seeds in their mouths and spit them as far as they can. Results are measured in inches and are treated with the reverence of the shot-put results at the Olympics. You want to compare the watermelon seed-spitting distances of female and male adults. Your data set includes ten people from each group.

You can see the results of the t-test in Figure 9-1. The mean spitting distance for females was 47.8 inches; the mean for males was 56.5 inches; and the difference (females – males) is –8.71 inches, meaning the females in the sample spit seeds at shorter distances, on average, than the males. The t-statistic for the difference in the two means (females – males) is $t = -2.23$, which has a $p$-value of 0.039 (see the last line of the output in Figure 9-1). At a level of $\alpha = 0.05$, this difference is significant (because $0.039 < 0.05$). You conclude that males and females differ with respect to their mean watermelon seed-spitting distance. And you can say males are likely spitting farther because their sample mean was higher.

**Figure 9-1:**
A *t*-test
comparing
mean
watermelon
seed-
spitting
distances
for females
versus
males.

```
Two-sample T for females vs males

            N    Mean    StDev    SE Mean
females     10   47.80   9.02     2.9
males       10   56.50   8.45     2.7


Difference = mu (females) - mu (males)
Estimate for difference:  -8.70000
95% CI for difference:  (-16.90914, -0.49086)
T-Test of difference = 0 (vs not =): T-Value = -2.23 P-Value = 0.039 DF = 18
```

# Evaluating More Means with ANOVA

When you can compare two independent populations inside and out, at some point two populations will not be enough. Suppose you want to compare more than two populations regarding some response variable $(y)$. This idea kicks the $t$-test up a notch into the territory of ANOVA. The ANOVA procedure is built around a hypothesis test called the *F-test*, which compares how much the groups differ from each other compared to how much variability is within each group. In this section, I set up an example of when to use ANOVA and show you the steps involved in the ANOVA process. You can then apply the ANOVA steps to the following example throughout the rest of the chapter.

## Spitting seeds: A situation just waiting for ANOVA

Before you can jump into using ANOVA, you must figure out what question you want answered and collect the necessary data.

Suppose you want to compare the watermelon seed-spitting distances for four different age groups: 6–8 years old, 9–11, 12–14, and 15–17. The hypotheses for this example are Ho: $\mu_1 = \mu_2 = \mu_3 = \mu_4$ versus Ha: At least two of these means are different, where the population means $\mu$ represent those from the age groups, respectively.

Over the years of this contest, you've collected data on 200 children from each age group, so you have some prior ideas about what the distances typically look like. This year, you have 20 entrants, 5 in each age group. You can see the data from this year, in inches, in Table 9-1.

| Table 9-1 | Watermelon Seed-Spitting Distances for Four Age Groups of Children (Measured in Inches) | | |
|---|---|---|---|
| *6–8 Years* | *9–11 Years* | *12–14 Years* | *15–17 Years* |
| 38 | 38 | 44 | 44 |
| 39 | 39 | 43 | 47 |
| 42 | 40 | 40 | 45 |
| 40 | 44 | 44 | 45 |
| 41 | 43 | 45 | 46 |

Do you see a difference in distances for these age groups based on this data? If you were to just combine all the data, you would see quite a bit of difference (the range of the combined data goes from 38 inches to 47 inches). And you may suspect that older kids can spit farther.

Perhaps accounting for which age group each contestant is in does explain at least some of what's going on. But don't stop there. The next section walks you through the official steps you need to perform to answer your question.

## Walking through the steps of ANOVA

You've decided on the quantitative response variable ($y$) you want to compare for your $k$ various population (or treatment) means, and you've collected a random sample of data from each population (refer to the preceding section). Now you're ready to conduct ANOVA on your data to see whether the population means are different for your response variable, $y$.

The characteristic that distinguishes these populations is called the *treatment variable, x.* Statisticians use the word *treatment* in this context because one of the biggest uses of ANOVA is for designed experiments where subjects are randomly assigned to treatments, and the responses are compared for the various treatment groups. So statisticians often use the word *treatment* even when the study isn't an experiment and they're comparing regular populations. Hey, don't blame me! I'm just following the proper statistical terminology.

Here are the general steps in a one-way ANOVA:

1. **Check the ANOVA conditions, using the data collected from each of the $k$ populations.**

   See the next section, "Checking the Conditions" for the specifics on these conditions.

2. **Set up the hypotheses Ho: $\mu_1 = \mu_2 = \ldots = \mu_k$ versus Ha: At least two of the population means are different.**

   Another way to state your alternative hypothesis is by saying Ha: At least two of $\mu_1, \mu_2, \ldots \mu_k$ are different.

3. **Collect data from $k$ random samples, one from each population.**

4. **Conduct an $F$-test on the data from step three, using the hypotheses from step two, and find the $p$-value.**

   See the section "Doing the $F$-Test" later in this chapter for these instructions.

5. **Make your conclusions: If you reject Ho (when your $p$-value is less than 0.05 or your predetermined $\alpha$ level), you conclude that at least two of the population means are different; otherwise, you conclude that you didn't have enough evidence to reject Ho (you can't say the means are different).**

If these steps seem like a foreign language to you, don't fear — I describe each in detail in the sections that follow.

# Checking the Conditions

Step one of ANOVA is checking to be sure all necessary conditions are met before diving into the data analysis. The conditions for using ANOVA are just an extension of the conditions for a $t$-test (see the section "Comparing Two Means with a $t$-Test"). The following conditions all need to hold in order to conduct ANOVA:

✔ The $k$ populations are independent. In other words, their outcomes don't affect each other.

✔ The $k$ populations each have a normal distribution.

✔ The variances of the $k$ normal distributions are equal.

## Verifying independence

To check the first condition, examine how the data were collected from each of the separate populations. In order to maintain independence, the outcomes from one population can't affect the outcomes of the other populations. If the data have been collected by using a separate random sample from each population (*random* here meaning that each individual in the population had an equal chance of being selected), this factor ensures independence at the strongest level.

In the watermelon seed-spitting data (see Table 9-1), the data aren't randomly sampled from each age group because the data represent everyone who participated in the contest. But, you can argue that in most cases the seed-spitting distances from one age group don't affect the seed-spitting distances from the other age groups, so the independence assumption is relatively okay.

## Looking for what's normal

The second ANOVA condition is that each of the $k$ populations has a normal distribution. To check this condition, make a separate histogram of the data from each group and see whether it resembles a normal distribution. Data from a normal distribution should look symmetric (in other words, if you split the histogram down the middle, it looks the same on each side) and have a bell shape. Don't expect the data in each histogram to follow a normal distribution exactly (remember, it's only a sample), but it shouldn't be extremely different from a normal, bell-shaped distribution.

Because the seed-spitting data contain only five children per age group, checking conditions can be iffy. But in this case, you have past years' data for 200 children in each age group, so you can use that to check the conditions. The histograms and descriptive statistics of the seed-spitting data for the four age groups are shown in Figure 9-2, all in one panel, so you can easily compare them to each other on the same scale.

Looking at the four histograms in Figure 9-2, you can see that each graph resembles a bell shape; the normality condition isn't being severely violated here. (Red flags should come up if you see two peaks in the data, a skewed shape where the peak is off to one side, or a flat histogram, for example.)

You can use Minitab to make histograms for each of your samples and have all of them appear on one large panel, all using the same scale. To do this, go to Graph>Histogram and click OK. Choose the variables that represent data from each sample by highlighting them in the left-hand box and clicking Select. Then click on Multiple Graphs, and a new window opens. Under the Show Graph Variables option, check the following box: In separate panels of the same graph. On the Same Scales for Graphs option, check the box for *x* and the box for *y*. This option gives you the same scale on both the *x* and *y* axes for all the histograms. Then click OK.

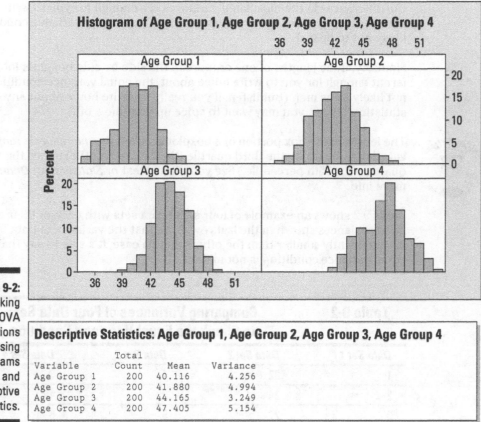

**Figure 9-2:** Checking ANOVA conditions by using histograms and descriptive statistics.

```
Descriptive Statistics: Age Group 1, Age Group 2, Age Group 3, Age Group 4

                  Total
Variable          Count    Mean     Variance
Age Group 1        200    40.116    4.256
Age Group 2        200    41.880    4.994
Age Group 3        200    44.165    3.249
Age Group 4        200    47.405    5.154
```

# *Taking note of spread*

The third condition for ANOVA is that the variance in each of the *k* populations is the same; statisticians call this the *equal variance condition*. You have two ways to check this condition on your data:

✔ Calculate each of the variances from each sample and see how they compare.

✔ Create one graph showing all the boxplots of each sample sitting side by side. This type of graph is called a *side-by-side boxplot*. (See your Stats I text or my book *Statistics For Dummies* (Wiley) for more information on boxplots.)

If one or more of the calculated variances is significantly different from the others, the equal variance condition is not likely to be met. What does *significantly different* mean? A hypothesis test for equal variances is the statistical tool used to handle this question; however, it falls outside the scope of most Stats II courses, so for now you can make a judgment call. I always say that if the differences in the calculated variances are enough for you to write home about (say they differ by 10 percent or more), the equal variance condition is likely not to be met.

Similarly, if the lengths of one or more of the side-by-side boxplots looks different enough for you to write home about, the equal variance condition is not likely to be met. (But listen, if you really do write home about any of your statistical issues, you may want to spice up your life a bit.)

The length of the box portion of a boxplot is called the *interquartile range*. You calculate it by taking the third quartile (the 75th percentile) minus the first quartile (the 25th percentile.) See your Stats I text or *Statistics For Dummies* for more info.

Table 9-2 shows an example of four small data sets with each of their calculated variances shown in the last row. Note that the variance of Data Set 4 is significantly smaller than the others. In this case, it's safe to say that the equal variance condition is not met.

| Table 9-2 | Comparing Variances of Four Data Sets to Check the Equal Variance Condition | | |
|---|---|---|---|
| **Data Set 1** | **Data Set 2** | **Data Set 3** | **Data Set 4** |
| 1 | 32 | 4 | 3 |
| 2 | 24 | 3 | 4 |
| 3 | 27 | 5 | 5 |
| 4 | 32 | 10 | 5 |
| 5 | 31 | 7 | 6 |
| 6 | 28 | 4 | 6 |
| 7 | 30 | 8 | 7 |
| 8 | 26 | 12 | 7 |
| 9 | 31 | 9 | 8 |
| 10 | 24 | 10 | 9 |
| Variance = 9.167 | Variance = 9.833 | Variance = 9.511 | Variance = 3.333 |

Figure 9-3 shows the side-by-side boxplots for these same four data sets. You see that the boxplot for Data Set 4 has an interquartile range (length of the box) that's significantly smaller than the others. I calculated the actual interquartile ranges for these four data sets; they're 5.50, 5.75, 6.00, and 2.50, respectively. These findings confirm the conclusion that the equal variance condition is not met, due to group 4's much smaller variability.

**Side-By-Side Boxplots of Data Sets 1–4**

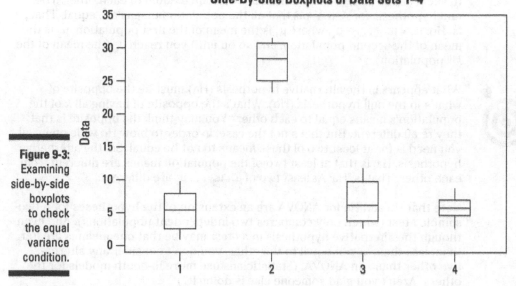

**Figure 9-3:**
Examining side-by-side boxplots to check the equal variance condition.

To find descriptive statistics (including the variance and interquartile range) for each sample, go to Stat>Basic Statistics>Display Descriptive Statistics. Click on each variable in the left-hand box for which you want the descriptive statistics, and click Select. Click on the Statistics option, and a new window appears with tons of different types of statistics. Click on the ones you want and click off the ones you don't want. Click OK. Then click OK again. Your descriptive statistics are calculated.

To find side-by-side boxplots in Minitab, go to Graph>Boxplot. A window appears. Click on the picture for Multiple Y's, Simple, and then click OK. Highlight the variables from the left-hand side that you want to compare, and click Select. Then click OK.

Note that you don't need the sample sizes in each group to be equal to carry out ANOVA; however, in Stats II, you'll typically see what statisticians call a *balanced design,* where each sample from each population has the same sample size. (As I explain in Chapter 3, for more precision in your data, the larger the sample sizes, the better.)

For the seed-spitting data, the variances for each age group are listed in Figure 9-2. These variances are close enough to say the equal variance condition is met.

# Setting Up the Hypotheses

Step two of ANOVA is setting up the hypotheses to be tested. You're testing to see if all the population means can be deemed equal to each other. The null hypothesis for ANOVA is that all the population means are equal. That is, Ho: $\mu_1 = \mu_2 = \ldots = \mu_k$, where $\mu_1$ is the mean of the first population, $\mu_2$ is the mean of the second population, and so on until you reach $\mu_k$ (the mean of the $k^{th}$ population).

What appears in the alternative hypothesis (Ha) must be the opposite of what's in the null hypothesis (Ho). What's the opposite of having all $k$ of the population's means equal to each other? You may think the opposite is that they're all different. But that's not the case. In order to blow Ho wide open, all you need is for at least two of those means to not be equal. So, the alternative hypothesis, Ha, is that at least two of the population means are different from each other. That is, Ha: At least two of $\mu_1, \mu_2, \ldots \mu_k$ are different.

Note that Ho and Ha for ANOVA are an extension of the hypotheses for a two-sample *t*-test (which only compares two independent populations). And even though the alternative hypothesis in a *t*-test may be that one mean is greater than, less than, or not equal to the other, you don't consider any alternative other than $\neq$ in ANOVA. (Statisticians use more in-depth models for the others. Aren't you glad someone else is doing it?)

You only want to know whether or not the means are equal — at this stage of the game anyway. After you reach the conclusion that Ho is rejected in ANOVA, you can proceed to figure out how the means are different, which ones are bigger than others, and so on, using multiple comparisons. Those details appear in Chapter 10.

# Doing the F-Test

Step three of ANOVA is collecting the data, and it includes taking $k$ random samples, one from each population. Step four of ANOVA is doing the *F*-test on this data, which is the heart of the ANOVA procedure. This test is the actual hypothesis test of Ho: $\mu_1 = \mu_2 = \ldots = \mu_k$ versus Ha: At least two of $\mu_1, \mu_2, \ldots \mu_k$ are different.

You have to carry out three major steps in order to complete the $F$-test. *Note:* Don't get these steps confused with the main ANOVA steps; consider the $F$-test a few steps within a step:

1. **Break down the variance of $y$ into sums of squares.**

2. **Find the mean sums of squares.**

3. **Put the mean sums of squares together to form the $F$-statistic.**

I describe each step of the $F$-test in detail and apply it to the example of comparing watermelon seed-spitting distances (see Table 9-1) in the following sections.

Data analysts rely heavily on computer software to conduct each step of the $F$-test, and you can do the same. All computer software packages organize and summarize the important information from the $F$-test into a table format for you.

This table of results for ANOVA is called (what else?) the *ANOVA table*. Because the ANOVA table is a critical part of the entire ANOVA process, I start the following sections out by describing how to run ANOVA in Minitab to get the ANOVA table, and I continue to reference this section as I describe each step of the ANOVA process.

## Running ANOVA in Minitab

In using Minitab to run ANOVA, you first have to enter the data from the $k$ samples. You can enter the data in one of two ways:

- **Stacked data:** You enter all the data into two columns. Column one includes the number indicating what sample the data value is from (1 to $k$), and the responses $(y)$ are in column two. To analyze this data, go to Stat>ANOVA>One-Way Stacked. Highlight the response $(y)$ variable, and click Select. Highlight the factor (population) variable, and click Select. Click OK.

- **Unstacked data:** You enter data from each sample into a separate column. To analyze the data entered this way, go to Stat>ANOVA>One-Way Unstacked. Highlight the names of the columns where your data are located, and click OK.

I typically use the unstacked version of data entry just because I think it helps visualize the data. However, the choice is up to you, and the results come out the same no matter which method you choose, as long as you're consistent.

# Breaking down the variance into sums of squares

Step one of the *F*-test is splitting up the variability in the *y* variable into portions that define where the variability is coming from. Each portion of variability is called a *sum of squares*. The term *analysis of variance* is a great description for exactly how you conduct a test of *k* population means. With the overall goal of testing whether *k* population (or treatment) means are equal, you take a random sample from each of the *k* populations.

You first put all the data together into one big group and measure how much total variability there is; this variability is called the *sums of squares total*, or SSTO. If the data are really diverse, SSTO is large. If the data are very similar, SSTO is small.

You can split the total variability in the combined data set (SSTO) into two parts:

  ✔ **SST:** The variability between the groups, known as the *sums of squares for treatment*

  ✔ **SSE:** The variability within the groups, known as the *sums of squares for error*

Splitting up the variability in your data results in one of the most important equalities in ANOVA:

SSTO = SST + SSE

The formula for SSTO is the numerator of the formula for $s^2$, the variance of a single data set, so SSTO $= \sum_i \sum_j \left(x_{ij} - \bar{x}\right)^2$, where *i* and *j* represent the $j^{th}$ value in the sample from the $i^{th}$ population and $\bar{x}$ is the *overall sample mean* (the mean of the entire data set). So, in terms of ANOVA, SSTO is the total squared distance between the data values and their overall mean.

The formula for SST is $\sum_i n_i \left(\bar{x}_i - \bar{x}\right)^2$, where $n_i$ is the size of the sample coming from the $i^{th}$ population and $\bar{x}$ is the overall sample mean. SST represents the total squared distance between the means from each sample and the overall mean.

The formula for SSE is $\sum_i \sum_j \left(x_{ij} - \bar{x}_i\right)^2$, where $x_{ij}$ is the $j^{th}$ value in the sample from the $i^{th}$ population and $\bar{x}_i$ is the mean of the sample coming from the $i^{th}$ population. This formula represents the total squared distance between the values in each sample and their corresponding sample means. Using algebra, you can confirm (with some serious elbow grease) that SSTO = SST + SSE.

The Minitab output for the watermelon seed-spitting contest for the four age groups is shown in Figure 9-4. Under the Source column of the ANOVA table, you see Factor listed in row one. The factor variable (as described by Minitab) represents the treatment or population variable. In column three of the Factor row, you see the SST, which is equal to 89.75. In the Error row (row two), you locate the SSE in column three, which equals 56.80. In column three of the Total row (row three), you see the SSTO, which is 146.55. Using the values of SST, SSE, and SSTO from the Minitab output, you can verify that SST + SSE = SSTO.

**Figure 9-4:**
ANOVA
Minitab out-
put for the
watermelon
seed-
spitting
example.

```
One-Way ANOVA: Age Group 1, Age Group 2, Age Group 3, Age Group 4

Source           DF       SS       MS       F       P
Factor            3    89.75    29.92    8.43   0.001
Error            16    56.80     3.55
Total            19   146.55

S = 1.884    R-Sq = 61.24%    R-Sq(adj) = 53.97%
```

Now you're ready to use these sums of squares to complete the next step of the *F*-test.

## Locating those mean sums of squares

After you have the sums of squares for treatment, SST, and the sums of squares for error, SSE (see the preceding section for more on these), you want to compare them to see whether the variability in the *y*-values due to the model (SST) is large compared to the amount of error left over in the data after the groups have been accounted for (SSE). So you ultimately want a ratio that somehow compares SST to SSE.

To make this ratio form a statistic that they know how to work with (in this case, an *F*-statistic), statisticians decided to find the means of SST and SSE and work with that. Finding the mean sums of squares is the second step of the *F*-test, and the mean sums are as follows:

✔ **MST** is the *mean sums of squares for treatments,* which measures the mean variability that occurs between the different treatments (the different samples in the data). What you're looking for is the amount of variability in the data as you move from one sample to another. A great deal of variability between samples (treatments) may indicate that the populations are different as well.

You can find MST by taking SST and dividing by $k - 1$ (where $k$ is the number of treatments).

✔ **MSE** is the *mean sums of squares for error,* which measures the mean within-treatment variability. The *within-treatment variability* is the amount of variability that you see within each sample itself, due to chance and/or other factors not included in the model.

You can find MSE by taking SSE and dividing by $n - k$ (where $n$ is the total sample size and $k$ is the number of treatments). The values of $k - 1$ and $n - k$ are called the *degrees of freedom* (or df) for SST and SSE, respectively.

Minitab calculates and posts the degrees of freedom for SST, SSE, MST, and MSE in the ANOVA table in columns two and four, respectively.

From the ANOVA table for the seed-spitting data in Figure 9-4, you can see that column two has the heading DF, which stands for degrees of freedom. You can find the degrees of freedom for SST in the Factor row (row two); this value is equal to $k - 1 = 4 - 1 = 3$. The degrees of freedom for SSE is found to be $n - k = 20 - 4 = 16$. (Remember, you have four age groups and five children in each group for a total of $n = 20$ data values.) The degrees of freedom for SSTO is $n - 1 = 20 - 1 = 19$ (found in the Total row under DF). You can verify that the degrees of freedom for SSTO = degrees of freedom for SST + degrees of freedom for SSE.

The values of MST and MSE are shown in column four of Figure 9-4, with the heading MS. You can see the MST in the Factor row, which is 29.92. This value was calculated by taking SST = 89.75 and dividing it by degrees of freedom, 3. You can see MSE in the Error row, equal to 3.55. MSE is found by taking SSE = 56.80 and dividing it by its degrees of freedom, 16.

By finding the mean sums of squares, you've completed step two of the F-test, but don't stop here! You need to continue to the next section in order to complete the process.

## *Figuring the F-statistic*

The test statistic for the test of the equality of the $k$ population means is $F = \dfrac{MST}{MSE}$. The result of this formula is called the *F-statistic.* The *F*-statistic has an *F*-distribution, which is equivalent to the square of a *t*-test (when the numerator degrees of freedom is 1; see more on this interesting connection between the *t*- and *F*-distributions in Chapter 12). All *F*-distributions start at zero and are skewed to the right. The degree of curvature and the height of the curvature of each $F$-distribution is reflected in two degrees of freedom, represented by $k - 1$ and $n - k$. (These come from the denominators of MST and MSE, respectively, where $n$ is the total sample size and $k$ is the total number of treatments or populations.) A shorthand way of denoting the $F$-distribution for this test is $F_{(k-1, n-k)}$.

In the watermelon seed-spitting example, you're comparing four means and have a sample size of five from each population. Figure 9-5 shows the corresponding $F$-distribution, which has degrees of freedom $4 - 1 = 3$ and $20 - 4 = 16$; in other words $F_{(3, 16)}$.

You can see the $F$-statistic on the Minitab ANOVA output (see Figure 9-4) in the Factor row, under the column indicated by F. For the seed-spitting example, the value of the $F$-statistic is 8.43. This number was found by taking MST = 29.92 divided by MSE = 3.55. Then locate 8.43 on the $F$-distribution in Figure 9-5 to see where it stands in terms of its $p$-value. (Turns out it's waaay out there; more on that in the next section.)

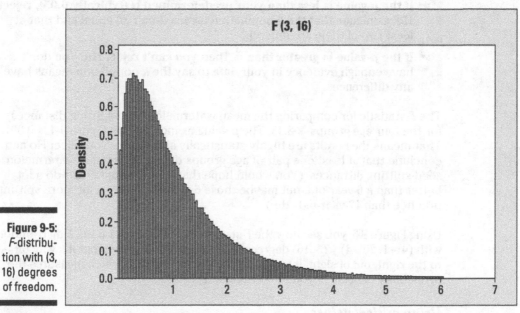

Figure 9-5: $F$-distribution with (3, 16) degrees of freedom.

Be sure to not to exchange the order of the degrees of freedom for the $F$-distribution. The difference between $F_{(3, 16)}$ and $F_{(16, 3)}$ is a big one.

# *Making conclusions from ANOVA*

If you've completed the *F*-test and found your *F*-statistic (step four in the ANOVA process), you're ready for step five of ANOVA: making conclusions for your hypothesis test of the *k* population means. If you haven't already done so, you can compare the *F*-statistic to the corresponding *F*-distribution with $(k - 1, n - k)$ degrees of freedom to see where it stands and make a conclusion. You can make the conclusion in one of two ways: the *p*-value approach or the critical-value approach. The approach you use depends primarily on whether you have access to a computer, especially during exams. I describe these two approaches in the following sections.

### *Using the p-value approach*

On Minitab ANOVA output (see Figure 9-4), the value of the *F*-statistic is located in the Factor row, under the column noted by F. The associated *p*-value for the *F*-test is located in the Factor row under the column headed by P. The *p*-value tells you whether or not you can reject Ho.

✔ **If the *p*-value is less than your predetermined α (typically 0.05), reject Ho.** Conclude that the *k* population means aren't all equal and that at least two of them are different.

✔ **If the *p*-value is greater than α, then you can't reject Ho.** You don't have enough evidence in your data to say the *k* population means have any differences.

The *F*-statistic for comparing the mean watermelon seed-spitting distances for the four age groups is 8.43. The *p*-value as indicated in Figure 9-4 is 0.001. That means the results are highly statistically significant. You reject Ho and conclude that at least one pair of age groups differs in its mean watermelon seed-spitting distances. (You would hope that a 17-year-old could do a lot better than a 6-year-old, but maybe those 6-year-olds have a lot more spitting practice than 17-year-olds do.)

Using Figure 9-5, you see how the *F*-statistic of 8.43 stands on the *F*-distribution with $(4 - 1, 20 - 4) = (3, 16)$ degrees of freedom. You can see that it's way off to the right, out of sight. It makes sense that the *p*-value, which measures the probability of being beyond that *F*-statistic, is 0.001.

### *Using critical values*

If you're in a situation where you don't have access to a computer (as is still the case in many statistics courses today when it comes to taking exams), finding the exact *p*-value for the *F*-statistic isn't possible using a table. You just choose the *p*-value of the *F*-statistic that's closest to yours. However, if you do have access to a computer while doing homework or an exam, statistical software packages automatically calculate all *p*-values exactly, so you can see them on any computer output.

To approximate the *p*-value from your *F*-statistic in the event you don't have a computer or computer output available, you find a cutoff value on the *F*-distribution with $(k – 1, n – k)$ degrees of freedom that draws a line in the sand between rejecting Ho and not rejecting Ho. This cutoff, also known as the *critical value,* is determined by your predetermined $\alpha$ (typically 0.05). You choose the critical value so that the area to its right on the *F*-distribution is equal to $\alpha$.

*F*-distribution tables are available in various statistics textbooks and Web sites for other values of $\alpha$; however, $\alpha = 0.05$ is by far the most common $\alpha$ level used for the *F*-distribution and is sufficient for your purposes.

This table of values for the *F*-distribution is called the *F-table,* and students typically receive these with their exams. For the seed-spitting example, the *F*-statistic has an *F*-distribution with degrees of freedom (3, 16), where $3 = k – 1$, and $16 = n – k$. To find the critical value, consult an *F*-table (Table A-5 in the appendix). Look up the degrees of freedom (3, 16), and you'll find that the critical value is 3.2389 (or 3.24). Your *F*-statistic for the seed-spitting example is 8.43, which is well beyond this critical value (you can see how 8.43 compares to 3.24 by looking at Figure 9-5). Your conclusion is to reject Ho at level $\alpha$. At least two of the age groups differ on mean seed-spitting distances.

With the critical value approach, any *F*-statistic that lies beyond the critical value results in rejecting Ho, no matter how far from or close to the line it is. If your *F*-statistic is beyond the value found in the *F*-table you consult, then you reject Ho and say at least two of the treatments (or populations) have different means.

## What's next?

After you've rejected Ho in the *F*-test and concluded that not all the populations means are the same, your next question may be: Which ones are different? You can answer that question by using a statistical technique called *multiple comparisons.* Statisticians use many different multiple comparison procedures to further explore the means themselves after the *F*-test has been rejected. I discuss and apply some of the more common multiple comparison techniques in Chapter 10.

# Checking the Fit of the ANOVA Model

As with any other model, you must determine how well the ANOVA model fits before you can use its results with confidence. In the case of ANOVA, the model basically boils down to a treatment variable (also known as the population you're in) plus an error term. To assess how well that model fits the data, see the values of $R^2$ and $R^2$ adjusted on the last line of the ANOVA output below the ANOVA table. For the seed-spitting data, you see those values at the bottom of Figure 9-4.

✔ **The value of $R^2$ measures the percentage of the variability in the response variable ($y$) explained by the explanatory variable ($x$).** In the case of ANOVA, the $x$ variable is the factor due to treatment (where the treatment can represent a population being compared). A high value of $R^2$ (say, above 80 percent) means this model fits well.

✔ **The value of $R^2$ adjusted, the preferred measure, takes $R^2$ and adjusts it for the number of variables in the model.** In the case of one-way ANOVA, you have only one variable, the factor due to treatment, so $R^2$ and $R^2$ adjusted won't be very far apart. For more on $R^2$ and $R^2$ adjusted, see Chapter 6.

For the watermelon seed-spitting data, the value of $R^2$ adjusted (as found in the last row of Figure 9-4) is only 53.97 percent. That means age group (shown to be statistically significant by the $F$-test; see the section "Making conclusions from ANOVA") explains just over half of the variability in the watermelon seed-spitting distances. Because of that connection, you may find other variables you can examine in addition to age group, making an even better model for predicting how far those seeds will go.

As you see in Figure 9-1, the results of the $t$-test done to compare the spitting distances of males and females in the section "Comparing Two Means with a $t$-Test" show that males and females were significantly different on mean seed-spitting distances ($p$-value = 0.039 < 0.05). So I would venture a guess that if you include gender as well as age group, thereby creating what statisticians call a *two-factor ANOVA* (or *two-way ANOVA*), the resulting model would fit the data even better, resulting in higher values of $R^2$ and $R^2$ adjusted. (Chapter 11 walks you through the two-way ANOVA.)

# Upfront rejection is the best policy for most refusal letters

Many medical and psychological studies use designed experiments to compare the responses of several different treatments, looking for differences. A *designed experiment* is a study in which subjects are randomly assigned to treatments (experimental conditions) and their responses are recorded. The results are used to compare treatments to see which one(s) work best, which ones work equally well, and so on.

Ohio State University researchers conducted one such experiment using ANOVA to determine the most effective way to write a rejection letter. (Is there really a best way to say "no" to someone? Turns out the answer is "yes.") The experiment tested three traditional principles of writing refusal letters:

✔ Using a buffer, which is a neutral or positive sentence that delays the negative information

✔ Placing the reason before the refusal

✔ Ending the letter on a positive note as a way of reselling the business

Subjects were randomly assigned to treatments, and their responses to the rejection letters were compared (likely on some sort of scale such as 1 = very negative to 7 = very positive with 4 being a neutral response).

You can analyze this scenario by using ANOVA because it compares three treatments (forms of the rejection letters) on some quantitative variable (response to the letter). You can argue that response to the letter isn't a continuous variable, however it has enough possible values that ANOVA isn't unreasonable. The data were also shown to have a bell shape.

The null hypothesis would be Ho: Mean responses to the three types of rejection letters are equal versus Ha: At least two forms of the rejection letter resulted in different mean responses.

In the end, the researchers did find some significant results; the different ways the rejection letter was written affected the participants differently (so the *F*-test was rejected). Using multiple comparison procedures (see Chapter 10), you could go in and determine which forms of the rejection letters gave different responses and how the responses differed.

In case you have to write a rejection letter at some point, the researchers recommend the following guidelines:

✔ Don't use buffers to begin negative messages.

✔ Give a reason for the refusal when it makes the sender's boss look good.

✔ Present the negative positively but clearly; offer an alternative or compromise if possible.

✔ A positive ending isn't necessary.

# Chapter 10

# Sorting Out the Means with Multiple Comparisons

. . . . . . . . . . . . . . . . . . . . . . . . . . . . . . . . . . . . .

### In This Chapter

▶ Knowing when and how to follow up ANOVA with multiple comparisons

▶ Comparing two well-known multiple comparison procedures

▶ Taking additional procedures into consideration

. . . . . . . . . . . . . . . . . . . . . . . . . . . . . . . . . . . . .

*I*magine this: You're comparing the means of not two, but $k$ independent populations, and you find out (using ANOVA; see Chapter 9) that you reject Ho: All the population means are equal, and you conclude Ha: At least two of the population means are different. Now you gotta know — which of those populations are different? Answering this question requires a follow-up procedure to ANOVA called *multiple comparisons,* which makes sense because you want to compare the multiple means you have to see which ones are different.

In this chapter, you figure out when you need to use a multiple comparison procedure. Two of the most well-known multiple comparison procedures are Fisher's LSD (least significant difference) and Tukey's test. They can help you answer that burning question: So some of the means are different, but which ones are different? In this chapter, I also tell you about other comparison procedures that you may encounter or want to try.

*Note:* For those individuals who come up with new multiple comparison procedures, the procedures are generally named after them. (It's like having a star named after you but less romantic and a whole lot more work.)

# Following Up after ANOVA

The main reason folks use ANOVA to analyze data is to find out whether there are any differences in a group of population means. Your null hypothesis is that there are no differences, and the alternative hypothesis is that there's at least one difference somewhere between two of the means. (Note it doesn't say that all the means have to be different.)

If it's established that at least two of the population means are different, the next natural question is: "Okay, which ones are different?" Although this is a very simple-sounding question, it doesn't have a simple answer. The concept of means being different can be interpreted in hundreds of ways. Is one larger than all the others? Are three pairs of them different from each other and the rest all the same? Statisticians have worked long and hard to come up with a wide range of choices of procedures to explore and find differences of all types in two or more population means. This family of procedures is called *multiple comparisons*.

This section starts off with an example in which the ANOVA procedure was used and Ho was rejected, leading you to the next step: multiple comparisons. You then get an overview of how and why multiple comparison procedures work.

## Comparing cellphone minutes: An example

Suppose you want to compare the average number of cellphone minutes used per month for various age groups, where the age groups are defined as the following:

✔ Group 1: 19 years old and under

✔ Group 2: 20–39 years old

✔ Group 3: Adult males 40–59 years old

✔ Group 4: Adult females 60 years old and over

You collect data on a random sample of ten people from each group (where no one knows anyone else to keep independence), and you record the number of minutes each person used their cellphone in one month. The first ten lines of a hypothetical data set are shown in Table 10-1.

| Table 10-1 | Cellphone Minutes Used in One Month | | |
|---|---|---|---|
| 19 and Under (Group 1) | 20–39 (Group 2) | 40–59 (Group 3) | 60 and Over (Group 4) |
| 800 | 250 | 700 | 200 |
| 850 | 350 | 700 | 120 |
| 800 | 375 | 750 | 150 |
| 650 | 320 | 650 | 90 |
| 750 | 430 | 550 | 20 |
| 680 | 380 | 580 | 150 |
| 800 | 325 | 700 | 200 |
| 750 | 410 | 700 | 130 |
| 690 | 450 | 590 | 160 |
| 710 | 390 | 650 | 30 |

The means and standard deviations of the sample data are shown in Figure 10-1, as well as confidence intervals for each of the population means separately (see Chapter 3 for info on confidence intervals). Looking at Figure 10-1, it appears that all four means are different, with the 19-and-under group heading the pack, 40- to 59-year-olds not far behind, with 20- to 39-year-olds and those over 60 bringing up the rear (in that order).

Knowing that you can't live by sample results alone, you decide that ANOVA is needed to see whether any differences that appear in the samples can be extended to the population (see Chapter 9). By using the ANOVA procedure, you test whether the average cell minutes used is the same across all groups. The results of the ANOVA, using the data from Table 10-1, are shown in Figure 10-2.

**Figure 10-1:**
Basic statistics and confidence intervals for the cellphone data.

```
                       Individual 95% CIs For Mean Based on
                       Pooled StDev
Level    N    Mean    StDev   ------+---------+---------+---------+---
Group 1  10   748.00  64.60                                    (-*-)
Group 2  10   368.00  59.08                  (-*-)
Group 3  10   657.00  64.99                            (-*-)
Group 4  10   125.00  62.41   (-*-)
                              ------+---------+---------+---------+---
                                  200       400       600       800
```

Looking at Figure 10-2, the $F$-test for equality of all four population means has a $p$-value of 0.000, meaning it's less than 0.001. That says at least two of these age groups have a significant difference in their cellphone use (see Chapter 9 for info on the $F$-test and its results).

Figure 10-2:
ANOVA
results for
comparing
cellphone
use for four
age groups.

```
One-way ANOVA: Group 1, Group 2, Group 3, Group 4

Source    DF        SS       MS       F      P
Factor     3   2416010   805337  204.13  0.000
Error     36    142030     3945
Total     39   2558040

S = 62.81    R-Sq = 94.5%    R-Sq(adj) = 93.99%
```

Okay, so what's your next question? You just found out that the average number of cellphone minutes used per month isn't the same across these four groups. This doesn't mean all four groups are different (see Chapter 9), but it does mean that at least two groups are significantly different in their cellphone use. So your questions are

✔ Which groups are different?

✔ How are they different?

## Setting the stage for multiple comparison procedures

Determining which populations have differing means after the ANOVA $F$-test has been rejected involves a new data-analysis technique called *multiple comparisons*. The basic idea of multiple comparison procedures is to compare various means and report where and what the differences are. For example, you may conclude from a multiple comparison procedure that the first population had a mean that was statistically lower than the second population, but it was statistically higher than the mean of the third population.

There are myriad different multiple comparison procedures out there; how do you know which one you should use when? Two basic elements distinguish multiple comparison procedures from each other. I call them purpose and price.

✔ **Purpose:** When you know that a group of means aren't all equal, you zoom in to explore the relationships between them, depending on the purpose of your research. Maybe you just want to figure out which means are equivalent and which are not. Maybe you want to sort them into statistically equivalent groups from smallest to largest. Or it may be important to compare the average of one group of means to the average of another group of means. Different multiple comparison procedures were built for different purposes; for the most part, if you use them for their designed purposes, you have a better chance of finding specific differences you're looking for, if those differences are actually there.

✔ **Price:** Any statistical procedure you use comes with a price: the probability of making a Type I error in your conclusions somewhere during the procedure, due to chance. (A *Type I error* is committed when Ho is rejected when it shouldn't be; in other words, you think two means are different but they really aren't. See your Stats I textbook or my book *Statistics For Dummies* (Wiley) for more info.) This probability of making at least one Type I error during a multiple comparisons procedure is called the *overall error rate* (also known as the experimentwise error rate (EER), or the familywise error rate). Small overall error rates are of course desirable. Each multiple comparison procedure has its own overall error rate; generally the more specific the relationships are that you're trying to find, the smaller your overall error rate is, assuming you're using a procedure that was designed for your purpose.

In the next section, I describe two all-purpose multiple comparison procedures: Fisher's LSD and Tukey's test.

Don't attempt to explore the data with a multiple comparison procedure if the test for equality of the populations isn't rejected. In this case, you must conclude that you don't have enough evidence to say the population means aren't all equal, so you must stop there. Always look at the *p*-value of the *F*-test on the ANOVA output before moving on to conduct any multiple comparisons.

# Pinpointing Differing Means with Fisher and Tukey

You've conducted ANOVA to see whether a group of *k* populations has the same mean, and you rejected Ho. You conclude that at least two of those populations have different means. But you don't have to stop there; you can go on to find out how many and which means are different by conducting multiple comparison tests.

In this section, you see two of the most well-known multiple comparison procedures: *Fisher's LSD* (also known as *Fisher's protected LSD* or *Fisher's test*) and *Tukey's test* (also known as *Tukey's simultaneous confidence intervals*).

Although I only discuss two procedures in detail in this chapter, tons of other multiple comparison procedures exist (see "So Many Other Procedures, So Little Time!" at the end of this chapter). Although the other procedures' methods differ a great deal, their overall goal is the same: to figure out which population means differ by comparing their sample means.

## Fishing for differences with Fisher's LSD

In this section, I outline the original least significant difference procedure (LSD) and R. A. Fisher's improvement on it (aptly called *Fisher's least significant difference procedure,* or *Fisher's LSD*). The LSD and Fisher's LSD procedures both compare pairs of means using some form of *t*-tests, but they do so in different ways (see Chapter 3 or your Stats I textbook for more on the *t*-test). You also see Fisher's LSD applied to the cellphone example from earlier in this chapter (see the section "Following Up after ANOVA").

### The original LSD procedure

To use the original (pre-Fisher) LSD (short for *least significant difference*) simply choose certain pairs of means in advance and conduct a *t*-test on each pair at level $\alpha = 0.05$ to look for differences. LSD doesn't require an ANOVA test first (which is a problem that R. A. Fisher later noticed). If $k$ population means are all to be compared to each other in pairs using LSD, the number of *t*-tests performed would be represented by $\frac{k(k-1)}{2}$.

Here's how to count the number of *t*-tests when all means are compared. To start, you compare the first mean and the second mean, the first mean and the third mean, and so on until you compare the first mean and the $k^{th}$ mean. Then compare the second and third, second and fourth, and so on all the way down to the $(k-1)^{th}$ mean and the $k^{th}$ mean. The total number of pairs of means to compare equals $k * (k-1)$. Because comparing the two means in either order (mean one and mean two versus mean two and mean one), gives you the same result regarding which one is largest, you divide the total by 2 to avoid double counting. For example, if you have four populations labeled A, B, C, and D, you have $\frac{4(4-1)}{2} = 6$ *t*-tests to perform: A versus B; A versus C; A versus D; B versus C; B versus D; and C versus D.

The original LSD procedure is very straightforward, easy to conduct, and easy to understand. However, the procedure has some issues. Because each *t*-test is conducted at $\alpha$ level 0.05, each test done has a 5 percent chance of making a Type I error (rejecting Ho when you shouldn't have, as I explain in Chapter 3).

Although a 5 percent error rate for each test doesn't seem too bad, the errors have a multiplicative effect as the number of tests increases. For example, the chance of making at least one Type I error with six *t*-tests, each at level $\alpha$ = 0.05, is 26.50 percent, which is your overall error rate for the procedure.

If you want or need to know how I arrived at the number 26.50 percent as the overall error rate in that last example, here it goes: The probability of making a Type I error for each test is 0.05. The chance of making at least one error in six tests equals 1 – the probability of making no errors in six tests. The chance of not making an error in one test is $1 - \alpha = 0.95$. The chance of no error in six tests is this quantity times itself six times, or $(0.95)^6$, which equals 0.735. Now take 1 – this quantity to get 1 – 0.735 = 0.2650 or 26.50 percent.

# Using Fisher's new and improved LSD

R. A. Fisher suggested an improvement over the regular LSD procedure, and his procedure is called *Fisher's LSD*, or *Fisher's protected LSD*. It adds the requirement that an ANOVA *F*-test must be performed first and must be rejected before any pairs of means can be compared individually or collectively. By requiring the *F*-test to be rejected, you're concluding that at least one difference exists in the means. Adding this requirement, the overall error rate of Fisher's LSD is somewhere in the area of $\alpha$, which is much lower than what you get from the regular LSD procedure.

The downside of Fisher's LSD is that because each *t*-test is made at level $\alpha$ and the overall error rate is also near $\alpha$, it's good at finding differences that really do exist, but it also makes some false alarms in the process (mainly saying there's a difference when there really isn't).

To conduct Fisher's LSD in Minitab, go to Stat>ANOVA>One-way or One-way unstacked. (If your data appear in two columns with Column 1 representing the population number and Column 2 representing the response, just click One-way because your data are stacked. If your data are shown in *k* columns, one for each of the *k* populations, click One-way unstacked.) Highlight the data for the groups you're comparing, and click Select. Then click on Comparisons, and then Fisher's. The individual error rate is listed at 5 (percent), which is typical. If you want to change it, type in the desired error rate (between 0.5 and 0.001), and click OK. You may type in your error rate as a decimal, 0.05, or as a number greater than 1, such as 5. Numbers greater than 1 are interpreted as a percentage.

An ANOVA procedure was done on the cellphone data presented in Table 10-1 to compare the mean number of minutes used for four age groups. Looking at the output in Figure 10-2, you see Ho (all the population means are equal) was rejected. The next step is to conduct multiple comparisons by using Fisher's LSD to see which population means differ. Figure 10-3 shows the Minitab output for those tests.

The first block of results shows "Group 1 subtracted from" where Group 1 = age 19 and under. Each line after that represents the other age groups (Group 2 = 20- to 39-year-olds, Group 3 = 40- to 59-year-olds, and Group 4 = age 60 and over). Each line shows the results of comparing the mean for some other group minus the mean for Group 1.

For example, the first row shows Group 2 being compared with Group 1. Moving to the right in that same row, you see the confidence interval for the difference in these two means, which turns out to be −436.97 to −323.03. Because zero isn't contained in this interval, you conclude that these two means are different in the populations also. Because this difference ($\mu_2 - \mu_1$) is negative, you also can say that $\mu_2$ is less than $\mu_1$. Or, a better way to think of it may be that $\mu_1$ is greater than $\mu_2$. That is, Group 1's mean is greater than Group 2's mean.

**Figure 10-3:**
Output
showing
Fisher's LSD
applied to
the cell-
phone data.

```
Fisher 95% Individual Confidence Intervals
All Pairwise Comparisons
Simultaneous confidence level = 80.32%

Group 1 subtracted from:
          Lower   Center   Upper  ---------+---------+---------+---------+
Group 2  -436.97  -380.00  -323.03         (*-)
Group 3  -147.97   -91.00   -34.03                      (*-)
Group 4  -679.97  -623.00  -566.03    (*-)
                                   ---------+---------+---------+---------+
                                         -350        0       350      700

Group 2 subtracted from:
          Lower   Center   Upper  ---------+---------+---------+---------+
Group 3   232.03   289.00   345.97                            (*-)
Group 4  -299.97  -243.00  -186.03        (-*-)
                                   ---------+---------+---------+---------+
                                         -350        0       350      700

Group 3 subtracted from:
          Lower   Center   Upper  ---------+---------+---------+---------+
Group 4  -588.97  -532.00  -475.03    (-*)
                                   ---------+---------+---------+---------+
                                         -350        0       350      700
```

If two means are equal, their difference equals zero, and a confidence interval for the difference should contain zero. If zero isn't included, you say the means are different.

In this case, each subsequent row in the "Group 1 subtracted from" section of Figure 10-3 shows similar results. None of the confidence intervals contain zero, so you conclude that the mean cellphone use for Group 1 is different from the mean cellphone use for any other group.

Moreover, because all confidence intervals are in negative territory, you can conclude that the mean cellphone use for those users age 19 and under is greater than all the others. (Remember, the mean for this group is subtracted from the others, so a negative difference means its mean is greater).

This process continues as you move down through the output until all six pairs of means are compared to each other. Then you put them all together into one conclusion. For example, in the second portion of the output, Group 2 is subtracted from Groups 3 and 4. You see the confidence interval for the "Group 3" line is (232.03, 345.97); this gives possible values for Group 3's mean minus Group 2's mean. The interval is entirely positive, so conclude that Group 3's mean is greater than Group 2's mean (according to this data).

On the next line, the interval for Group 4 minus Group 2 is –299.97 to –186.03. All these numbers are negative, so conclude Group 4's mean is less than Group 2's. Combine conclusions to say that Group 3's mean is greater than Group 2's, which is greater than Group 4's.

In the cellphone example, none of the means are equal to each other, and based on the signs of confidence intervals and the results of all the individual pairwise comparisons, the following order of cellphone mean usage prevails: $\mu_1 > \mu_3 > \mu_2 > \mu_4$. (Hypothetical data aside, it may be the case that 40- to 59-year-olds use a lot of cellphone time because of their jobs.) Comparing these results to the sample means in Figure 10-1, this ordering makes sense and the means are separated enough to be declared statistically significant.

Notice near the top of Figure 10-3 that you see "Simultaneous confidence level = 80.32%." That means the overall error rate for this procedure is $1 - 0.8032 = 0.1968$, which is close to 20 percent, a bit on the high side.

## Separating the turkeys with Tukey's test

The basic idea behind Tukey's test is to provide a series of simultaneous tests for differences in the means. It still examines all possible pairs of means and keeps the overall error rate at $\alpha$ and also keeps the individual Type I error rate for each pair of means at $\alpha$. Its distinguishing feature is that it performs the tests all at the same time.

Although the details of the formulas used for Tukey's test are beyond the scope of this book, they're not based on the $t$-test but rather something called a *studentized range statistic,* which is based on the highest and lowest means in the group and their difference. The individual error rates are held at 0.05 because Tukey developed a cutoff value for his test statistic that's based on all pairwise comparisons (no matter how many means are in each group).

To conduct Tukey's test, go to Stat>ANOVA>One-way or One-way unstacked. (If your data appear in two columns with Column 1 representing the population number and Column 2 representing the response, just click One-way because your data are stacked. If your data are shown in $k$ columns, one for each of the $k$ populations, click One-way unstacked.) Highlight the data for the groups you're comparing, and click Select. Then click on Comparisons, and then Tukey's. The familywise (overall) error rate is listed at 5 (percent), which is typical. If you want to change it, type in the desired error rate (between 0.5 and 0.001), and click OK. You may type in your error rate as a decimal, such as 0.05, or as a number greater than 1, such as 5. Numbers greater than 1 are interpreted as a percentage.

The Minitab output for comparing the groups regarding cellphone use by using Tukey's test appears in Figure 10-4. You can interpret its results in the same ways as those in Figure 10-3. Some of the numbers in the confidence intervals are different, but in this case, the main conclusions are the same: Those age 19 and under use their cellphones most, followed by 40- to 59-year-olds, then 20- to 39-year-olds, and finally those age 60 and over.

The results of Fisher and Tukey don't always agree, usually because the overall error rate of Fisher's procedure is larger than Tukey's (except when only two means are involved). Most statisticians I know prefer Tukey's procedure over Fisher's. That doesn't mean they don't have other procedures they like even better than Tukey's, but Tukey's is a commonly used procedure, and many people like to use it.

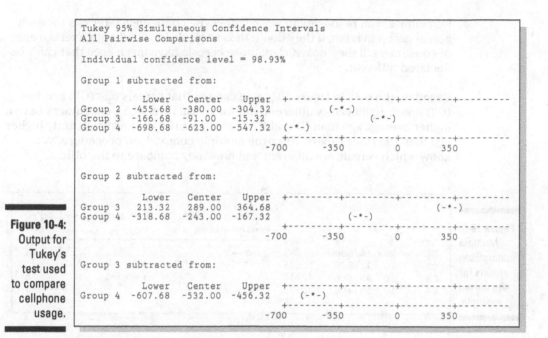

```
Tukey 95% Simultaneous Confidence Intervals
All Pairwise Comparisons

Individual confidence level = 98.93%

Group 1 subtracted from:

            Lower    Center    Upper    +---------+---------+---------+---------
Group 2   -455.68   -380.00   -304.32               (-*-)
Group 3   -166.68    -91.00    -15.32                          (-*-)
Group 4   -698.68   -623.00   -547.32      (-*-)
                                        +---------+---------+---------+---------
                                      -700       -350         0        350


Group 2 subtracted from:

            Lower    Center    Upper    +---------+---------+---------+---------
Group 3    213.32    289.00    364.68                                  (-*-)
Group 4   -318.68   -243.00   -167.32                     (-*-)
                                        +---------+---------+---------+---------
                                      -700       -350         0        350


Group 3 subtracted from:

            Lower    Center    Upper    +---------+---------+---------+---------
Group 4   -607.68   -532.00   -456.32         (-*-)
                                        +---------+---------+---------+---------
                                      -700       -350         0        350
```

**Figure 10-4:**
Output for
Tukey's
test used
to compare
cellphone
usage.

# Examining the Output to Determine the Analysis

Sometimes, the process of answering questions is flipped around in your stats courses. Instead of asking you a question that you use computer output to answer, your professor may give you computer output and ask you to determine the question that the analysis answers. (Kind of like *Jeopardy.*) To work your way backward to the question, you look for clues that tell you what type of analysis was done, and then fill in the details using what you already know about that particular type of analysis.

For example, your professor gives you computer output comparing the ages of ten consumers of each of four cereal brands, labeled C1–C4 (see Figure 10-5). On the analysis, you can see the mean consumer ages for the four cereals being compared to each other, and the analysis also shows and compares the confidence intervals for the averages. The comparison of confidence intervals tells you that you're dealing with a multiple comparison procedure.

Remember, you're looking to see whether the confidence intervals for each cereal group overlap; if they don't, those cereals have different average ages of consumers. If they do overlap, those cereals have mean ages that can't be declared different.

Based on the data in Figure 10-5, you can see that cereals one (C1) and two (C2) aren't significantly different, but for cereal three (C3), consumers have a higher average age than C1 and C2. Cereal four (C4) has a significantly higher age than the three others. After the multiple comparison procedure, you know which cereals are different and how they compare to the others.

**Figure 10-5:**
Multiple
comparison
results for
the cereal
example.

```
                               Individual 95% CIs For Mean Based on
                               Pooled StDev
Level   N     Mean    StDev   -------+---------+---------+---------+--
C1      10    8.800   1.687   (--*--)
C2      10   11.800   1.033     (--*--)
C3      10   36.500   7.735                        (--*--)
C4      10   55.400  10.309                                  (--*--)
```

Sometimes multiple comparison procedures give you groups of means that are equivalent to each other, different from each other, or overlapping. In this case, the final result is $\mu_{C1} = \mu_{C2} < \mu_{C3} < \mu_{C4}$.

# So Many Other Procedures, So Little Time!

Many more multiple comparison procedures exist beyond Fisher's and Tukey's imaginations. Those that I discuss in this section are a little more specialized in what they were designed to look for, compared to Tukey's and Fisher's. For example, you may want to know whether a certain combination of means is larger than another combination of means; or you may want to only compare specific means to each other, not all the pairs of means.

One thing to note, however, is that in many cases you don't know exactly what you're looking for when comparing means — you're just looking for differences, period. If that's the case, one of the more general procedures, like Fisher's or Tukey's, is the way to go. They're built for general exploration and do a better job of it than more-specialized procedures.

This section provides an overview of other multiple comparison procedures that exist and tells a little bit about each one, including the people who developed them. Given the dates of when these procedures were developed, I think you'll agree with me that the 1950s was the golden age of the multiple comparison procedure.

## Controlling for baloney with the Bonferroni adjustment

The *Bonferroni adjustment* (or *Bonferroni correction*) is a technique used in a host of situations, not just for multiple comparisons. It was basically created to stop people from over-analyzing data. There's a limit to what you should do when analyzing data; there's a line that, when crossed, results in something statisticians call *data snooping*. And the Bonferroni adjustment curbs that.

Data snooping is when someone analyzes her data over and over again until she gets a result that she can say is *statistically significant* (meaning the result is said to have been unlikely to have happened by chance; see Chapter 3). Because the number of tests completed by the data snooper is so high, she's likely to find something significant just by chance. And that result is highly likely to be bogus.

For example, suppose a researcher wants to find out what variable is related to sales of bedroom slippers. He collects data on everything he can think of, including the size of people's feet, the frequency with which they go out to get the paper in their slippers, and their favorite colors. Not finding anything significant, he goes on to examine marital status, age, and income.

Still coming up short, he goes out on a limb and looks at hair color, whether or not the subjects have seen a circus, and where they like to sit on an airplane (aisle or window, sir?). Then wouldn't you know, he strikes gold. Turns out that, according to his data, people who sit on the aisles on planes are more likely to buy bedroom slippers than those who sit by the window or in the middle of a row.

What's wrong with this picture? Too many tests. Each time the researcher examines one variable and conducts a test on it, he chooses an $\alpha$ level at which to conduct the test. (Recall that the $\alpha$ level is the amount of chance you're willing to take of rejecting the null hypothesis and making a false alarm.) As the number of tests increases, the $\alpha$'s pile up.

Suppose α is chosen to be 0.05. The researcher then has a 5 percent chance of being wrong in finding a significant conclusion, just by chance. So if he does 100 tests, each with a 5 percent chance of an error, on average 5 of those 100 tests will result in a statistically significant result, just by chance. However, researchers who don't know that (or who know and go ahead regardless) find results that they claim are significant even though they're really bogus.

An Italian mathematician named Carlo Emilio Bonferroni (1892–1960) said "enough already" and created something statisticians call the Bonferroni adjustment in 1950 to control the madness. The *Bonferroni adjustment* simply says that if you're doing $k$ tests of your data, you can't do each one at level α = 0.05, you need to have an α level for each test equal to 0.05 ÷ $k$.

For example, someone who conducts 20 tests on one data set needs to do each one at level α = 0.05 ÷ 20 = 0.0025. This adjustment makes it harder to find a conclusion that's significant because the *p*-value for any test must be less than 0.0025. The Bonferroni adjustment curbs the chance of data snooping until you find something bogus.

The downside of Bonferroni's adjustment is that it's very conservative. Although it reduces the chance of concluding two means differ when they really don't, it fails to catch some differences that really are there. In statistical terms, Bonferroni has power issues. (See your Stats I text or *Statistics For Dummies* for a discussion on power.)

## Comparing combinations by using Scheffe's method

*Scheffe's method* was developed in 1953 by Henry Scheffe (1907–1977). This method doesn't just compare two means at time, like Tukey's and Fisher's tests do; it compares all different combinations (called *contrasts*) of the means. For example, if you have the means from four populations, you may want to test to see if their sum equals a certain value, or if the average of two of them equals the average of the two others.

## Finding out whodunit with Dunnett's test

Dunnett's test was developed in 1955 by Charles Dunnett (1921–1977). *Dunnett's test* is a special multiple comparison procedure used in a designed experiment that contains a control group. The test compares each treatment group to the control group and determines which treatments do better than others.

Compared to other multiple comparison procedures, Dunnett's test is better able to find real differences in this situation because it focuses only on the differences between each treatment and the control — not on the differences between every single pair of treatments in the entire study.

## Staying cool with Student Newman-Keuls

Student Newman-Keuls test is a different approach from Tukey and Fisher in comparing pairs of means in a multiple comparison procedure. This test comes from the work of three people: "Student," Newman, and Keuls.

The *Student Newman-Keuls procedure* is based on a stepwise or layer approach. You order sample means from the smallest to the largest and then examine the differences between the ordered means.

You first test the largest minus smallest difference, and if that turns out to be statistically significant, you conclude that their two respective populations are different in terms of their means. Of the remaining means, the ones that are farthest apart in the order are tested for a significant difference, and so on. You stop when you don't find any more differences.

## Duncan's multiple range test

David B. Duncan designed the *Duncan's multiple range test* (MRT) in 1955. The test is based on the Student Newman-Keuls test but has increased power in its ability to detect when the null hypothesis is not true (see Chapter 3) because it increases the value of $\alpha$ at each step of the Student Newman-Keul's test. Duncan's test is used especially in agronomy (crop and farm land management) and other types of agricultural research. One of the neatest things about being a statistician is you never know what kinds of problems you'll be working on or who will use your methods and results.

Although Duncan won the favor of many researchers who used his test (and still do), he wasn't without his critics. Both John Tukey (who developed Tukey's test) and Henry Scheffe (who developed Scheffe's test) accused Duncan's test of being too liberal by not controlling the rate of an overall error (called a *familywise error rate* in the big leagues). But Duncan stood his ground. He said that means are usually never equal anyway, so he wanted to err on the side of making a false alarm (Type I error) rather than missing an opportunity (Type II error) to find out when means are different.

Every procedure in statistics has some chance of making the wrong conclusion, not because of an error in the process but because results vary from data set to data set. You just have to know your situation and choose the procedure that works best for that situation. When in doubt, consult a statistician for help in sorting it all out.

# The secret lives of statisticians

Sometimes it's hard to imagine famous people having real lives, and it may be especially hard to picture statisticians doing anything but sitting in the back room calculating numbers. But the truth is, famous statisticians are interesting folks with interesting lives, just like you and me. Consider these stellar statisticians:

✔ **Henry Scheffe:** Scheffe was a very distinguished statistician at University of California, Berkeley. One of his five books *The Analysis of Variance,* written in 1959, is the classic book on the subject and is still used today. (I used it in grad school and still have a copy in my office.) Scheffe enjoyed backpacking, swimming, cycling, reading, and music, having learned to play the recorder during his adult life. Sadly, he died from a bicycle accident on his way to the university in 1977.

✔ **Charles Dunnett:** Nicknamed "Charlie" (did you ever think of famous statisticians as having nicknames?), Dunnett was a distinguished, award-winning professor in the Departments of Mathematics, Statistics, Clinical Epidemiology and Biostatistics at McMaster University in Ontario, Canada.

He wrote many papers, two of which were so important that they made it onto the list of the top 25 most-cited statistical papers of all time.

✔ **William Sealy Gosset, or "Student":** The first name included on the Student Newman-Keuls test is a story in itself. "Student" is a pseudonym of the English statistician William Sealy Gosset (1876–1937). Gosset was a statistician working for the Guinness brewery in Dublin, Ireland, when he became famous for developing the *t*-test, also known as the Student *t*-distribution (see Chapter 3), one of the most commonly used hypothesis tests in the statistical world. Gosset devised the *t*-test as a way to cheaply monitor the quality of beer. He published his work in the best of statistical journals, but his employer regarded his use of statistics in quality control to be a trade secret and wouldn't let him use his real name on his publications (although all his cronies knew exactly who "Student" was). So if not for Guinness beer, the Student's *t*-test would have been called the Gossett *t*-test (or you'd be drinking "Gosset beer").

# Going nonparametric with the Kruskal-Wallis test

The Kruskal-Wallis test was developed in 1952 by American statisticians William Kruskal (1919–2005) and W. Allen Wallis (1912–1998). The *Kruskal-Wallis test* is the nonparametric version of a multiple comparison procedure. Nonparametric procedures don't have nearly as many conditions to meet as their traditional counterparts. All the other procedures described in this chapter require normal distributions from the populations and often the same variance as well.

The Kruskal-Wallis test doesn't use the actual values of the data; it's based on ranks (orderings of the data from smallest to largest). The test ranks all the data together, and then looks at how those ranks are distributed out amongst the samples that represent separate populations. If one sample gets all the small ranks, that population is concluded to have a smaller mean than the others, and so on. (Turn to Chapter 16 for the full story on nonparametric statistics and Chapter 19 for all the details of the Kruskal-Wallis test.)

# Chapter 11

# Finding Your Way through Two-Way ANOVA

......................................

## In This Chapter

▶ Building and carrying out ANOVA with two factors

▶ Getting familiar with (and looking for) interaction effects and main effects

▶ Putting the terms to the test

▶ Demystifying the two-way ANOVA table

......................................

*A*nalysis of variance (ANOVA) is often used in experiments to see whether different levels of an explanatory variable *(x)* get different results on some quantitative variable *y*. The *x* variable in this case is called a *factor,* and it has certain levels to it, depending on how the experiment is set up.

For example, suppose you want to compare the average change in blood pressure on certain dosages of a drug. The factor is drug dosage. Suppose it has three levels: 10mg per day, 20mg per day, or 30mg per day. Suppose someone else studies the response to that same drug and examines whether the times taken per day (one time or two times) has any effect on blood pressure. In this case, the factor is number of times per day, and it has two levels: once and twice.

Suppose you want to study the effects of dosage *and* number of times taken together because you believe both may have an effect on the response. So what you have is called a *two-way ANOVA,* using two factors together to compare the average response. It's an extension of one-way ANOVA (refer to Chapter 9) with a twist, because the two factors you use may operate on the response differently together than they would separately.

In this chapter, first I give you an example of when you'd need to use a two-way ANOVA. Then I show you how to set up the model, make your way through the ANOVA table, take the *F*-tests, and draw the appropriate conclusions.

# Setting Up the Two-Way ANOVA Model

The two-way ANOVA model extends the ideas of the one-way ANOVA model and adds an interaction term to examine how various combinations of the two factors affect the response. In this section, you see the building blocks of a two-way ANOVA: the treatments, main effects, the interaction term, and the sums of squares equation that puts everything together.

## Determining the treatments

The two-way ANOVA model contains two factors, A and B, and each factor has a certain number of levels — say *i* levels of Factor A and *j* levels of Factor B.

In the drug study example from the chapter intro, you have A = drug dosage with *i* = 1, 2, or 3, and B = number of times taken per day with *j* = 1 or 2. Each person involved in the study is subject to one of the three different drug dosages and will take the drug in one of the two methods given. That means you have 3 * 2 = 6 different combinations of Factors A and B that you can apply to the subjects, and you can study these combinations and their effects on blood pressure changes in the two-way ANOVA model.

Each different combination of levels of Factors A and B is called a *treatment* in the model. Table 11-1 shows the six treatments in the drug study. For example, Treatment 4 is the combination of 20mg of the drug taken in two doses of 10mg each per day.

| Table 11-1 | Six Treatment Combinations for the Drug Study Example | |
|---|---|---|
| *Dosage Amount* | *One Dose Per Day* | *Two Doses Per Day* |
| 10mg | Treatment 1 | Treatment 2 |
| 20mg | Treatment 3 | Treatment 4 |
| 30mg | Treatment 5 | Treatment 6 |

If Factor A has *i* levels and Factor B has *j* levels, you have *i* * *j* different combinations of treatments in your two-way ANOVA model.

## Stepping through the sums of squares

The two-way ANOVA model contains the following three terms:

- ✔ **The main effect A:** Term for the effect of Factor A on the response
- ✔ **The main effect B:** Term for the effect of Factor B on the response
- ✔ **The interaction of A and B:** The effect of the combination of Factors A and B (denoted AB)

The sums of squares equation for the one-way ANOVA (which I cover in Chapter 9) is SSTO = SST + SSE, where SSTO is the total variability in the response variable, *y*; SST is the variability explained by the treatment variable (call it factor A); and SSE is the variability left over as error.

The purpose of a one-way ANOVA model is to test to see whether the different levels of Factor A produce different responses in the *y* variable. The way you do it is by using Ho: $\mu_1 = \mu_2 = \ldots = \mu_i$, where *i* is the number of levels of Factor A (the treatment variable). If you reject Ho, Factor A (which separates the data into the groups being compared) is significant. If you can't reject Ho, you can't conclude that Factor A is significant.

In the two-way ANOVA, you add another factor to the mix (B) plus an interaction term (AB). The sums of squares equation for the two-way ANOVA model is SSTO = SSA + SSB + SSAB + SSE. Here, SSTO is the total variability in the *y*-values; SSA is the sums of squares due to Factor A (representing the variability in the *y*-values explained by Factor A); and similarly for SSB and Factor B. SSAB is the sums of squares due to the interaction of Factors A and B, and SSE is the amount of variability left unexplained, and deemed error.

Although the mathematical details of all the formulas for these terms are unwieldy and beyond the focus of this book, they just extend the formulas for one-way ANOVA found in Chapter 9. ANOVA handles the calculations for you, so you don't have to worry about that part.

To carry out a two-way ANOVA in Minitab, enter your data in three columns.

- ✔ Column 1 contains the responses (the actual data).
- ✔ Column 2 represents the level of Factor A (Minitab calls it the *row factor*).
- ✔ Column 3 represents the level of Factor B (Minitab calls it the *column factor*).

Go to Stat>Anova>Two-way. Click on Column 1 in the left-hand box, and it appears in the Response box on the right-hand side. Click on Column 2, and it appears in the row factor box; click on Column 3, and it appears in the column factor box. Click OK.

For example, suppose you have six data values in Column 1: 11, 21, 38, 14, 15, and 62. Suppose Column 2 contains 1, 1, 1, 2, 2, 2, and Column 3 contains 1, 2, 3, 1, 2, 3. This means that Factor A has two levels (1, 2), and Factor B has three levels (1, 2, 3). Table 11-2 shows a breakdown of the data values and which combinations of levels and factors are affiliated with them.

| Table 11-2 | Data and Its Respective Levels from Two Factors | |
|---|---|---|
| Data Value | Level of Factor A | Level of Factor B |
| 11 | 1 | 1 |
| 21 | 1 | 2 |
| 38 | 1 | 3 |
| 14 | 2 | 1 |
| 15 | 2 | 2 |
| 62 | 2 | 3 |

Suppose Factor A has $i$ levels and Factor B has $j$ levels, with a sample of size $m$ collected on each combination of A and B. The degrees of freedom for Factor A, Factor B, and the interaction term AB are $(i-1)$, $(j-1)$, and $(i-1) * (j-1)$, respectively. This formula is just an extension of the degrees of freedom for the one-way model for Factors A and B. The degrees of freedom for SSTO is $(i * j * m) - 1$, and the degrees of freedom for SSE is $i * j * (m-1)$. (See Chapter 9 for details on degrees of freedom.)

# Understanding Interaction Effects

The interaction effect is the heart of the two-way ANOVA model. Knowing that the two factors may act together in a different way than they would separately is important and must be taken into account. In this section, you see the many ways in which the interaction term AB and the main effects of Factors A and B affect the response variable in a two-way ANOVA model.

## *What is interaction, anyway?*

*Interaction* is when two factors meet, or interact with each other, on the response in a way that's different from how each factor affects the response separately.

For example, before you can test to see whether dosage of medicine (Factor A) or number of times taken (Factor B) are important in explaining changes in blood pressure, you have to look at how they operate together to affect blood pressure. That is, you have to examine the interaction term.

Suppose you're taking one type of medicine for cholesterol and another medicine for a heart problem. Suppose researchers only looked at the effects of each drug alone, saying each one was good for managing the problem for which it was designed with little or no side effects. Now you come along and mix the two drugs in your system. As far as the individual study results are concerned, all bets are off. With only those separate studies to go on, no one knows how the drugs will interact with each other, and you can find yourself in a great deal of trouble very quickly if you take them together.

Fortunately, drug companies and medical researchers do a great deal of work studying drug interactions, and your pharmacist knows which drugs interact as well. You can bet a statistician was involved in this work from day one!

Baking is another good example of how interaction works. Slurp down one raw egg, drink a cup of milk, and eat a cup of sugar, a cup of flour, and a stick of margarine. Then eat a cup of chocolate chips. Each one of these items has a certain taste, texture, and effect on your taste buds that, in most cases, isn't all that great. But mix them all together in a bowl and voilà! You have a batch of chocolate chip cookie dough, thanks to the magical effects of interaction.

 In any two-way ANOVA, you must check out the interaction term first. If A and B interact with each other and the interaction is statistically significant, you can't examine the effects of either factor separately. Their effects are intertwined and can't be separated.

## *Interacting with interaction plots*

In the two-way ANOVA model, you're dealing with two factors and their interaction. A number of results could come out of this model in terms of significance of the individual terms, as you can see in the following list:

- ✔ Factors A and B are both significant.
- ✔ Factor A is significant but not Factor B.
- ✔ Factor B is significant but not Factor A.
- ✔ Neither Factors A nor B are significant.
- ✔ The interaction term AB is significant, so you don't examine A or B separately.

Figure 11-1 depicts each of these five situations in terms of a diagram using the drug study example. Plots that show how Factors A and B react separately and together on the response variable *y* are called *interaction plots*. In the following sections, I describe each of these five situations in detail in terms of what the plots tell you and what the results mean in the context of the drug study example.

### Factors A and B are significant

Figure 11-1a shows the situation when both A and B are significant in the model and no interaction is present. The lines represent the levels of the times-per-day factor (B); the *x*-axis represents the levels of the dosage factor (A); and the *y*-axis represents the average value of the response variable *y*, which is change in blood pressure, at each combination of treatments.

In order to interpret these interaction plots, you first look at the general trends each line is making. The top line in Figure 11-1a is moving uphill from left to right, meaning that when the drug is taken two times per day, the changes in blood pressure increase as dosage level increases. The bottom line shows a similar result when the drug is taken once per day; blood pressure changes increase as dosage level increases. Assuming these differences are large enough, you conclude that dosage level (Factor A) is significant.

Now you look at how the lines compare to each other. Note that the lines, although parallel, are quite far apart. In particular, the amounts of blood pressure changes are higher overall when taking the drug twice per day (top line) than they are when taking the drug once per day (bottom line). Again, assuming these differences are large enough, you conclude that times per day (Factor B) is significant.

In this case, the different combinations of Factors A and B don't affect the overall trends in blood pressure changes in opposite ways (that is, the lines don't cross each other) so there's no interaction effect between dosage level and times per day.

Two parallel lines in an interaction plot indicate a lack of an interaction effect. In other words, the effect of Factor A on the response doesn't change as you move across different levels of Factor B. In the drug study example, the levels of A don't change blood pressure differently for different levels of B.

### Factor A is significant but not Factor B

Figure 11-1b shows that blood pressure changes increase across dosage levels for people taking the drug once or twice a day. However, the two lines are so close together that it makes no difference whether you take the drug once or twice a day. So Factor A (dosage) is significant, and Factor B (times per day) isn't. Parallel lines indicate no interaction effect.

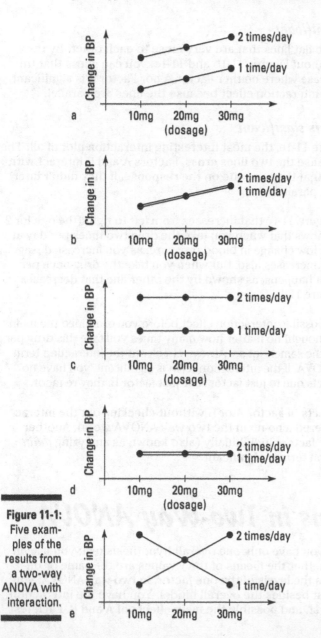

Figure 11-1:
Five exam-
ples of the
results from
a two-way
ANOVA with
interaction.

### Factor B is significant but not Factor A

Figure 11-1c shows where Factor B (times per day) is significant but Factor A (dosage level) isn't. The lines are flat across dosage levels, indicating that dosage has no effect on blood pressure. However, the two lines for times per day are spread apart, so their effect on blood pressure is significant. Parallel lines indicate no interaction effect.

### Neither factor is significant

Figure 11-1d shows two flat lines that are very close to each other. By the previous discussions about Figures 11-1b and 11-1c, you can guess that this figure represents the case where neither Factor A nor Factor B is significant, and you don't have an interaction effect because the lines are parallel.

### Interaction term AB is significant

Finally you get to Figure 11-1e, the most interesting interaction plot of all. The big picture is that because the two lines cross, Factors A and B interact with each other in the way that they operate on the response. If they didn't interact, the lines would be parallel.

Start with the line in Figure 11-1e that increases from left to right (the one for 2 times/day). This line shows that when you take the drug two times per day at the low dose, you get a low change in blood pressure; as you increase dosage, blood pressure change increases also. But when you take the drug once per day, the opposite result happens, as shown by the other line that decreases from left to right in Figure 11-1e.

If you didn't look for a possible interaction effect before you examined the main effects, you may have thought no matter how many times you take this drug per day, the effects will be the same. Not so! Always check out the interaction term first in any two-way ANOVA. If the interaction term is significant, you have no way to pull out the effects due to just factor A or just factor B; they're moot.

Checking the main effects of Factor A or B without checking out the interaction AB term is considered a no-no in the two-way ANOVA world. Another taboo is examining the factors individually (also known as analyzing *main effects*) if the interaction term is significant.

# Testing the Terms in Two-Way ANOVA

In a one-way ANOVA, you have only one overall hypothesis test; you use an *F*-test to determine whether the means of the *y* values are the same or different as you go across the levels of the one factor. In two-way ANOVA, you have more items to test besides the overall model. You have the interaction term AB to examine first, and possibly the main effects of A and B. Each test in a two-way ANOVA is an *F*-test based on the ideas of one-way ANOVA (see Chapter 9 for more on this).

To conduct the *F*-tests for these terms, you basically want to see whether more of the total variability in the *y*'s can be explained by the term you're testing compared to what's left in the error term. A large value of *F* means that the term you're testing is significant.

First, you test whether the interaction term AB is significant. To do this, you use the test statistic $F = \frac{MS_{AB}}{MSE}$, which has an *F*-distribution with $(i-1) * (j-1)$ degrees of freedom from $MS_{AB}$ (mean sum of squares for the interaction term of A and B) and $i * j * (m-1)$ degrees of freedom from MSE (mean sum of squares for error), respectively. (Recall that $i$ and $j$ are the number of levels of A and B, and $m$ is the sample size at each combination of A and B.)

If the interaction term isn't significant, you take the AB term out of the model, and you can explore the effects of Factors A and B separately regarding the response variable *y*.

The test for Factor A uses the test statistic $F = \frac{MS_A}{MSE}$, which has an *F*-distribution with $i-1$ degrees of freedom from $MS_A$ (mean sum of squares for Factor A) and $i * j * (m-1)$ degrees of freedom from MSE (mean sum of squares for error), respectively.

Testing for Factor B uses the test statistic $F = \frac{MS_B}{MSE}$, which has an *F*-distribution with $j-1$ and $i * j * (m-1)$ degrees of freedom. (See Chapter 9 for all the details on *F*-tests, MSE, and degrees of freedom.)

The results you can get from testing the terms of the ANOVA model are the same as those represented in Figure 11-1. They're all provided in Minitab output outlined in the next section, including their sum of squares, degrees of freedom, mean sum of squares, and *p*-values for their appropriate *F*-tests.

# *Running the Two-Way ANOVA Table*

The ANOVA table for two-way ANOVA includes the same elements as the ANOVA table for one-way ANOVA (see Chapter 9). But where in the one-way ANOVA you have one line for Factor A's contributions, now you add lines for the effects of Factor B and the interaction term AB. Minitab calculates the ANOVA table for you as part of the output from running a two-way ANOVA.

In this section, you figure out how to interpret the results of a two-way ANOVA, assess the model's fit, and use a multiple comparisons procedure, all using the drug data study.

# *Interpreting the results: Numbers and graphs*

The drug study example involves four people in each treatment combination of three possible dosage levels (10mg, 20mg, and 30mg per day) and two possible times for taking the drug (one time per day and two times per day). The total sample size is 4 * 3 * 2 = 24. I made up five different data sets in which the analyses represent each of the five scenarios shown in Figure 11-1. Their ANOVA tables, as created by Minitab, are shown in Figure 11-2.

Notice that each ANOVA table in Figure 11-2 shows the degrees of freedom for dosage is 3 − 1 = 2; the degrees of freedom for times per day is 2 − 1 = 1; the degrees of freedom for the interaction term is (3 − 1) * (2 − 1) = 2; the degrees of freedom for total is 3 * 2 * 4 − 1 = 23; and the degrees of freedom for error is 3 * 2 * (4 − 1) = 18.

The order of the graphs in Figure 11-1 and the ANOVA tables in Figure 11-2 isn't the same. Can you match them up? (I promise to give you the answers, so keep reading.)

Here's how the graphs from Figure 11-1 match up with the output in Figure 11-2:

✔ In the ANOVA table for Figure 11-2a, you see that the interaction term isn't significant (*p*-value = 0.526), so the main effects can be studied. The *p*-values for dosage (Factor A) and times taken (Factor B) are 0.000 and 0.001, indicating both Factors A and B are significant; this matches the plot in Figure 11-1a.

✔ In Figure 11-2b, you see that the *p*-value for interaction is significant (*p*-value = 0.000), so you can't examine the main effects of Factors A and B (in other words, don't look at their *p*-values). This represents the situation in Figure 11-1e.

✔ Figure 11-2c shows nothing is significant. The *p*-value for the interaction term is 0.513; *p*-values for main effects of Factors A (dosage) and B (times taken) are 0.926 and 0.416, respectively. These results coincide with Figure 11-1d.

✔ Figure 11-2d matches Figure 11-1b. It has no interaction effect (*p*-value = 0.899); dosage (Factor A) is significant (*p*-value = 0.000), and times per day (Factor B) isn't (*p*-value = 0.207).

✔ Figure 11-2e matches Figure 11-1c. The interaction term, dosage * times per day, isn't significant (*p*-value = 0.855); times per day is significant with *p*-value 0.000, but dosage level isn't significant (*p*-value = 0.855).

**a**

```
Two-way ANOVA: BP versus Dosage, Times
Source        DF      SS        MS        F       P
Dosage         2   56.3333   28.1667   112.67   0.000
Times          1    4.1667    4.1667    16.67   0.001
Interaction    2    0.3333    0.1667     0.67   0.526
Error         18    4.5000    0.2500
Total         23   65.3333

S = 0.5        R-Sq = 93.11%    R-Sq(adj) = 91.20%
```

**b**

```
Two-way ANOVA: BP versus Dosage, Times
Source        DF      SS        MS        F       P
Dosage         2    0.0833   0.04167    0.16   0.855
Times          1    0.3750   0.37500    1.42   0.249
Interaction    2   16.7500   8.37500   31.74   0.000
Error         18    4.7500   0.26389
Total         23   21.9583

S = 0.5137     R-Sq = 78.37%    R-Sq(adj) = 72.36%
```

**c**

```
Two-way ANOVA: BP versus Dosage, Times
Source        DF      SS         MS         F       P
Dosage         2    0.0833   0.041667    0.08   0.926
Times          1    0.3750   0.375000    0.69   0.416
Interaction    2    0.7500   0.375000    0.69   0.513
Error         18    9.7500   0.541667
Total         23   10.9583

S = 0.7360     R-Sq = 11.03%    R-Sq(adj) = 0.00%
```

**d**

```
Two-way ANOVA: BP versus Dosage, Times
Source        DF      SS         MS        F       P
Dosage         2   36.7500   18.3750    47.25   0.000
Times          1    0.6667    0.6667     1.71   0.207
Interaction    2    0.0833    0.0417     0.11   0.899
Error         18    7.0000    0.3889
Total         23   44.5000

S = 7.6236     R-Sq = 84.27%    R-Sq(adj) = 79.90%
```

**Figure 11-2:**
ANOVA
tables
for the
interaction
plots from
Figure 11-1.

**e**

```
Two-way ANOVA: BP versus Dosage, Times
Source        DF      SS         MS        F       P
Dosage         2    0.0833    0.0417     0.16   0.855
Times          1   12.0417   12.0417    45.63   0.000
Interaction    2    0.0833    0.0417     0.16   0.855
Error         18    4.7500    0.2639
Total         23   16.9583

S = 0.5137     R-Sq = 71.99%    R-Sq(adj) = 64.21%
```

### Assessing the fit

To assess the fit of the two-way ANOVA models, you can use the $R^2$ adjusted (see Chapter 6). The higher this number is, the better (the maximum is 100 percent or 1.00). Notice that all the ANOVA tables in Figure 11-2 show a fairly high $R^2$ adjusted except for Figure 11-2c. In this table, none of the terms were significant.

### Multiple comparisons

In the case where you find that an interaction effect is statistically significant, you can conduct multiple comparisons to see which combinations of Factors A and B create different results in the response. The same ideas hold here as do for multiple comparisons (covered in Chapter 10), except the tests can be performed on all $i * j$ interactions.

To perform multiple comparisons for a two-way ANOVA by using Minitab, enter your responses (data) in Column 1 (C1), your levels of Factor A in Column 2 (C2), and your levels of factor B in Column 3 (C3). Choose Stat>ANOVA>General Linear Model. In the Responses box, enter your Column 1 variable. In Model, enter C1 <space> C2 <space> C1*C2 (for the main effects and the interaction effect, respectively; here, <space> means leave a space). Click on Comparisons. In Terms, enter Columns 2 and 3. Check the Method you want to use for your multiple comparisons (see Chapter 10), and click OK.

# Are Whites Whiter in Hot Water? Two-Way ANOVA Investigates

You use two-way ANOVA when you want to compare the means of $n$ populations that are classified according to two different categorical variables (factors). For example, suppose you want to see how four brands of detergent (Brands A, B, C, D) and water temperature (1 = cold, 2 = warm, 3 = hot) work together to affect the whiteness of dirty t-shirts being washed. (Product-testing groups can use this information as well as the detergent companies to investigate or advertise how a detergent measures up to its competitors.)

Because this question involves two different factors and their effects on some numerical (quantitative) variable, you know that you need to do a two-way ANOVA. You can't assume that water temperature affects whiteness of clothes in the same way for each brand, so you need to include an interaction effect of brand and temperature in the two-way ANOVA model. Because brand of detergent has four possible types (or levels) and water temperature has three possible values (or levels), you have 4 * 3 = 12 different combinations to examine in terms of how brand and temperature interact. Those combinations are: Brand A in cold water, Brand A in warm water, Brand A in hot water, Brand B in cold water, Brand B in warm water, Brand B in hot water, and so on.

The resulting two-way ANOVA model looks like this: $y = b_i + w_j + bw_{ij} + e$, where $b$ represents the brand of detergent, $w$ represents the water temperature, $y$ represents the whiteness of the clothes after washing, and $bw_{ij}$ represents the interaction of brand $i$ of detergent ($i$ = A, B, C, D) and temperature $j$ of the water ($j$ = 1, 2, 3). (Note that $e$ represents the amount of variation in the $y$ values (whiteness) that isn't explained by either brand or temperature.)

Suppose you decide to run the experiment five times on each of the 12 combinations, which means 60 observations. (That's 60 t-shirts to wash — hey, it's a dirty job but someone's got to do it!) The results of the two-way ANOVA are shown in Figure 11-3.

**Figure 11-3:** ANOVA table for the clothing example.

```
ANOVA Table: Clothing Example
Source          DF        SS        MS        F         P
Brand           3      22.983    7.6611    20.89     0.000
Water           2       1.433    0.7167     1.95     0.153
Interaction     6     308.167   51.3611   140.08     0.000
Error          48      17.600    0.3667
Total          59     350.183

S = 0.6055     R-Sq = 94.97%        R-Sq(adj) = 93.82%
```

Note that the degrees of freedom (DF) for Brand, Water, Interaction, Error, and Total were arrived at from the following:

✔ DF for brand: $4 - 1 = 3$

✔ DF for water temperature: $3 - 1 = 2$

✔ DF for interaction term: $(4 - 1) * (3 - 1) = 6$

✔ DF for error: $60 - (4 * 3) = 48$

✔ DF for total: $n - 1 = 60 - 1 = 59$

Looking at the ANOVA table in Figure 11-3, you can see that the model fits the data very well, with $R^2$ *adjusted* equal to 93.82 percent. The interaction term (brand of detergent interacting with water temperature) is significant, with a $p$-value of 0.000. This means you can't look separately at the effect of brand of detergent or water temperature separately. One brand of detergent isn't always best, and one water temperature is not always best; it's the combination of the two that has different effects.

Your next question may be: Okay, which combination of detergent brand and water temperature is best? To answer this question, I did multiple comparisons on the means from all 12 combinations. (To do this, I followed the Minitab directions from the previous section.) Luckily, Tukey gives me an overall error rate of only 5 percent, so doing this many tests doesn't lead to making a lot of incorrect conclusions.

Because of the high number of combinations to compare, making sense of all the results on Tukey's output was a little difficult. Instead, I opted to first make box-plots of the data for each combination of brand and water temperature to help me see what was going on. The results of my boxplots are shown in Figure 11-4.

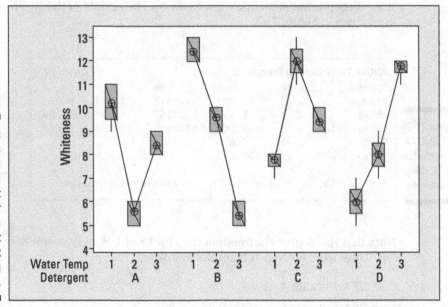

**Figure 11-4:**
Boxplots showing how brand of detergent and water temperature interact to affect clothing whiteness.

To create one set of boxplots for the data from each of the combinations in a two-way ANOVA, first ask Minitab to conduct a two-way ANOVA (you can find directions in the earlier section "Stepping through the sums of squares"). In that same Minitab window for two-way ANOVA, click Graphs, and a new window comes up. Click Boxplots of Data, and then OK. Finally, click OK to run the analysis and get the boxplots with it.

Figure 11-4 shows four groups of three connected boxes; each group of three represents data from one brand of detergent, tested under each of the three water temperatures (1 = cold, 2 = warm, and 3 = hot). For example, the first group of three shows the data from Brand A under each of the three water temperatures 1, 2, and 3, respectively. Each boxplot shows the results of the whiteness levels for the five shirts washed under that combination of detergent and water temp.

Looking at these plots you can see that each detergent reacts differently with different water temperatures. For example, Brand A does best in cold water (water temp level 1) and worst in warm water (water temp level 2), while Brand C is just the opposite, having the highest scores in warm water and the lowest in cold water. Each detergent does best/worst under a different combination of water temperatures. You can really see why the interaction term in this model is significant!

Now which combination of detergent and water temperature does the best? If you look at the plots, Brand B in cold water looks really good, and so does Brand C in warm water, closely followed perhaps by Brand D in hot water. This is where Tukey's multiple comparisons come in.

Running multiple comparisons on all 12 combinations of detergent and water temperature, you confirm that the three top combinations identified are all significantly higher than all the others (because their sample means were higher and their differences from all the other means had $p$-values less than 0.05). But the top three can't be distinguished from each other (because the $p$-values for the differences between them all exceed 0.05). Tukey also tells you that the three worst combinations are Brand A in warm water, Brand B in hot water, and Brand D in cold water. And they're all at the bottom of the barrel together (their means are significantly lower than all the rest but can't be distinguished from each other). So no single combination can claim all the bragging rights or shoulder all the blame.

You can imagine the many other comparisons that you could make from here to put the other combinations in some sort of order, but I think the best and worst are the most interesting for this case. It's like the fashion police commenting on what the stars wear on awards night. (Whatever they do wear, let's hope their statistician told them which brand of detergent to use and what water temperature to wash it in!)

# Chapter 12

# Regression and ANOVA: Surprise Relatives!

. . . . . . . . . . . . . . . . . . . . . . . . . . . . . . . . . . . . . . . . . . .

## In This Chapter
▶ Rewriting a regression line as an ANOVA model
▶ Connecting regression equations to the ANOVA table

. . . . . . . . . . . . . . . . . . . . . . . . . . . . . . . . . . . . . . . . . . .

**S**o you're motoring on in your Stats II course, working your way through regression (where you estimate *y* using one or more *x* variables; see Chapter 4). Then you hit a new topic, ANOVA, which stands for *analysis of variance* and refers to comparing the means of several populations (see Chapter 9). That seems to be no problem. But wait a minute; now your professor starts talking about how ANOVA is related to regression, and suddenly everything starts to spin out of control. How do you reconcile two techniques that appear to be as different as apples and oranges? That's what this chapter is all about.

Think of this chapter as your bridge across the gap between simple linear regression and ANOVA, allowing you to walk smoothly across, answering any questions that a professor may throw your way. Keep in mind that you don't actually apply these two techniques in this chapter (you can find that information in Chapters 4 and 9); the goal of this chapter is to determine and describe the relationship between regression and ANOVA so they don't look quite so much like an apple and an orange.

# Seeing Regression through the Eyes of Variation

Every basic statistical model tries to explain why the different outcomes (*y*) are what they are. It tries to figure out what factors or explanatory variables (*x*) can help explain that variability in those *y*'s. In this section, you start with the *y*-values by themselves and see how their variability plays a central role in the regression model. This is the first step toward applying ANOVA to the regression model.

No matter what *y* variable you're interested in predicting, you'll always have variability in those *y*-values. If you want to predict the length of a fish, for example, you know that fish have many different lengths (indicating a great deal of variability). Even if you put all the fish of the same age and species together, you still have some variability in their lengths (it's less than before but still there nonetheless). The first step in understanding the basic ideas of regression and ANOVA is to understand that variability in the *y*'s is to be expected, and your job is to try to figure out what can explain most of it.

## Spotting variability and finding an "x-planation"

Both regression and ANOVA work to get a handle on explaining the variability in the *y* variable using an *x* variable. After you collect your data, you can find the standard deviation in the *y* variable to get a sense of how much the data varies within the sample. From there, you collect data on an *x* variable and see how much it contributes to explaining that variability.

Suppose you notice that people spend different amounts of time on the Internet, and you want to explore why that may be. You start by taking a small sample of 20 people and record how many hours per month they spend on the Internet. The results (in hours) are 20, 20, 22, 39, 40, 19, 20, 32, 33, 29, 24, 26, 30, 46, 37, 26, 45, 15, 24, and 31. The first thing you notice about this data is the large amount of variability in it. The *standard deviation* (average distance from the data values to their mean) of this data set is 8.93 hours, which is quite large given the size of the numbers in the data set.

So you figured out that the *y*-values — the amount of time someone uses the Internet — have a great deal of variability in them. What can help explain this? Part of the variability is due to chance. But you suspect some variable is out there (call it *x*) that has some connection to the *y* variable, and that *x* variable can help you make more sense out of this seemingly wide range of *y*-values.

Suppose you have a brainstorm that number of years of education could possibly be related to Internet use. In this case, the explanatory variable (input variable, $x$) is years of education, and you want to use it to try to estimate $y$, the number of hours spent on the Internet in a month. You ask a larger random sample of 250 Internet users how many years of education they have (so $n = 250$). You can check out the first ten observations from your data set containing the $(x, y)$ pairs in Table 12-1. If a significant connection of some sort exists between the $x$-values and the $y$-values, then you can say that $x$ is helping to explain some of the variability in the $y$'s. If it explains enough variability, you can place $x$ into a simple regression model and use it to estimate $y$.

| Table 12-1 | First Ten Observations from the Education and Internet Use Example |
|---|---|
| **Years of Education** | **Hours Spent on Internet (In One Month)** |
| 15 | 41 |
| 15 | 32 |
| 11 | 33 |
| 10 | 42 |
| 10 | 28 |
| 10 | 21 |
| 10 | 17 |
| 10 | 14 |
| 9 | 18 |
| 9 | 14 |

## Getting results with regression

After you have a possible $x$ variable picked, you collect pairs of data $(x, y)$ on a random sample of individuals from the population, and you look for a possible linear relationship between them. Looking at the small snippet of 10 out of the 250-person data set in Table 12-1, you can begin to see that you may have a pattern between education and Internet use. It looks like as education increases so does Internet use.

To delve deeper, you make a scatterplot of the data and calculate the correlation $(r)$. If the data appear to follow a straight line (as shown on the scatterplot), go ahead and perform a simple linear regression of the response variable $y$ based on the $x$ variable. The $p$-value of the $x$ variable in the simple linear regression analysis tells you whether or not the *x* variable does a significant job in predicting *y*. (For the details on simple linear regression, see Chapter 4.)

To do a simple linear regression using Minitab, enter your data in two columns: the first column for your *x* variable and the second column for your *y* variable (as in Table 12-1). Go to Stat>Regression>Regression. Click on your *y* variable in the left-hand box; the *y* variable then appears in the Response box on the right-hand side. Click on your *x* variable in the left-hand box; the *x* variable then appears in the Predictor box in the right-hand side. Click OK, and your regression analysis is done. As part of every regression analysis, Minitab also provides you with the corresponding ANOVA results, found at the bottom of the output.

The simple linear regression output that Minitab gives you for the education and Internet example is in Figure 12-1. (Notice the ANOVA output at the bottom; you can see the connection in the upcoming section "Regression and ANOVA: A Meeting of the Models.")

**Figure 12-1:**
Output for simple linear regression applied to education and Internet use data.

```
Regression Analysis: Internet versus Education

The regression equation is
Internet = -8.29 + 3.15 Education

Predictor      Coef   SE Coef       T       P
Constant     -8.290     2.665   -3.11   0.002
Education    3.1460    0.2387   13.18   0.000

S = 7.23134      R-Sq = 41.2%      R-Sq(adj) = 41.0%

Analysis of Variance

Source            DF       SS      MS       F       P
Regression         1   9085.6  9085.6  173.75   0.000
Residual Error   248  12968.5    52.3
Total            249  22054.0
```

Looking at Figure 12-1, you see that the *p*-value on the row marked *Education* is 0.000, which means the *p*-value's less than 0.001. Therefore the relationship between years of education and Internet use is statistically significant. A scatterplot of the data (not shown here) also indicates that the data appear to have a positive linear relationship, so as you increase number of years of education, Internet use also tends to increase (on average).

# Assessing the fit of the regression model

Before you go ahead and use a regression model to make predictions for $y$ based on an $x$ variable, you must first assess the fit of your model. You can do this with a scatterplot and correlation or $R^2$.

## Using a scatterplot and correlation

One way to get a rough idea of how well your regression model fits is by using a *scatterplot*, which is a graph showing all the pairs of data plotted in the $x$-$y$ plane. Use the scatterplot to see whether the data appear to fall in the pattern of a line. If the data appear to follow a straight-line pattern (or even something close to that — anything but a curve or a scattering of points that has no pattern at all), you calculate the *correlation, r*, to see how strong the linear relationship between $x$ and $y$ is. The closer $r$ is to +1 or –1, the stronger the relationship; the closer $r$ is to zero, the weaker the relationship. Minitab can do scatterplots and correlations for you; see Chapter 4 for more on simple linear regression, including making a scatterplot and finding the value of $r$.

If the data don't have a significant correlation and/or the scatterplot doesn't look linear, stop the analysis; you can't go further to find a line that fits a relationship that doesn't exist.

## Using $R^2$

The more general way of assessing not only the fit of a simple linear regression model but many other models too is to use $R^2$, also known as the *coefficient of determination*. (For example, you can use this method in multiple, nonlinear, and logistic regression models in Chapters 5, 7, and 8, to name a few.) In simple linear regression, the value of $R^2$ (as indicated by Minitab and statisticians as a capital $R$ squared) is equal to the square of the Pearson correlation coefficient, $r$ (indicated by Minitab and statisticians by a small $r$). In all other situations, $R^2$ provides a more general measure of model fit. (Note that $r$ only measures the fit of a straight-line relationship between one $x$ variable and one $y$ variable; see Chapter 4.) An even better statistic, $R^2$ *adjusted*, modifies $R^2$ to account for the number of variables in the model. (For more information on $R^2$ and its use and interpretation, see Chapter 6.)

The value of $R^2$ *adjusted* for the model of using education to estimate Internet use (see Figure 12-1) is equal to 41 percent. This value reflects the percentage of variability in Internet use that can be explained by a person's years of education. This number isn't close to one, but note that $r$, the square root of 41 percent, is 0.64, which in the case of linear regression indicates a moderate relationship.

This evidence gives you the green light to use the results of the regression analysis to estimate number of hours of Internet use in a month by using years of education. The regression equation as it appears in the top part of the Figure 12-1 output is Internet use = –8.29 + 3.15 * years of education. So if you have 16 years of education, for example, your estimated Internet use is –8.29 + 3.15 * 16 = 42.11, or about 42 hours per month (about 10.5 hours per week).

But wait! Look again at Figure 12-1 and zoom in on the bottom part. I didn't ask for anything special to get this info on the Minitab output, but you can see an ANOVA table there. That seems like a fish out of water doesn't it? The next section connects the two, showing you how an ANOVA table can describe regression results (albeit it in a different way).

# Regression and ANOVA: A Meeting of the Models

After you've broken down the regression output into all its pieces and parts, the next step toward understanding the connection between regression and ANOVA is to apply the sums of squares from ANOVA to regression (something that's typically not done in a regression analysis). Before you start, think of this process as going to a 3-D movie, where you have to wear special glasses in order to see all the special effects!

In this section, you see the sums of squares in ANOVA applied to regression and how the degrees of freedom work out. You build an ANOVA table for regression and discover how the *t*-test for a regression coefficient is related to the *F*-test in ANOVA.

## Comparing sums of squares

*Sums of squares* is a term you may remember from ANOVA (see Chapter 9), but it certainly isn't a term you normally use when talking about regression (see Chapter 4). Yet, you can break down both types of models into sums of squares, and that similarity gets at the true connection between ANOVA and regression.

In step-by-step terms, you first partition out the variability in the *y* variable by using formulas for sums of squares from ANOVA (sums of squares for total, treatment, and error). Then you find those same sums of squares for regression — this is the twist on the process. You compare the two procedures through their sums of squares. This section explains how this comparison is done.

## Partitioning variability by using SSTO, SSE, and SST for ANOVA

ANOVA is all about partitioning the total variability in the $y$-values into sums of squares (find all the info you ever need on one-way ANOVA in Chapter 9). The key idea is that SSTO = SST + SSE, where SSTO is the total variability in the $y$-values; SST measures the variability explained by the model (also known as the treatment, or $x$ variable in this case); and SSE measures the variability due to error (what's left over after the model is fit).

Following are the corresponding formulas for SSTO, SSE, and SST, where $\bar{y}$ is the mean of the $y$'s, $y_i$ is each observed value of $y$, and $\hat{y}_i$ is each predicted value of $y$ from the ANOVA model:

$$SSTO = \sum (y_i - \bar{y})^2$$
$$SSE = \sum (y_i - \hat{y}_i)^2$$
$$SST = \sum (\hat{y}_i - \bar{y})^2$$

Use these formulas to calculate the sums of squares for ANOVA. (Minitab does this for you when it performs ANOVA.) Keep these values of SSTO, SST, and SSE. You'll use them to compare to the results from regression.

## Finding sums of squares for regression

In regression, you measure the deviations in the $y$-values by taking each $y_i$ minus its mean, $\bar{y}$. Square each result and add them all up, and you have SSTO. Next, take the residuals, which represent the difference between each $y_i$ and its estimated value from the model, $\hat{y}_i$. Square the residuals and add them up, and you get the formula for SSE.

After you calculate SSTO and SSE, you need the bridge between them — that is, you need a formula that connects the variability in the $y_i$'s (SSTO) and the variability in the residuals after fitting the regression line (SSE). That bridge is called the *sum of squares for regression*, or SSR (equivalent to SST in ANOVA). In regression, $\hat{y}_i$ represents the predicted value of $y_i$ based on the regression model. These are the values on the regression line. To assess how much this regression line helps to predict the $y$-values, you compare it to the model you'd get without any $x$ variable in it.

Without any other information, the only thing you can do to predict $y$ is look at the average, $\bar{y}$. So, SSR compares the predicted value from the regression line to the predicted value from the flat line (the mean of the $y$'s) by subtracting them. The result is $(\hat{y}_i - \bar{y})$. Square each result and sum them all up, and you get the formula for SSR, which is the same as the formula for SST in ANOVA. *Voilà!*

Instead of calling the sum of squares for the regression model SST as is done in ANOVA, statisticians call it SSR for *sum of squares regression*. Consider SSR to be equivalent to the SST from ANOVA. You need to know the difference because computer output lists the sums of squares for the regression model as SSR, not SST.

To summarize the sums of squares as they apply to regression, you have SSTO = SSR + SSE where

- ✔ **SSTO** measures the variability in the observed *y*-values around their mean. This value represents the variance of the *y*-values.

- ✔ **SSE** represents the variability between the predicted values for *y* (the values on the line) and the observed *y*-values. SSE represents the variability left over after the line has been fit to the data.

- ✔ **SSR** measures the variability in the predicted values for *y* (the values on the line) from the mean of *y*. SSR is the sum of squares due to the regression model (the line) itself.

Minitab calculates all the sums of squares for you as part of the regression analysis. You can see this calculation in the section "Bringing regression to the ANOVA table."

# Dividing up the degrees of freedom

In ANOVA, you test a model for the treatment (population) means by using an $F$-test, which is $F = \dfrac{MST}{MSE}$. To get MST (the mean sum of squares for treatment), you take SST (the sum of squares for treatment) and divide by its degrees of freedom. You do the same with MSE (that is, take SSE, the sum of squares for error, and divide by its degrees of freedom). The questions now are, what do those degrees of freedom represent, and how do they relate to regression?

## Degrees of freedom in ANOVA

In ANOVA, the degrees of freedom for SSTO is $n - 1$, which represents the sample size minus one. In the formula for SSTO, $\sum \left( y_i - \bar{y} \right)^2$, you see there are $n$ observed *y*-values minus one mean. In a very general way, that's where the $n - 1$ comes from.

Note that if you divide SSTO by $n - 1$, you get $\dfrac{\sum \left( y_i - \bar{y} \right)^2}{n - 1}$, the variance in the *y*-values. This calculation makes good sense because the variance measures the *total* variability in the *y*-values.

### Degrees of freedom in regression

The degrees of freedom for SST in ANOVA equal the number of treatments minus one. How does the degrees of freedom idea relate to regression? The number of treatments in regression is equivalent to the number of parameters in a model (a *parameter* being an unknown constant in the model that you're trying to estimate).

When you test a model, you're always comparing it to a different (simpler) model to see whether it fits the data better. In linear regression, you compare your regression line $y = a + bx$, to the horizontal line $y = \bar{y}$. This second, simpler model just uses the mean of $y$ to predict $y$ all the time, no matter what $x$ is. In the regression line, you have two coefficients: one to estimate the parameter for the $y$-intercept ($a$) and one to estimate the parameter for slope ($b$) in the model. In the second, simpler model, you have only one parameter: the value of the mean. The degrees of freedom for SSR in simple linear regression is the difference in the number of parameters from the two models: $2 - 1 = 1$.

The degrees of freedom for SSE in ANOVA is $n - k$. In the formula for SSE, $\sum(\hat{y}_i - \bar{y})^2$, you see there are $n$ predicted $y$-values, and $k$ is the number of treatments in the model. In regression, the number of parameters in the model is $k = 2$ (the slope and the $y$-intercept). So you have degrees of freedom $n - 2$ associated with SSE when you're doing regression.

Putting all this together, the degrees of freedom for regression must add up for the equation SSTO = SSR + SSE. The degrees of freedom corresponding to this equation are $(n - 1) = (2 - 1) + (n - 2)$, which is true if you do the math. So the degrees of freedom for regression, using the ANOVA approach, all check out. Whew!

In Figure 12-1, you can see the degrees of freedom for each sum of squares listed under the DF column of the ANOVA part of the output. You see SSR has $2 - 1 = 1$ degree of freedom, SSE has $250 - 2 = 248$ degrees of freedom (because $n = 250$ observations were in the data set and $k = 2$ and you find $n - k$ to get degrees of freedom for SSE). The degrees of freedom for SSTO is $250 - 1 = 249$.

## Bringing regression to the ANOVA table

In ANOVA, you test your model Ho: All $k$ population means are equal versus Ha: At least two population means are different by using a $F$-test. You build your $F$-test statistic by relating the sums of squares for treatment to the sum of squares for error. To do this, you divide SSE and SST by their degrees of freedom ($n - k$ and $k - 1$, respectively, where $n$ is the sample size and $k$ is the number of treatments) to get the mean sums of squares for error (MSE) and mean sums of squares for treatment (MST). In general, you want MST to be large compared to MSE, indicating that the model fits well. The results of all these statistical gymnastics are summarized by Minitab in a table called (cleverly) the ANOVA table.

The ANOVA table shown in the bottom part of Figure 12-1 for the Internet use data example represents the ANOVA table you get from using the regression line as your model. Under the Source column, you may be used to seeing treatment, error, and total. For regression, the treatment is the regression line, so you see *regression* instead of treatment. The error term in ANOVA is labeled *residual error*, because in regression, you measure error in terms of residuals. Finally you see *total*, which is the same the world around.

The SS column represents the sums of squares for the regression model. The three sums of squares listed in the SS column are SSR (for regression), SSE (for residuals), and SST (total). These sums of squares are calculated using the formulas from the previous section; the degrees of freedom, DF in the table, are found by using the formulas from the previous section also.

The MS column takes the value of SS[you fill in the blank] and divides it by the respective degrees of freedom, just like ANOVA. For example in Figure 12-1, SSE is 12968.5, and the degrees of freedom is 248. Take the first value divided by the second one to get 52.29 or 52.3, which is listed in the ANOVA table for MSE.

The value of the $F$-statistic, using the ANOVA method, is $F = \dfrac{MST}{MSE} = \dfrac{9085.6}{52.3} = 173.7$ in the Internet use example, which you can see in column five of the ANOVA part of Figure 12-1 (subject to rounding). The $F$-statistics's $p$-value is calculated based on an $F$-distribution with $k - 1 = 2 - 1 = 1$ and $n - k = 250 - 2 = 248$ degrees of freedom, respectively. (In the Internet use example, the $p$-value listed in the last column of the ANOVA table is 0.000, meaning the regression model fits.) But remember, in regression you don't use an $F$-statistic and an $F$-test. You use a $t$-statistic and a $t$-test. (Whoa. . .)

## Relating the F- and t-statistics: The final frontier

In regression, one way of testing whether the best-fitting line is statistically significant is to test Ho: Slope = 0 versus Ha: Slope $\neq$ 0. To do this, you use a $t$-test (see Chapter 3). The slope is the heart and soul of the regression line, because it describes the main part of the relationship between $x$ and $y$. If the slope of the line equals zero (you can't reject Ho), you're just left with $y = a$, a horizontal line, and your model $y = a + bx$ isn't doing anything for you.

In ANOVA, you test to see whether the model fits by testing Ho: The means of the populations are all equal versus Ha: At least two of the population means aren't equal. To do this you use an $F$-test (taking MST and dividing it by MSE; see Chapter 9).

REMEMBER

The sets of hypotheses in regression and ANOVA seem totally different, but in essence, they're both doing the same general thing: testing whether a certain model fits. In the regression case, the model you want to see fit is the straight line, and in the ANOVA case, the model of interest is a set of (normally distributed) populations with at least two different means (and the same variance). Here each population is labeled as a treatment by ANOVA.

But more than that, you can think of it this way: Suppose you took all the populations from the ANOVA and lined them up side by side on an $x$-$y$ plane (see Figure 12-2). If the means of those distributions are all connected by a flat line (representing the mean of the $y$'s), then you have no evidence against Ho in the $F$-test, so you can't reject it — your model isn't doing anything for you (simply put, it doesn't fit). This idea is similar to the idea of fitting a flat horizontal line through the $y$-values in regression; a straight-line model with a nonzero slope. This also indicates no relationship between $x$ and $y$.

The big thing is that statisticians can prove (so you don't have to) that an $F$-statistic is equivalent to the square of a $t$-statistic and that the $F$-distribution is equivalent to the square of a $t$-distribution when the SSR has df $= 2 - 1 = 1$. And when you have a simple linear regression model, the degrees of freedom is exactly 1! (Note that $F$ is always greater than or equal to zero, which is needed if you're making it the square of something.) So there you have it! The $t$-statistic for testing the regression model is equivalent to an $F$-statistic for ANOVA when the ANOVA table is formed for the simple regression model.

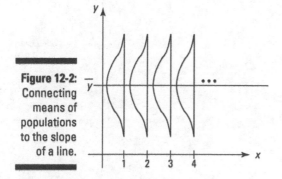

**Figure 12-2:**
Connecting
means of
populations
to the slope
of a line.

Indeed (the stats professor's way of saying "and this is the *really* cool part . . ."), if you look at the value of the *t*-statistic for testing the slope of the education variable in Figure 12-1, you see that it's 13.18 (look at the row marked Education and the column marked T). Square that value, and you get 173.71, the *F*-statistic in the ANOVA table of Figure 12-1. The *F*-statistic from ANOVA and the square of the *t*-statistic from regression are equal to each other in Figure 12-2, subject to a little round-off error done by Minitab on the output. (Just like magic! I still get chills just thinking about it.)

# Part IV
# Building Strong Connections with Chi-Square Tests

The 5th Wave — By Rich Tennant

"Is it just me or did the whole '50% satisfaction' statistic seem a little unimpressive?"

# In this part . . .

**H**ave you ever wondered if the percentage of M&M'S of each color is the same in every bag? Or whether someone's vote in an election is related to gender? Have you ever wondered if banks really have a case for denying loans based on a low credit score? This part answers all those questions and more using the Chi-square distribution. In particular, you get to use Chi-square to test for independence and to run goodness-of-fit tests.

# Chapter 13

# Forming Associations with Two-Way Tables

. . . . . . . . . . . . . . . . . . . . . . . . . . . . . . . . . . . . . . . . . . . . . . . . .

### In This Chapter

▶ Reading and interpreting two-way tables

▶ Figuring probabilities and checking for independence

▶ Watching out for Simpson's Paradox

. . . . . . . . . . . . . . . . . . . . . . . . . . . . . . . . . . . . . . . . . . . . . . . . .

*L*ooking for relationships between two categorical variables is a very common goal for researchers. For example, many medical studies center on how some characteristic about a person either raises or lowers his chance of getting some disease. Marketers ask questions like, "Who's more likely to buy our product: males or females?" Sports stats freaks wonder about things like, "Does winning the coin toss at the beginning of a football game increase a team's chance of winning the game?" ( I believe it does!)

To answer each of the preceding questions, you must first collect data (from a random sample) on the two categorical variables being compared — call them *x* and *y*. Then you organize that data into a table that contains columns and rows, showing how many individuals from the sample appear in each combination of *x* and *y*. Finally, you use the information in the table to conduct a hypothesis test (called the *Chi-square test*). Using the Chi-square test, you can determine whether you can see a relationship between *x* and *y* in the population from which the data were drawn. You need the machinery from Chapter 14 to accomplish this last step.

The goals of this chapter are to help you to understand what it means for two categorical variables (*x* and *y*) to be associated and to discover how to use percentages to determine whether a sample data set appears to show a relationship between *x* and *y*.

Suppose you're collecting data on cellphone users, and you want to find out whether more females use cellphones for personal use than males. A study of 508 randomly selected male cellphone users and 508 randomly selected female cellphone users conducted by a wireless company found that women tend to use their phones for personal calls more than men (big shocker). The survey showed that 427 of the women said they used their wireless phones primarily to talk with friends and family, while only 325 of the men admitted to doing so.

But you can't stop there. You need to break down this information, calculate some percentages, and compare those percentages to see how close they really are. Sample results vary from sample to sample, and differences can appear by chance.

In this chapter, you find out how to organize data from categorical variables (that's data based on categories rather than measurements) into a table format. This skill is especially useful when you're looking for relationships between two categorical variables, such as using a cellphone for personal calls (a yes or no category) and gender (male or female). You also summarize the data to answer your questions. And, finally, you get to figure out, once and for all, what's going on with that Simpson's Paradox thing.

# Breaking Down a Two-Way Table

A *two-way table* is a table that contains rows and columns that helps you organize data from categorical variables in the following ways:

- ✔ **Rows** represent the possible categories for one categorical variable, such as males and females.

- ✔ **Columns** represent the possible categories for a second categorical variable, such as using your cellphone for personal calls, or not.

## Organizing data into a two-way table

To organize your data into a two-way table, first set up the rows and columns. Table 13-1 shows the setup for the cellphone data example that I set up in the chapter introduction.

| Table 13-1 | Two-Way Table for the Cellphone Data | |
|---|---|---|
| | *Personal Calls: Yes* | *Personal Calls: No* |
| *Males* | | |
| *Females* | | |

Notice that Table 13-1 has four empty cells inside of it (not counting the empty space in the upper-left corner). Because gender has two choices (male or female) and personal cellphone use has two choices (yes or no), the resulting two-way table has 2 * 2 = 4 cells.

To figure out the number of cells in any two-way table, multiply the number of possible categories for the row variable times the number of possible categories for the column variable.

## Filling in the cell counts

After you set up the table with the appropriate number of rows and columns, you need to fill in the appropriate numbers in each of the cells of the two-way table. The number in each cell of a two-way table is called the *cell count* for that cell. Of the four cells in the two-way table shown in Table 13-1, the upper-left cell represents the number of males who use their cellphones for personal calls. With the information you have in the cellphone example, the cell count for this cell is 325. You also know that 427 females use their cellphones for personal calls, and this number goes into the lower-left cell.

To figure out the numbers in the remaining two cells, you do a bit of subtraction. You know from the information given that the total number of male cellphone users in the survey is 508. Each male either uses his cellphone for personal calls (falling into the *yes* group) or doesn't use it for personal calls (falling into the *no* group). Because 325 males fall into the *yes* group, and you have 508 males total, 183 males (508 – 325 = 183) don't use their cellphones for personal calls. This number is the cell count for the upper-right cell of the two-way table. Finally, because 508 females took the survey, and 427 of them use their cellphones for personal calls, you know that the rest of them (508 – 427 = 81) don't. Therefore, 81 is the cell count for the lower-right cell of the table. Table 13-2 shows the completed table for the cellphone user example, with the four cell counts filled in.

| Table 13-2 | Completed Two-Way Table for the Cellphone Data | |
|---|---|---|
| | **Personal Calls: Yes** | **Personal Calls: No** |
| *Males* | 325 | 183 = (508 − 325) |
| *Females* | 427 | 81 = (508 − 427) |

Just to save you a little time, if you have the total number in a group and the number of individuals who fall into one of the categories of the two-way table, you can determine the number falling into the remaining category by subtracting the total number in the group minus the number in the given category. You can complete this process for each remaining group in the table.

## Making marginal totals

One of the most important characteristics of a two-way table is that it gives you easy access to all the pertinent totals. Because every two-way table is made up of rows and columns, you can imagine that the totals for each row and the totals for each column are important. Also, the grand total is important to know.

If you take a single row and add up all the cell counts in the cells of that row, you get a *marginal row total* for that row. Where does this marginal row total go on the table? You guessed it — out in the margin at the end of that row. You can find the marginal row totals for every row in the table and put them into the margins at the end of the rows. This group of marginal row totals for each row represents what statisticians call the *marginal distribution* for the row variable.

The marginal row totals should add up to the *grand total*, which is the total number of individuals in the study. (The individuals may be people, cities, dogs, companies, and so on, depending on the scenario of the problem at hand.)

Similarly, if you take a single column and add up all the cell counts in the cells of that column, you get the *marginal column total* for that column. This number goes in the margin at the bottom of the column. Follow this pattern for each column in the table, and you have the marginal distribution for the column variable. Again, the sum of all the marginal column totals equals the grand total. The grand total is always located in the lower-right corner of the two-way table.

The marginal row total, marginal column totals, and the grand total for the cellphone example are shown in Table 13-3.

| Table 13-3 | Marginal and Grand Totals for the Cellphone Data | | |
|---|---|---|---|
| | **Personal Calls: Yes** | **Personal Calls: No** | **Marginal Row Totals** |
| *Males* | 325 | 183 = (508 − 325) | 508 |
| *Females* | 427 | 81 = (508 − 427) | 508 |
| *Marginal Column Totals* | 752 | 264 | 1,016 (Grand Total) |

The marginal row totals add the cell counts in each row; yet the marginal row totals show up as a column in the two-way table. This phenomenon occurs because when summing the cell counts in a row, you put the result in the margin at the end of the row, and when you do this for each row, you're stacking the row totals into a column. Similarly, the marginal column totals add the cell counts in each column; yet they show up as a row in the two-way table. Don't let this result be a source of confusion when you're trying to navigate or set up a two-way table. I recommend that you label your totals as marginal row, marginal column, or grand total to help keep it all clear.

# Breaking Down the Probabilities

In the context of a two-way table, a percentage can be interpreted in one of two ways — in terms of a group or an individual. Regarding a group, a percentage represents the portion of the group that falls into a certain category. However, a percentage also represents the probability that an individual selected at random from the group falls into a certain category.

A two-way table gives you the opportunity to find many different kinds of probabilities, which help you to find the answers to different questions about your data or to look at the data another way. In this section, I cover the three most important types of probabilities found in a two-way table: marginal probabilities, joint probabilities, and conditional probabilities. (For more complete coverage of these types of probabilities, check out *Probability For Dummies,* by yours truly and published by Wiley.)

When you find probabilities based on a sample, as you do in this chapter, you have to realize that those probabilities pertain to that sample only. They don't transfer automatically to the population being studied. For example, if you take a random sample of 1,000 adults and find that 55 percent of them watch reality TV, this study doesn't mean that 55 percent of all adults in the entire population watch reality TV. (The media makes this mistake every day.) You need to take into account the fact that sample results vary; in Chapters 14 and 15, you do just that. But this chapter zeros in on summarizing the information in your sample, which is the first step toward that end (but not the last step in terms of making conclusions about your corresponding population).

## Marginal probabilities

A *marginal probability* makes a probability out of the marginal total, for either the rows or the columns. A marginal probability represents the proportion of the entire group that belongs in that single row or column category. Each marginal probability represents only one category for only one variable — it doesn't consider the other variable at all. In the cellphone example, you have four possible marginal probabilities (refer to Table 13-3):

✔ Marginal probability of female $\left(\frac{508}{1,016} = 0.50\right)$, meaning that 50 percent of all the cellphone users in this sample were females

✔ Marginal probability of male $\left(\frac{508}{1,016} = 0.50\right)$, meaning that 50 percent of all the cellphone users in this sample were males

✔ Marginal probability of using a cellphone for personal calls $\left(\frac{752}{1,016} = 0.74\right)$, meaning that 74 percent of all cellphone users in this sample make personal calls with their cellphones

✔ Marginal probability of not using a cellphone for personal calls $\left(\frac{264}{1,016} = 0.26\right)$, meaning that 26 percent of all the cellphone users in this sample don't make personal calls with their cellphones

Statisticians use shorthand notation for all probabilities. If you let M = male, F = female, Yes = personal cellphone use, and No = no personal cellphone use, then the preceding marginal probabilities are written as follows:

✔ P(F) = 0.50

✔ P(M) = 0.50

✔ P(Yes) = 0.74

✔ P(No) = 0.26

Notice that P(F) and P(M) add up to 1.00. This result is no coincidence because these two categories make up the entire gender variable. Similarly, P(Yes) and P(No) sum up to 1.00 because those choices are the only two for the personal cellphone use variable. Everyone has to be classified somewhere.

Be advised that some probabilities aren't useful in terms of discovering information about the population in general. For example, P(F) = 0.50 because the researchers determined ahead of time that they wanted exactly 508 females and exactly 508 males. The fact that 50 percent of the sample is female and 50 percent of the sample is male doesn't mean that in the entire population of cellphone users 50 percent are males and 50 percent are females. If you want to study what proportion of cellphone users are females and males, you need to take a combined sample instead of two separate ones and see how many males and females appear in the combined sample.

## Joint probabilities

A *joint probability* gives the probability of the intersection of two categories, one from the row variable and one from the column variable. It's the probability that someone selected from the whole group has two particular characteristics at the same time. In other words, both characteristics happen jointly, or together. You find a joint probability by taking the cell count for those having both characteristics and dividing by the grand total.

Here are the four joint probabilities in the cellphone example:

- The probability that someone from the entire group is male and uses his cellphone for personal calls is $\frac{325}{1,016} = 0.32$, meaning that 32 percent of all the cellphone users in this sample are males using their cellphones for personal calls.

- The probability that someone from the entire group is male and doesn't use his cellphone for personal calls is $\frac{183}{1,016} = 0.18$.

- The probability that someone from the entire group is female and makes personal calls with her cellphone is $\frac{427}{1,016} = 0.42$.

- The probability that someone from the entire group is female and doesn't make personal calls with her cellphone is $\frac{81}{1,016} = 0.08$.

The notation for the joint probabilities listed is as follows, where ∩ represents the intersection of the two categories listed:

- P(M ∩ Yes) = 0.32
- P(M ∩ No) = 0.18
- P(F ∩ Yes) = 0.42
- P(F ∩ No) = 0.08

The sum of all the joint probabilities for any two-way table should be 1.00, unless you have a little round-off error, which makes it very close to 1.00 but not exactly. The sum is 1.00 because everyone in the group is classified somewhere with respect to both variables. It's like dividing the entire group into four parts and showing which proportion falls into each part.

## Conditional probabilities

A *conditional probability* is what you use to compare subgroups in the sample. In other words, if you want to break down the table further, you turn to a conditional probability. Each row has a conditional probability for each cell within the row, and each column has a conditional probability for each cell within that column.

*Note:* Because conditional probability is one of the sticking points for a lot of students, I spend extra time on it. My goal with this section is for you to have a good understanding of what a conditional probability really means and how you can use it in the real world (something many statistics textbooks neglect to mention, I have to say).

### Figuring conditional probabilities

To find a conditional probability, you first look at a single row or column of the table that represents the known characteristic about the individuals. The marginal total for that row (column) now represents your new grand total, because this group becomes your entire universe when you examine it. Then take the cell counts from that row (column) and divide the sum by the marginal total for that row (column).

Consider the cellphone example in Table 13-3. Suppose you want to look at just the males who took the survey. The total number of males is 508. You can break down this group into two subgroups by using conditional probability: You can find the probability of using cellphones for personal calls (males only) and the probability of not using cellphones for personal calls (males only). Similarly, you can break down the females into those females who use cellphones for personal calls and those females who don't.

In the cellphone example, you have the following conditional probabilities when you break down the table by gender:

✔ The conditional probability that a male uses a cellphone for personal calls is $\frac{325}{508} = 0.64$.

✔ The conditional probability that a male doesn't use a cellphone for personal calls is $\frac{183}{508} = 0.36$.

✔ The conditional probability that a female uses a cellphone for personal calls is $\frac{427}{508} = 0.84$.

✔ The conditional probability that a female doesn't use a cellphone for personal calls is $\frac{81}{508} = 0.16$.

To interpret these results, you say that within this sample, if you're male, you're more likely than not to use your cellphone for personal calls (64 percent compared to 36 percent). However, the percentage of personal-call makers is higher for females (84 percent versus 16 percent).

Notice that for the males in the previous example, the two conditional probabilities (0.64 and 0.36) add up to 1.00. This is no coincidence. The males have been broken down by cellphone use for personal calls, and because everyone in the study is a cellphone user, each male has to be classified into one group or the other. Similarly, the two conditional probabilities for the females sum to 1.00.

## Notation for conditional probabilities

You denote conditional probabilities with a straight vertical line that lists and separates the event that's known to have happened (what's given) and the event for which you want to find the probability. You can write the notation like this: P(XX|XX). You place the given event to the right of the line and the event for which you want to find the probability to the left of the line. For example, suppose you know someone is female (F) and you want to find out the chance she's a Democrat (D). In this case, you're looking for P(D|F). On the other hand, say you know a person is a Democrat and you want the probability that person is female — you're looking for P(F|D).

The vertical line in the conditional probability notation isn't a division sign; it's just a line separating events A and B. Also, be careful of the order in which you place A and B into the conditional probability notation. In general, P(A|B) ≠ P(B|A).

Following is the notation used for the conditional probabilities in the cellphone example:

✔ **P(Yes|M) = 0.64.** You can say it this way: "The probability of Yes given Male is 0.64."

✔ **P(No|M) = 0.36.** In human terms, say, "The probability of No given Male is 0.36."

✔ **P(Yes|F) = 0.84.** Say this one with gusto: "The probability of Yes given Female is 0.84."

✔ **P(No|F) = 0.16.** You translate this notation by saying, "The probability of No given Female is 0.16."

You can see that P(Yes|M) + P(No|M) = 1.00 because you're breaking all males into two groups: those using cellphones for personal calls (Y) and those not (N). Notice, however, that P(Yes|M) + P(Yes|F) doesn't sum to 1.00. In the first case, you're looking only at the males, and in the second case, only at the females.

### Comparing two groups with conditional probabilities

One of the most common questions regarding two categorical variables is this: Are they related? To answer this question, you compare their conditional probabilities.

To compare the conditional probabilities, follow these steps:

1. **Take one variable and find the conditional probabilities based on the other variable.**

2. **Repeat step one for each category of the first variable.**

3. **Compare those conditional probabilities (you can even graph them for the two groups) and see whether they're the same or different.**

   If the conditional probabilities are the same for each group, the variables aren't related in the sample. If they're different, the variables are related in the sample.

4. **Generalize the results to the entire population by using the sample results to draw a conclusion from the overall population involved by doing a Chi-square test (see Chapter 14).**

Revisiting the cellphone example from the previous section, you can ask specifically: Is personal use related to gender? You know that you want to compare cellphone use for males and females to find out whether use is related to gender. However, it's very difficult to compare cell counts; for example, 325 males use their phones for personal calls, compared to 427 females. In fact, it's impossible to compare these numbers without using some total for perspective. 325 out of what?

You have no way of comparing the cell counts in two groups without creating percentages (achieved by dividing each cell count by the appropriate total). Percentages give you a means of comparing two numbers on equal terms. For example, suppose you gave a one-question opinion survey (yes, no, and no opinion) to a random sample of 1,099 people; 465 respondents said yes, 357 said no, and 277 had no opinion. To truly interpret this information, you're probably trying to compare these numbers to each other in your head. That's what percentages do for you. Showing the percentage in each group in a side-by-side fashion gives you a relative comparison of the groups with each other.

But first, you need to bring conditional probabilities into the mix. In the cell-phone example, if you want the percentage of females who use their cellphones for personal calls, you take 427 divided by the total number of females (508) to get 84 percent. Similarly, to get the percentage of males who use their cellphones for personal calls, take the cell count (325) and divide it by that row total for males (508), which gives you 64 percent. This percentage is the conditional probability of using a cellphone for personal calls, given the person is male.

Now you're ready to compare the males and females by using conditional probabilities. Take the percentage of females who use their cellphones for personal calls and compare it to the percentage of males who use their cellphones for personal calls. By finding these conditional probabilities, you can easily compare the two groups and say that in this sample at least, more females use their cellphones (84 percent) for personal calls than men (64 percent).

### Using graphs to display conditional probabilities

One way to highlight conditional probabilities as a tool for comparing two groups is to use graphs, such as a pie chart comparing the results of the other variable for each group or a bar chart comparing the results of the other variable for each group. (For more info on pie charts and bar charts, see my book *Statistics For Dummies* (Wiley) or your Stats I textbook.)

You may be wondering how close the two pie charts need to look (in terms of how close the slice amounts are for one pie compared to the other) in order to say the variables are independent. This question isn't one you can answer completely until you conduct a hypothesis test for the proportions themselves (see the Chi-square test in Chapter 14). For now, with respect to your sample data, if the difference in the appearance of the slices for the two graphs is enough that you would write a newspaper article about it, then go for dependence. Otherwise, conclude independence.

Figures 13-1a and 13-1b use two pie charts to compare cellphone use of males and females. Figure 13-1a shows the conditional distribution of cellphone use for (given) males. Figure 13-1b shows the conditional distribution of cellphone use for (given) females. A comparison of Figures 13-1a and 13-1b reveals that the slices for cellphone use aren't equal (or even close) for males compared to females, meaning that gender and cellphone use for personal calls are dependent in this sample. This confirms the previous conclusions.

Figure 13-1:
Pie charts
comparing
male versus
female
personal
cellphone
use.

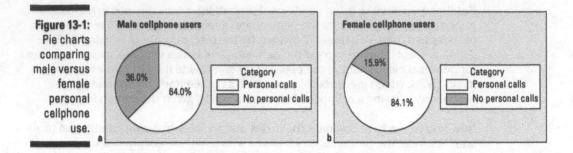

Another way you can make comparisons is to break down the two-way table by the column variable. (You don't always have to use the row variable for comparisons.) In the cellphone example (Table 13-3), you can compare the group of personal-call makers to the group of nonpersonal-call makers and see what percentage in each group is male and female. This type of comparison puts a different spin on the information because you're comparing the behaviors to each other in terms of gender.

With this new breakdown of the two-way table, you get the following:

- ✔ The conditional probability of being male, given you use your cellphone for personal calls, is $P(M \mid \text{Yes}) = \frac{325}{752} = 0.43$. *Note:* The denominator is 752, the total number of people who make personal calls with their cellphones.

- ✔ The conditional probability of being female, given you use your cellphone for personal calls, is $P(F \mid \text{Yes}) = \frac{427}{752} = 0.57$.

Again, these two probabilities add up to 1.00 because you're breaking down the personal-call makers according to gender (male or female). The conditional probabilities for the nonpersonal cellphone users are $P(M \mid \text{No}) = \frac{183}{264} = 0.69$ and $P(F \mid \text{No}) = \frac{81}{264} = 0.31$. These two probabilities also sum to 1.00 because you're breaking down the nonpersonal-call makers by gender (male and female).

The overall conclusions are similar to those found in the previous section, but the specific percentages and the interpretation are different. Interpreting the data this way, if you use your cellphone for personal calls, you're more likely to be female than male (57 percent compared to 43 percent). And if you don't use your cellphone to make personal calls, you're more likely to be male (69 percent compared to 31 percent).

## What should you divide by? That is the question!

To get the correct answer for any probability in a two-way table, here's the trick: Always identify the group being examined. What's the probability "out of"? In the cellphone example (refer to Table 13-3),

✔ If you want the percentage of *all* users who are males using their phones for personal calls, you take the cell count 325 divided by 1,016, the grand total.

✔ If you want the percentage of *males* who are using their cellphones for personal calls, you take 325 divided by 508, the total number of males.

✔ If you want the percentage of *personal-call makers* who are male, you take 325 divided by 752, the total number of people who make personal calls with their cellphones.

In each of these three cases, the numerator is the same but the denominators are different, leading you to very different answers. Deciding which number to divide by is a very common source of confusion for people, and this trick can really give you an edge on keeping it straight.

# Trying To Be Independent

*Independence* is a big deal in statistics. The term generally means that two items have outcomes whose probabilities don't affect each other. The items could be events A and B, variables $x$ and $y$, survey results from two people selected at random from a population, and so on. If the outcomes of the two items do affect each other, statisticians call those two items *dependent* (or not independent). In this section, you check for and interpret independence of individual categories, one from each categorical variable in a sample, and you check for and interpret independence of two categorical variables in a sample.

## Checking for independence between two categories

Statistics instructors often have students check to see whether two categories (one from a categorical variable $x$ and the other from a categorical variable $y$) are independent. I prefer to just compare the two groups and talk about how similar or different the percentages are, broken down by another variable. However, to cover all the bases and make sure you can answer this very popular question, here's the official definition of independence, straight from the statistician's mouth: Two categories are *independent* if their joint probability equals the product of their marginal probabilities. The only caveat here is that neither of the categories can be completely empty.

For example, if being female is independent of being a Democrat, then $P(F \cap D) = P(F) * P(D)$, where D = Democrat and F = Female. So, to show that two categories are independent, find the joint probability and compare it to the product of the two marginal probabilities. If you get the same answer both times, the categories are independent. If not, then the categories are dependent.

You may be wondering: Don't all probabilities work this way, where the joint probability equals the product of the marginals? No, they don't. For example, if you draw a card from a standard 52-card deck, you get a red card with probability $\frac{1}{2}$. You draw a heart with probability $\frac{1}{4}$. The chance of drawing both a heart and a red card with one draw is still $\frac{1}{4}$ (because all hearts are red). However, the product of the individual probabilities for red and heart comes out to $\frac{1}{4} * \frac{1}{2} = \frac{1}{8}$ which is not equal to $\frac{1}{4}$. This tells you that the categories "red" and "heart" aren't independent (that is, they're dependent).

Now the joint probability of a red two is $\frac{2}{52}$, or $\frac{1}{52}$. This equals the probability of a red card, $\frac{1}{2}$, times the probability of a two (because $\frac{1}{2} * \frac{4}{52} = \frac{1}{26}$). This tells you that the categories "red" and "two" are independent.

Another way to check for independence is to compare the conditional probability to the marginal probability. Specifically, if you want to check whether being female is independent of being Democrat, check either of the following two situations (they'll both work if the variables are independent):

- **Is $P(F|D) = P(F)$?** That is, if you know someone is a Democrat, does that affect the chance that they'll also be female? If yes, F and D are independent. If not, F and D are dependent.

- **Is $P(D|F) = P(D)$?** This question is asking whether being female changes your chances of being a Democrat. If yes, D and F are independent. If not, D and F are dependent.

Is knowing that you're in one category going to change the probability of being in another category? If so, the two categories aren't independent. If knowing doesn't affect the probability, then the two categories are independent.

# Checking for independence between two variables

The previous section focuses on checking whether two specific categories are independent in a sample. If you want to extend this idea to showing that two entire categorical variables are independent, you must check the independence conditions for every combination of categories in those variables. All of them must work, or independence is lost. The first case where dependence is found between two categories means that the two variables are dependent. If you find that the first case shows independence, you must continue checking all the combinations before declaring independence.

Suppose a doctor's office wants to know whether calling patients to confirm their appointments is related to whether they actually show up. The variables are $x$ = called the patient (called or didn't call) and $y$ = patient showed up for his appointment (showed or didn't show). Here are the four conditions that need to hold before you declare independence:

✔ P(showed) = P(showed | called)

✔ P(showed) = P(showed | didn't call)

✔ P(didn't show) = P(didn't show | called)

✔ P(didn't show) = P(didn't show | didn't call)

If any one of these conditions isn't met, you stop there and declare the two variables to be dependent in the sample. Only if all the conditions are met do you declare the two variables independent in the sample.

You can see the results of a sample of 100 randomly selected patients for this example scenario in Table 13-4.

| Table 13-4 | Confirmation Calls Related to Showing Up for the Appointment | | |
|---|---|---|---|
| | *Called* | *Didn't Call* | *Row Totals* |
| *Showed* | 57 | 33 | 90 |
| *Didn't Show* | 3 | 7 | 10 |
| *Column Totals* | 60 | 40 | 100 |

Checking the conditions for independence, you can start at the first condition and check to see whether P(showed) = P(showed|called). From the last column of Table 13-4, you can see that P(showed) is equal to $\frac{90}{100}$ = 0.90, or 90 percent. Next, look at the first column to find P(showed|called); this probability is $\frac{57}{60}$ = 95 percent. Because these two probabilities aren't equal (although they're close), you say that showing up and calling first are dependent in this sample. You can also say that people come a little more often when you call them first. (To determine whether these sample results carry through to the population, which also takes care of the question of how close the probabilities need to be in order to conclude independence, see Chapter 14.)

# Demystifying Simpson's Paradox

*Simpson's Paradox* is a phenomenon in which results appear to be in direct contradiction to one another, which can make even the best student's heart race. This situation can go unnoticed unless three variables (or more) are examined, in which case you organize the results into a *three-way table*, with columns within columns or rows within rows.

Simpson's Paradox is a favorite among statistics instructors (because it's so mystical and magical — and the numbers get so gooey and complex), but it's a nonfavorite among many students, mainly because of the following two reasons (in my opinion):

✔ Due to the way Simpson's Paradox is presented in most statistics courses, you can easily get buried in the details and have no hope of seeing the big picture. Simpson's Paradox draws attention to a big problem in terms of interpreting data, and you need to understand the paradox fully in order to avoid it.

✔ Most textbooks do a good job of showing you examples of Simpson's Paradox, but they fall short in explaining why it occurs, so it just looks like smoke and mirrors. Some even neglect to explain the why part at all!

This section helps you get a handle on what Simpson's Paradox is, better understand why and how it happens, and know how to watch for it.

## Experiencing Simpson's Paradox

Simpson's Paradox was discovered in 1951 by an American statistician named E. H. Simpson. He realized that if you analyze some data sets one way by breaking them down by two variables only, you can get one result, but when you break down the data further by a third variable, the results switch direction. That's why his result is called *Simpson's Paradox* — a paradox being an apparent contradiction in results.

### Simpson's Paradox in action: Video games and the gender gap

The best way to sort through Simpson's Paradox is to watch it play out in an example and explain all the whys along the way. Suppose I'm interested in finding out who's better at playing video games, men or women. I watch males and females choose and play a variety of video games, and I record whether the player wins or loses. Suppose I record the results of 200 video games, as seen in Table 13-5. (Note that the females played 120 games, and the males played 80 games.)

| Table 13-5 | Video Games Won and Lost for Males Versus Females | | |
|---|---|---|---|
| | **Won** | **Lost** | **Marginal Row Totals** |
| *Males* | 44 | 36 | 80 |
| *Females* | 84 | 36 | 120 |
| *Marginal Column Totals* | 128 | 72 | 200 (Grand Total) |

Looking at Table 13-5, you see the proportion of males who won their video games, P(Won | Male), is $\frac{44}{80} = 0.55$. The proportion of females who won their video games, P(Won | Female), is $\frac{84}{120} = 0.70$. So overall, the females won more of their video games than the males did. Does this finding mean that women are better than men at video games in general in the sample?

Not so fast, my friend. Notice that the people in the study were allowed to choose the video games they played. This factor blows the study wide open. Suppose females and males choose different types of video games: Can this affect the results? The answer may be yes. Considering other variables that could be related to the results but weren't included in the original study (or at least not in the original data analysis) is important. These additional variables that cloud the results are called *lurking variables*.

### Factoring in difficulty level

Many people may expect the video game results from the previous section to be turned around to indicate that men are better at playing video games than women. According to the research, men spend more time playing video games, on average, and are by far the primary purchasers of video games, compared to women. So what explains the eyebrow-raising results in this study? Is there another possible explanation? Is important information missing that's relevant to this case?

One of the variables that wasn't considered when I made Table 13-5 was the difficulty level of the video game being played. Suppose I go back and include the difficulty level of the chosen game each time, along with each result (won or lost). Level one indicates easy video games, comparable to the level of Ms. Pac Man (games that are my speed), and level two means more challenging video games (like war games or sophisticated strategy games).

Table 13-6 represents the results with the addition of this new information on difficulty level of games played. You have three variables now: level of difficulty (one or two), gender (male or female), and outcome (won or lost). That makes Table 13-6 a three-way table.

| Table 13-6 | A Three-Way Table for Gender, Game Level, and Game Outcome | | | |
| --- | --- | --- | --- | --- |
| | Level-One Games | | Level-Two Games | |
| | Won | Lost | Won | Lost |
| Males | 9 | 1 | 35 | 35 |
| Females | 72 | 18 | 12 | 18 |

Note in Table 13-6 that the number of level-one video games chosen was $9 + 1 + 72 + 18 = 100$, and the number of level-two video games chosen was $35 + 35 + 12 + 18 = 100$. In order to reevaluate the data based on the game level information, you need to look at who chose which level of game. The next section probes this very issue.

### Comparing success rates with conditional probabilities

To compare the success rates for males versus females using Table 13-6, you can figure out the appropriate conditional probabilities, first for level-one games and then for level-two games.

For level-one games (only), the conditional probability of winning given male is $P\left(\text{Won}|\text{Male}\right) = \frac{9}{10} = 0.90$. So for the level-one games, males won 90 percent of the games they played. For level-one games, the percentage of games won by the females is $P\left(\text{Won}|\text{Female}\right) = \frac{72}{90} = 0.80$, or 80 percent. These results mean that at level one, the males did 10 percent better than the females at winning their games. But this percentage appears to contradict the results found in Table 13-5. (Just wait — the contradictions don't end here!)

Now figure the conditional probabilities for the level-two video games won. For the men, the percentage of males winning level-two games was $\frac{35}{70} = 0.50$, or 50 percent. For the ladies, the percentage of women winning level-two games was $\frac{12}{30} = 0.40$, or 40 percent. Once again, the males outdid the females!

Step back and think about this scenario for a minute. Table 13-5 shows that females won a higher percentage of the video games they played overall. But Table 13-6 shows that males won more of the level-one games and more of the level-two games. What's going on? No need to check your math. No mistakes were made — no tricks were pulled. This inconsistency in results happens in real life from time to time in situations where an important third variable is left out of a study, a situation aptly named *Simpson's Paradox*. (See why it's called a paradox?)

## *Figuring out why Simpson's Paradox occurs*

Lurking variables are the underlying cause of Simpson's Paradox. A *lurking variable* is a third variable that's related to each of the other two variables and can affect the results if not accounted for.

In the video game example, when you look at the video game outcomes (won or lost) broken down by gender only (Table 13-5), females won a higher percentage of their overall games than males (70 percent overall winning percentage for females compared to 55 percent overall winning for males). Yet, when you split up the results by the level of the video game (level one or level two; see Table 13-6), the results reverse themselves, and you see that males did better than females on the level-one games (90 percent compared to 80 percent), and males also did better on the level-two games (50 percent compared to 40 percent).

To see why this seemingly impossible result happens, take a look at the marginal row *probabilities* versus the marginal row *totals* for the level-one games in Table 13-6. The percentage of times a male won when he played an easy video game was 90 percent. However, males chose level-one video games only 10 times out of 80 total level-one games played by men. That's only 12.5 percent.

To break this idea down further, the males' nonstellar performance on the challenging video games (50 percent — but still better than the females) coupled with the fact that the males chose challenging video games 87.5 percent of the time (that's 70 out of 80 times) really brought down their overall winning percentage (55 percent). And even though the men did really well on the level-one video games, they didn't play many of them (compared to the females), so their high winning percentage on level-one video games (90 percent) didn't count much toward their overall winning percentage.

Meanwhile, in Table 13-6, you see that females chose level-one video games 90 times (out of 120). Even though the females only won 72 out of the 90 games (80 percent, a lower percentage than the males, who won 9 out of 10 of their games), they chose to play many more level-one games, therefore boosting their overall winning percentage.

Now the opposite situation happens when you look at the level-two video games in Table 13-6. The males chose the harder video games 70 times (out of 80), while the females only chose the harder ones only 30 times out of 120. The males did better than the females on level-two video games (winning 50 percent of them versus 40 percent for the females). However, level-two video games are harder to win than level-one video games. This factor means that the males' winning percentage on level-two video games, being only 50 percent, doesn't contribute much to their overall winning percentage. However, the low winning percentage for females on level-two video games doesn't hurt them much, because they didn't play many level-two video games.

The bottom line is that the occurrence or nonoccurrence of Simpson's Paradox is a matter of weights. In the overall totals from Table 13-5, the males don't look as good as the females. But when you add in the difficulty of the games, you see that most of the males' wins came from harder games (which have a lower winning percentage). The females played many more of the easier games on average, and easy games carry a higher chance of winning no matter who plays them. So it all boils down to this: Which games did the males choose to play, and which games did the females choose to play? The males chose harder games, which contributed in a negative way to their overall winning percentage and made the females look better than they actually were.

## Keeping one eye open for Simpson's Paradox

Simpson's Paradox shows you the importance of including data about possible lurking variables when attempting to look at relationships between categorical variables.

Level of game wasn't included in the original summary, Table 13-5, but it should have been included because it's a variable that affected the results. Level of game, in this case, was the lurking variable. More men chose to play the more difficult games, which are harder to win, thereby lowering their overall success rate.

You can avoid Simpson's Paradox by making sure that obvious lurking variables are included in a study; that way, when you look at the data you get the relationships right the first time and there's a lower chance of reversing the results. And, as with all other statistical results, if it looks too good to be true or too simple to be correct, it probably is! Beware of someone who tried to oversimplify any result. While three-way tables are a little more difficult to examine, they're often worth using.

# Chapter 14

# Being Independent Enough for the Chi-Square Test

. . . . . . . . . . . . . . . . . . . . . . . . . . . . . . . . . . . . . . . . . . . . . .

## In This Chapter

▶ Testing for independence in the population (not just the sample)

▶ Using the Chi-square distribution

▶ Discovering the connection between the Z-test and the Chi-square test

. . . . . . . . . . . . . . . . . . . . . . . . . . . . . . . . . . . . . . . . . . . . . .

You've seen these hasty judgments before — people who collect one sample of data and try to use it to make conclusions about the whole population. When it comes to two categorical variables (where data fall into categories and don't represent measurements), the problem seems to be even more widespread.

For example, a TV news show finds that out of 1,000 presidential voters, 200 females are voting Republican, 300 females are voting Democrat, 300 males are voting Republican, and 200 males are voting Democrat. The news anchor shows the data and then states that 30 percent (300 + 1,000) of *all* presidential voters are females voting Democrat (and so on for the other counts).

This conclusion is misleading. It's true that in this sample of 1,000 voters, 30 percent of them are females voting Democrat. However, this result doesn't automatically mean that 30 percent of the entire population of voters is females voting Democrat. Results change from sample to sample.

In this chapter, you see how to move beyond just summarizing the sample results from a two-way table (discussed in Chapter 13) to using those results in a hypothesis test to make conclusions about an entire population. This process requires a new probability distribution called the *Chi-square distribution*. You also find out how to answer a very popular question among researchers: Are these two categorical variables independent (not related to each other) in the entire population?

# The Chi-square Test for Independence

Looking for relationships between variables is one of the most common reasons for collecting data. Looking at one variable at a time usually doesn't cut it. The methods used to analyze data for relationships are different depending on the type of data collected. If the two variables are quantitative (for example, study time and exam score), you use correlation and regression (see Chapter 4). If the two variables are categorical (for example, gender and political affiliation), you use a Chi-square test to examine relationships. In this section, you see how to use a Chi-square test to look for relationships between two categorical variables.

If two categorical variables don't have a relationship, they're deemed to be *independent*. If they do have a relationship, they're called *dependent variables*. Many folks get confused by these terms, so it's important to be clear about the distinction right up front.

To test whether two categorical variables are independent, you need a Chi-square test. The steps for the Chi-square test follow. (Minitab can conduct this test for you, from step three on down.)

1. **Collect your data, and summarize it in a two-way table.**

   These numbers represent the observed cell counts. (For more on two-way tables, see Chapter 13.)

2. **Set up your null hypothesis, Ho: Variables are independent; and the alternative hypothesis, Ha: Variables are dependent.**

3. **Calculate the expected cell counts under the assumption of independence.**

   The expected cell count for a cell is the row total times the column total divided by the grand total.

4. **Check the conditions of the Chi-square test before proceeding; each expected cell count must be greater than or equal to five.**

5. **Figure the Chi-square test statistic.**

   This statistic finds the observed cell count minus the expected cell count, squares the difference, and divides it by the expected cell count. Do these steps for each cell, and then add them all up.

6. **Look up your test statistic on the Chi-square table (Table A-3 in the appendix) and find the *p*-value (or one that's close).**

7. If your result is less than your predetermined cutoff (the α level), usually 0.05, reject Ho and conclude dependence of the two variables.

   If your result is greater than the α level, fail to reject Ho; the variables can't be deemed dependent.

To conduct a Chi-square test in Minitab, enter your data in the spreadsheet exactly as it appears in your two-way table (see Chapter 13 for setting up a two-way table for categorical data). Go to Stat>Tables>Chi-Square Test. Click on the two variable names in the left-hand box corresponding to your column variables in the spreadsheet. They appear in the box labeled Columns Contained in the Table. Then click OK.

# Collecting and organizing the data

The first step in any data analysis is collecting your data. In the case of two categorical variables, you collect data on the two variables at the same time for each individual in the study

A survey conducted by American Demographics asked men and women about the color of their next house. The results showed that 36 percent of the men wanted to paint their houses white, and 25 percent of the women wanted to paint their houses white. Keeping the data together in pairs (for example: male, white paint; female, nonwhite paint), you organize them into a two-way table where the rows represent the categories of one categorical variable (males and females for gender) and the columns represent the categories of the other categorical variable (white paint and nonwhite paint). Table 14-1 contains the results from a sample of 1,000 people (500 men and 500 women).

| Table 14-1 | Gender and House Paint Color Preference: Observed Cell Counts | | |
|---|---|---|---|
| | *White Paint* | *Nonwhite Paint* | *Marginal Row Totals* |
| *Men* | 180 | 320 | 500 |
| *Women* | 125 | 375 | 500 |
| *Marginal Column Totals* | 305 | 695 | 1,000 (Grand Total) |

The *marginal row totals* represent the total number in each row; the *marginal column totals* represent the total number in each column. (See Chapter 13 for more information on row and column marginal totals.)

Notice that of the males, the percentage that wants to paint the house white is 180 ÷ 500 = 0.36, or 36 percent, as stated previously. And the percentage of females that wants to paint the house white is 125 ÷ 500 = 0.25, or 25 percent. (Both of these percentages represent conditional probabilities as explained in Chapter 13.)

The American Demographics report concluded from this data that ". . . men and women generally agree on exterior house paint colors; the main exception being the top male choice, white (36 percent would paint their next house white versus 25 percent of women)." This type of conclusion is commonly formed, but it's an overgeneralization of the results at this point.

You know that in this sample, more men wanted to paint their houses white than women, but is 180 really that different from 125 when you're dealing with a sample size of 1,000 people whose results will vary the next time you do the survey? How do you know these results carry over to the population of all men and women? That question can't be answered without a formal statistical procedure called a *hypothesis test* (see Chapter 3 for the basics of hypothesis tests).

To show that men and women in the population differ according to favorite house color, first note that you have two categorical variables:

- ✔ Gender (male or female)
- ✔ Paint color (white or nonwhite)

Making conclusions about the population based on the sample (observed) data in a two-way table is taking too big of a leap. You need to conduct a Chi-square test in order to broaden your conclusions to the entire population. The media, and even some researchers, can get into trouble by ignoring the fact that sample results vary. Stopping with the sample results only and going merrily on your way can lead to conclusions that others can't confirm when they take new samples.

You keep the connection between the two pieces of information by organizing the data into one two-way table versus two individual tables — one for gender and one for house-paint preference. With one two-way table, you can look at the relationship between the two variables. (For the full details on organizing and interpreting the results from a two-way table, see Chapter 13.)

## Determining the hypotheses

Every hypothesis test (whether it be a Chi-square test or some other test) has two hypotheses:

- **Null hypothesis:** You have to believe this unless someone shows you otherwise. The notation for this hypothesis is Ho.

- **Alternative hypothesis:** You want to conclude this in the event that you can't support the null hypothesis anymore. The notation for this hypothesis is Ha.

In the case where you're testing for the independence of two categorical variables, the null hypothesis is when no relationship exists between them. In other words, they're independent. The alternative hypothesis is when the two variables are related, or dependent.

For the paint color preference example from the previous section, you write Ho: Gender and paint color preference are independent versus Ha: Gender and paint color preference are dependent. And there you have it — step two of the Chi-square test.

For a quick review of hypothesis testing, turn to Chapter 3. For a full discussion of the topic, see my other book *Statistics For Dummies* (Wiley) or your Stats I textbook.

## Figuring expected cell counts

When you've collected your data and set up your two-way table (for example, see Table 14-1), you already know what the observed values are for each cell in the table. Now you need something to compare them to. You're ready for step three of the Chi-square test — finding expected cell counts.

The null hypothesis says that the two variables $x$ and $y$ are independent. That's the same as saying $x$ and $y$ have no relationship. Assuming independence, you can determine which numbers should be in each cell of the table by using a formula for what's called the *expected cell counts.* (Each individual square in a two-way table is called a *cell,* and the number that falls into each cell is called the *cell count;* see Chapter 13.)

Table 14-1 shows the observed cell counts from the gender and paint color preference example. To find the expected cell counts you take the row total times the column total divided by the grand total, and do this for each cell in the table. Table 14-2 shows the calculations for the expected cell counts for the gender and paint color preference data.

| Table 14-2 | Gender and House Paint Color Preference: Expected Cell Counts | | |
|---|---|---|---|
| | **White Paint** | **Nonwhite Paint** | **Marginal Row Totals** |
| *Men* | (500 * 305) ÷ 1000 = 152.5 | (500 * 695) ÷ 1000 = 347.5 | 500 |
| *Women* | (500 * 305) ÷ 1000 = 152.5 | (500 * 695) ÷ 1000 = 347.5 | 500 |
| *Marginal Column Totals* | 305 | 695 | 1000 (Grand Total) |

Next you compare the observed cell counts in Table 14-1 to the expected cell counts in Table 14-2 by looking at their differences. The differences between the observed and expected cell counts shown in these tables are the following:

$$180 - 152.5 = 27.5$$

$$320 - 347.5 = -27.5$$

$$125 - 152.5 = -27.5$$

$$375 - 347.5 = 27.5$$

Next you do a Chi-square test for independence (see Chapter 15) to determine whether the differences found in the sample between the observed and expected cell counts are simply due to chance, or whether they carry through to the population.

Under independence, you conclude there is not a significant difference between what you observed and what you expected.

## Checking the conditions for the test

Step four of the Chi-square test is checking conditions. The Chi-square test has one main condition that must be met in order to test for independence on a two-way table: The expected count for each cell must be at least five — that is, greater than or equal to five. Expected cell counts that fall below five aren't reliable in terms of the variability that can take place.

In the gender and paint color preference example, Table 14-2 shows that all the expected cell counts are at least five, so the conditions of the Chi-square test are met.

If you're analyzing data and you find that your data set doesn't meet the expected cell count of at least five for one or more cells, you can combine some of your rows and/or columns. This combination makes your table smaller, but it increases the cell counts for the cells that you do have, which helps you meet the condition.

# Calculating the Chi-square test statistic

Every hypothesis test uses data to make the decision about whether or not to reject Ho in favor of Ha. In the case of testing for independence in a two-way table, you use a hypothesis test based on the Chi-square test statistic. In the following sections, you can see the steps for calculating and interpreting the Chi-square test statistic, which is step five of the Chi-square test.

### Working out the formula

A major component of the Chi-square test statistic is the expected cell count for each cell in the table. The formula for finding the expected cell count, $e_{ij}$, for the cell in row $i$, column $j$ is $e_{ij} = \dfrac{\text{row } i \text{ total} * \text{column } j \text{ total}}{\text{grand total}}$.

Note that the values of $i$ and $j$ vary for each cell in the table. In a two-way table, the upper-left cell of the table is in row one, column one. The cell in the upper-right corner is in row one, column two. The cell in the lower-left corner is in row two, column one, and the lower-right cell is in row two, column two.

The formula for the Chi-square test statistic is $\chi^2 = \sum_i \sum_j \dfrac{\left(o_{ij} - e_{ij}\right)^2}{e_{ij}}$, where $o_{ij}$ is the observed cell count for the cell in row $i$, column $j$, and $e_{ij}$ is the expected cell count for the cell in row $i$, column $j$.

When you calculate the expected cell count for some cells, you typically get a number that has some digits after the decimal point (in other words, the number isn't a whole number). Don't round this number off, despite the temptation to do so. This expected cell count is actually an overall-average expected value, so keep the count as it is, with decimal included.

### Calculating the test statistic

Here are the major steps of how to calculate the Chi-square test statistic for independence (Minitab does these steps for you as well):

1. **Subtract the observed cell count from the expected cell count for the upper-left cell in the table.**

2. **Square the result from step one to make the number positive.**

3. **Divide the result from step two by the expected cell count.**

4. **Repeat this process for all the cells in the table, and add up all the results to get the Chi-square test statistic.**

The reason you divide by the expected cell count in the Chi-square test statistic is to account for cell-count sizes. If you expect a big cell count, say 100, and are off by only 5 for the observed count of that cell, that difference shouldn't count as much as if you expected a small cell count (like 10) and the observed cell count was off by 5. Dividing by the expected cell count puts a more fair weight on the differences that go into the Chi-square test statistic.

To perform a Chi-square test in Minitab, you have to first enter the raw data (the data on each person) in two columns. The first column contains the values of the first variable in your data set. (For example, if your first variable is gender, go down the first Minitab column entering the gender of each person.) Then enter the data from your second variable in the second column, where each row represents a single person in the data set. (If your second variable is house paint color preference, for example, enter each person's paint color preference in column two, keeping the data from each person together in each row.) Go to Stat>Tables>Cross-tabulation and Chi-square.

Now Minitab needs to know which is your row variable and which is your column variable in your table. On the left-hand side, click on the variable that you want to represent the rows of your two-way table (you may click on the first variable). Click Select, and the variable name appears in the row variable portion of the table on the right. Now find the column variable blank on the right-hand side and click on it. Go to the left-hand side and click on the name of your second variable. Click Select. Then click on the Chi-square button and choose Chi-square analysis by checking the box. If you want the expected cell counts included, check that box also. Then click OK. Finally, click OK again to clear all the windows.

### Picking through the output

The Minitab output for the Chi-square analysis for the gender and house paint color preference example (from Table 14-1) is shown in Figure 14-1. You can pick out quite a few numbers from the output in Figure 14-1 that are especially important. The following three numbers are listed in each cell:

✔ The first (top) number is the observed cell count for that cell; this matches the observed cell count for each cell shown in Table 14-1. (Notice that the marginal row and column totals of Figure 14-1 also match those from Table 14-1.)

✔ The second number in each cell of Figure 14-1 is the expected cell count for that cell; you find it by taking the row total times the column total divided by the grand total (see the section "Figuring expected cell counts"). For example, the expected cell count for the upper-left cell (males who prefer white house paint) is (500 * 305) ÷ 1,000 = 152.50.

✔ The third number in each cell of Figure 14-1 is that part of the Chi-square test statistic that comes from that cell. (See steps one through three of the previous section "Working out the formula.") The sum of the third numbers in each cell equals the value of the Chi-square statistic listed in the last line of the output. (For the house paint color preference example, the Chi-square test statistic is 14.27.)

**Figure 14-1:**
Minitab
output for
the house
paint color
preference
data.

```
Chi-Square Test: Gender, House-Paint Preference

Expected counts are printed below observed counts
Chi-Square contributions are printed below expected counts

          White Paint  Nonwhite Paint    Total
     M            180             320       500
             152.50          347.50
              4.959           2.176

     F            125             375       500
             152.50          347.50
              4.959           2.176

Total            305             695      1000

Chi-Sq = 14.271, DF = 1, P-Value = 0.000
```

# Finding your results on the Chi-square table

The only way to make an assessment about your Chi-square test statistic is to compare it to all the possible Chi-square test statistics you would get if you had a two-way table with the same row and column totals, yet you distributed the numbers in the cells in every way possible. (You can do that in your sleep, right?) Some resulting tables give large Chi-square test statistics, and some give small Chi-square test statistics.

Putting all these Chi-square test statistics together gives you what's called a *Chi-square distribution*. You find your particular test statistic on that distribution (step six of the Chi-square test), and see where it stands compared to the rest.

If your test statistic is large enough that it appears way out on the right tail of the Chi-square distribution (boldly going where no test statistic has gone before), you reject Ho and conclude the two variables are not independent. If the test statistic isn't that far out, you can't reject Ho.

In the next sections, you find out more about the Chi-square distribution and how it behaves, so you can make a decision about the independence of your two variables based on your Chi-square statistic.

### Determining degrees of freedom

Each type of two-way table has its own Chi-square distribution, depending on the number of rows and columns it has, and each Chi-square distribution is identified by its *degrees of freedom*.

In general, a two-way table with *r* rows and *c* columns uses a Chi-square distribution with $(r-1)*(c-1)$ degrees of freedom. A two-way table with two rows and two columns uses a Chi-square distribution with one degree of freedom. Notice that $1 = (2-1)*(2-1)$. A two-way table with three rows and two columns uses a Chi-square distribution with $(3-1)*(2-1) = 2$ degrees of freedom.

Understanding *why* degrees of freedom are calculated this way is likely to be beyond the scope of your statistics class. But if you really want to know, the degrees of freedom represents the number of cells in the table that are flexible, or free, given all the marginal row and column totals.

For example, suppose that a two-way table has all row and column totals equal to 100 and the upper-left cell is 70. Then the upper-right cell must be 100 (row total) – 30 = 70. Because the column one total is 100, and the upper-left cell count is 70, the lower-left cell count must be 100 – 70 = 30. Similarly, the lower-right cell count must be 70.

So you have only one free cell in a two-way table after you have the marginal totals set up. That's why the degree of freedom for a two-way table is 1. In general, you always lose one row and one column because of knowing the marginal totals. That's because the last row and column values can be calculated through subtraction. That's where the formula $(r-1)*(c-1)$ comes from. (That's more than you wanted to know, isn't it?)

### Discovering how Chi-square distributions behave

Figure 14-2 shows pictures of Chi-square distributions with 1, 2, 4, 6, 8, and 10 degrees of freedom, respectively. Here are some important points to keep in mind about Chi-square distributions:

- For 1 degree of freedom, the distribution looks like a hyperbola (see Figure 14-2, top left); for more than 1 degree of freedom, it looks like a mound that has a long right tail (see Figure 14-2, lower right).

- All the values are greater than or equal to zero.

- The shape is always skewed to the right (tail going off to the right).

- As the number of degrees of freedom increases, the mean (the overall average) increases (moves to the right) and the variances increase (resulting in more spread).

- No matter what the degree of freedom is, the values on the Chi-square distribution (known as the *density*) approach zero for increasingly larger Chi-square values. That means that larger and larger Chi-square values are less and less likely to happen.

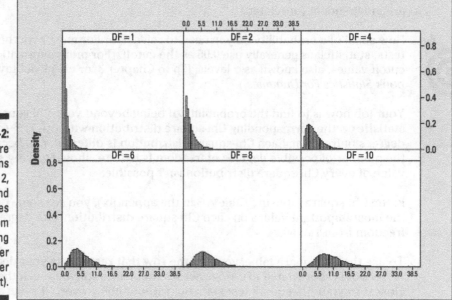

**Figure 14-2:** Chi-square distributions with 1, 2, 4, 6, 8, and 10 degrees of freedom (moving from upper left to lower right).

### Using the Chi-square table

After you find your Chi-square test statistic and its degrees of freedom, you want to determine how large your statistic is, relative to its corresponding distribution. (You're now venturing into step seven of the Chi-square test.)

If you think about it graphically, you want to find the probability of being beyond (getting a larger number than) your test statistic. If that probability is small, your Chi-square test statistic is something unusual — it's out there — and you can reject Ho. You then conclude that your two variables are not independent (they're related somehow).

In case you're following along at home, the Chi-square test statistic for the independent data from Table 14-2 is zero because the observed cell counts are equal to the expected cell counts for each cell, and their differences are always equal to zero. (This result never happens in real life!) This scenario represents a *perfectly independent* situation and results in the smallest possible value of a Chi-square test statistic.

If the probability of being to the right of your Chi-square test statistic (on a graph) isn't small enough, you don't have enough evidence to reject Ho. You then stick with Ho; you can't reject it. You conclude that your two variables are independent (unrelated).

How small of a probability do you need to reject Ho? For most hypothesis tests, statisticians generally use 0.05 as the cutoff. (For more information on cutoff values, also known as α levels, flip to Chapter 3, or check out my other book *Statistics For Dummies*.)

Your job now is to find the probability of being beyond your Chi-square test statistic on the corresponding Chi-square distribution with $(r-1) * (c-1)$ degrees of freedom. Each Chi-square distribution is different, and because the number of possible degrees of freedom is infinite, showing every single value of every Chi-square distribution isn't possible.

In the Chi-square table in (Table A-3 in the appendix), you see some of the most important values on each Chi-square distribution with degrees of freedom from 1 to 50.

To use the Chi-square table, you find the row that represents your degrees of freedom (abbreviated df). Move across that row until you reach the value closest to your Chi-square test statistic, without going over. (It's like a game show where you're trying to win the showcase by guessing the price.)

Then go to the top of the column you're in. That number represents the area to the right (above) of the Chi-square test statistic you saw in the table. The area above your particular Chi-square test statistic is less than or equal to this number. This result is the approximate *p*-value of your Chi-square test.

In the house paint color preference example (see Figure 14-1), the Chi-square test statistic is 14.27. You have $(2 - 1) * (2 - 1) = 1$ degree of freedom. In the Chi-square table, go to the row for df = 1, and go across to the number closest to 14.27 (without going over), which is 7.88.

## Drawing your conclusions

You have two alternative ways to draw conclusions from the Chi-square test statistic. You can look up your test statistic on the Chi-square table and see the probability of being greater than that. This method is known as *approximating the p-value*. (The *p*-value of a test statistic is the probability of being at or beyond your test statistic on the distribution to which the test statistic is being compared — in this case, the Chi-square distribution.) Or you can have the computer calculate the exact *p*-value for your test. (For a quick review of *p*-values and α levels, turn to Chapter 3. For a full review of these topics, see my other book *Statistics For Dummies*.)

Before you do anything though, set your α, the cutoff probability for your *p*-value, in advance. If your *p*-value is less than your α level, reject Ho. If it's more, you can't reject Ho.

### Approximating p-value from the table

For the house paint color preference example (see Figure 14-1), the Chi-square test statistic is 14.27 with $(2 - 1) * (2 - 1) = 1$ df (degree of freedom). The closest number in row one of the Chi-square table (see Table A-3 in the appendix), without going over, is 7.88 (in the last column).

The number at the top of that column is 0.005. This number is less than your typical α level of 0.05, so you reject Ho. You know that your *p*-value is less than 0.005 because your test statistic was more than 7.88. In other words, if 7.88 is the minimum evidence you need to reject Ho, you have more evidence than that with a value of 14.28. More evidence against Ho means a smaller *p*-value.

However, because Chi-square tables in general only give a few values for each Chi-square distribution, the best you can say using this table is that your p-value for this test is less than 0.005.

Here's the big news: Because your p-value is less than 0.05, you can conclude based on this data that gender and house paint color preference are likely to be related in the population (dependent), like the American Demographics Survey said (quoted at the beginning of this chapter). Only now, you have a formal statistical analysis that says this result found in the sample is also likely to occur in the entire population. This statement is much stronger!

If your data shows you can reject Ho, you only know at that point that the two variables have some relationship. The Chi-square test statistic doesn't tell you what that relationship is. In order to explore the relationship between the two variables, you find the conditional probabilities in your two-way table (see Chapter 13). You can use those results to give you some ideas as to what may be happening in the population.

For the gender and house paint color preference example, because paint color preference is related to gender, you can examine the relationship further by comparing the male versus female paint color preferences and describing how they're different. Start by finding the percentage of men that prefer white houses, which comes out to 180 ÷ 500 = 0.36, or 36 percent, calculated from Table 14-1. Now compare this result to the percentage of women who prefer white houses: 125 ÷ 500 = 0.25, or 25 percent. You can now conclude that in this population (not just the sample) men prefer white houses more than women do. Hence, gender and house paint color preference are dependent.

Dependent variables affect each other's outcomes, or cell counts. If the cell counts you actually observe from the sample data won't match the expected cell counts under Ho: The variables are independent, you conclude that the dependence relationship you found in the sample data carries over to the population. In other words, big differences between observed and expected cell counts mean that the variables are dependent.

### Extracting the p-value from computer output

After Minitab calculates the test statistic for you, it reports the exact p-value for your hypothesis test. The p-value measures the likelihood that your results were found just by chance while Ho is still true. It tells you how much strength you have against Ho. If the p-value is 0.001, for example, you have much more strength against Ho than if the p-value, say, is 0.10.

Looking at the Minitab output for the gender-paint color preference data in Figure 14-1, the *p*-value is reported to be 0.000. This means that the *p*-value is smaller than 0.001; for example, it may be 0.0009. That's a very small *p*-value! (Minitab only reports results to three decimal points, which is typical of many statistical software packages.)

I've seen situations where people get a result that isn't quite what they want (like a *p*-value of 0.068), and so they do some tweaking to get what they want. They change their α level from 0.05 to 0.10 after the fact. This change makes the *p*-value less than the α level, and they feel they can reject Ho and say that a relationship exists.

But what's wrong with this picture? They changed the α after they looked at the data, which isn't allowed. That's like changing your bet in blackjack after you find out what the dealer's cards look like. (Tempting, but a serious no-no.) Always be wary of large α levels, and make sure that you always choose your α before collecting any data — and stick to it.

The good news is that when *p*-values are reported, anyone reading them can make his or her own conclusion; no cut-and-dry rejection and acceptance region is set in stone. But setting an α level once and then changing it after the fact to get a better conclusion is never good!

## Putting the Chi-square to the test

If two variables turn out to be dependent, you can describe the relationship between them. But if two variables are independent, the results are the same for each group being compared. The following example illustrates this idea.

There has been much speculation and debate as to whether cellphone use should be banned while driving. You're interested in Americans' opinions on this issue, but you also suspect that the results may differ by gender. You decide to do a Chi-square test for independence to see if your theory plays out. Table 14-3 shows a two-way table of observed data from 60 men and 60 women regarding whether they agree with the policy (banning cellphone use while driving) or not. From Table 14-3 you see that 12 ÷ 60 = 20 percent of men agree with the policy of banning cellphones while driving, compared to 9 ÷ 60 = 15 percent of women. You see these percentages are different, but is this enough to say that gender and opinion on this issue are dependent? Only a Chi-square test for independence can help you decide.

| Table 14-3 | Gender and Opinion on Cellphone Ban: Observed Cell Counts | | |
|---|---|---|---|
| | Agree with Cellphone Ban | Disagree with Cellphone Ban | Marginal Row Totals |
| Men | 12 | 48 | 60 |
| Women | 9 | 51 | 60 |
| Marginal Column Totals | 21 | 99 | 120 (Grand Total) |

Table 14-4 shows the expected cell counts under Ho, along with their calculations.

| Table 14-4 | Gender and Opinion on Cellphone Ban: Expected Cell Counts | | |
|---|---|---|---|
| | Agree with Cellphone Ban | Disagree with Cellphone Ban | Marginal Row Totals |
| Men | $(60 * 21) \div 120$ $= 10.5$ | $(60 * 99) \div 120$ $= 49.5$ | 60 |
| Women | $(60 * 21) \div 120$ $= 10.5$ | $(60 * 99) \div 120$ $= 49.5$ | 60 |
| Marginal Column Totals | 21 | 99 | 120 (Grand Total) |

Running a Chi-square test in Minitab for this data, the degrees of freedom equals $(2 - 1) * (2 - 1) = 1$; the Chi-square test statistic can be shown to be equal to 0.519, and the $p$-value is 0.471. Because the $p$-value is greater than 0.05 (the typical cutoff), you can't reject Ho; therefore you conclude that gender and opinion on the banning of cellphones while driving are independent and therefore not related. Your theory that gender had something to do with it just doesn't pan out; there's not sufficient evidence for it.

In general, *independence* means that you can find no major difference in the way the rows look as you move down a column. Put another way, the proportion of the data falling into each column across the row is about the same for each row. Because Table 14-4 has the same number of men as women, the row totals are the same, and you get the same expected cell counts for men and women in both the Agree column (10.5) and the Disagree column (49.5).

# Comparing Two Tests for Comparing Two Proportions

You can use the Chi-square test to check whether two population proportions are equal. For example, is the proportion of female cellphone users the same as the proportion of male cellphone users?

You may be thinking, "But wait a minute, don't statisticians already have a test for two proportions? I seem to remember it from my Stats I course . . . I'm thinking . . . yeah, it's the Z-test for two proportions. What's that test got to do with a Chi-square test?" In this section, you get an answer to that question and practice using both methods to investigate a possible gender gap in cellphone use.

## Getting reacquainted with the Z-test for two population proportions

The way that most people figure out how to test the equality of two population proportions is to use a *Z-test for two population proportions*. With this test, you collect a random sample from each of the two populations, find and subtract their two sample proportions, and divide by their pooled standard error (see your Stats I textbook for details on this particular test).

This test is possible to do as long as the sample sizes from the two populations are large — at least five successes and five failures in each sample.

The null hypothesis for the Z-test for two population proportions is Ho: $p_1 = p_2$, where $p_1$ is the proportion of the first population that falls into the category of interest, and $p_2$ is the proportion of the second population that falls into the category of interest. And as always, the alternative hypothesis is one of the following choices, Ha: Not equal to, greater than, or less than.

Suppose you want to compare the proportion of male versus female cellphone users, where $p_1$ is the proportion of males who own a cellphone, and $p_2$ is the proportion of all females who own a cellphone. You collect data, find the sample proportions from each group, take their difference and make a

Z-statistic out of it using the formula $Z = \dfrac{\hat{p}_1 - \hat{p}_2}{\sqrt{\hat{p}(1-\hat{p})}\sqrt{\dfrac{1}{n_1}+\dfrac{1}{n_2}}}$, where $\hat{p} = \dfrac{x_1 + x_2}{n_1 + n_2}$.

Here, $x_1$ and $x_2$ are the number of individuals from samples one and two, respectively, with the desired characteristic; $n_1$ and $n_2$ are the two sample sizes.

Suppose that you collect data on 100 men and 100 women and find 45 male cellphone owners and 55 female cellphone owners. This means that $\hat{p}_1$ equals $45 \div 100 = 0.45$, and $\hat{p}_2$ equals $55 \div 100 = 0.55$. Your samples have at least five successes (having the desired characteristic; in this case, cellphone ownership) and five failures (not having the desired characteristic, which is cellphone ownership). So you compute the $Z$-statistic for comparing the two population proportions (males versus females) based on this data; it's $-1.41$, as shown on the last line of the Minitab output in Figure 14-3.

**Figure 14-3:**
Minitab
output
comparing
proportion
of male
and female
cellphone
owners.

```
Test Cellphone for Two Proportions

Sample   X     N    Sample p
M       45    100   0.450000
F       55    100   0.550000


Difference = p (1) − p (2)
Estimate for difference: −0.1
95% CI for difference:(−0.237896, 0.0378957)
Test for difference = 0 (vs not = 0): Z = −1.41 P-Value = 0.157
```

The $p$-value for the test statistic of $Z = -1.41$ is 0.157 (calculated by Minitab, or by looking at the area below the $Z$-value of $-1.41$ on a $Z$-table, which you should have in your Stats I text). This $p$-value (0.157) is greater than the typical $\alpha$ level (predetermined cutoff) of 0.05, so you can't reject Ho. You can't say that the two population proportions aren't equal, so you must conclude that the proportion of male cellphone owners is no different than females.

Even though the sample seemed to have evidence for a difference (after all, 45 percent isn't equal to 55 percent), you don't have enough evidence in the data to say that this same difference carries over to the population. So you can't lay claim to a gender gap in cellphone use, at least not with this sample.

## Equating Chi-square tests and Z-tests for a two-by-two table

Here's the key to relating the $Z$-test to a Chi-square test for independence. The $Z$-test for two proportions and the Chi-square test for independence in a two-by-two table (one with two rows and two columns) are equivalent if the sample sizes from the two populations are large enough — that is, when the number of successes and the number of failures in each cell of the two samples is at least five.

If you use the Z-test to see whether the proportion of male cellphone owners is equal to the proportion of female cellphone owners, you're really looking at whether you can expect the same proportion of cellphone owners despite gender (after you take the sample sizes into account). And that means you're testing whether gender (male or female) is independent of cellphone owner-ship (yes or no).

If the proportion of female cellphone owners equals the proportion of male cellphone owners, the proportion of cellphone owners is the same regardless of gender, so gender and cellphone ownership are independent. On the other hand, if you find the proportion of male cellphone owners to be unequal to the proportion of female cellphone owners, you can say that cellphone use differs by gender, so gender and cellphone ownership are dependent.

With the cellphone data, you have 45 males using cellphones (out of 100 males) and 55 females using cellphones (out of 100 females). The Minitab output for the Chi-square test for independence (complete with observed and expected cell counts, degrees of freedom, test statistic, and $p$-value) is shown in Figure 14-4. The $p$-value for this test is 0.157, which is greater than the typical $\alpha$ level 0.05, so you can't reject Ho.

Because the Chi-square test for independence and the Z-test tests are equivalent when you have a two-by-two table, the $p$-value from the Chi-square test for independence is identical to the $p$-value from the Z-test for two proportions. If you compare the $p$-values from Figures 14-3 and 14-4, you can see that for yourself.

**Figure 14-4:**
Minitab output testing independence of gender and cellphone ownership.

```
Chi-Square Test: Gender, Cellphone

Expected counts are printed below observed counts
Chi-Square contributions are printed below expected counts

              Y       N     Total
    M        45      55      100
          50.00   50.00
           0.500   0.500

    F        55      45      100
          50.00   50.00
           0.500   0.500

Total       100     100      200

Chi-Sq =   2.000,   DF =  1, P-Value = 0.157
```

Also, note that if you take the Z-test statistic for this example (from Figure 14-3), which is –1.41, and square it, you get 2.00, which is equal to the Chi-square test statistic for the same data (last line of Figure 14-4). It's also the case that the square of the Z-test statistic (when testing for the equality of two proportions) is equal to the corresponding Chi-square test statistic for independence.

The Chi-square test and Z-test are equivalent only if the table is a two-by-two table (two rows and two columns) and if the Z-test is two-tailed (the alternative hypothesis is that the two proportions aren't equal, instead of using Ha: One proportion is greater than or less than the other). If the Z-test isn't two-tailed, a Chi-square test isn't appropriate. If the two-way table has more than two rows or columns, use the Chi-square test for independence (because many categories mean you no longer have only two proportions, so the Z-test isn't applicable).

# The car accident–cellphone connection

Researchers are doing a great deal of study of the effects of cellphone use while driving. One study published in the *New England Journal of Medicine* observed and recorded data in 1997 on 699 drivers who had cellphones and were involved in motor vehicle collisions resulting in substantial property damage but no personal injury. Each person's cellphone calls on the day of the collision and during the previous week were analyzed through the use of detailed billing records. A total of 26,798 cellphone calls were made during the 14-month study period.

One conclusion the researchers made was that ". . . the risk of a collision when using a cellphone is four times higher than the risk of a collision when a cellphone was not being used." They basically conducted a Chi-square test to see whether cellphone use and having a collision are independent, and when they found out the events were not, the researchers were able to examine the relationship further using appropriate ratios. In particular, they found that the risk of a collision is four times higher for those drivers using cellphones than for those who aren't.

Researchers also found out that the relative risk was similar for drivers who differed in personal characteristics, such as age and driving experience. (This finding means that they conducted similar tests to see whether the results were the same for drivers of different age groups and drivers of different levels of experience, and the results always came out about the same. Therefore, age and the experience of the driver weren't related to the collision outcome.)

The research also shows that ". . . calls made close to the time of the collision were found to be particularly hazardous ($p < 0.001$). Hands-free cellphones offered no safety advantage over hand-held units (p-value not significant)." *Note:* The items in parentheses show the typical way that researchers report their results: using p-values. The p in both cases of parentheses represents the p-value of each test.

In the first case, the p-value is very tiny, less than 0.001, indicating strong evidence for a relationship between collisions and cellphone use at the time. The second p-value in parentheses was stated to be insignificant, meaning that it was substantially more than 0.05, the usual $\alpha$ level. This second result indicates that using hands-free equipment didn't affect the chances of a collision happening; the proportion of collisions using hands-free cellphones versus using regular cellphones were found to be statistically the same (they could have easily occurred by chance under independence). Whether you use a regular or hands-free cellphone, may this study be a lesson to you!

# Chapter 15

# Using Chi-Square Tests for Goodness-of-Fit (Your Data, Not Your Jeans)

---

### In This Chapter

▶ Understanding what goodness-of-fit really means

▶ Using the Chi-square model to test for goodness-of-fit

▶ Looking at the conditions for goodness-of-fit tests

---

**M**any phenomena in life may appear to be haphazard in the short term, but they actually occur according to some preconceived, preselected, or predestined model over the long term. For example, even though you don't know whether it will rain tomorrow, your local meteorologist can give you her model for the percentage of days that it rains, snows, is sunny, or cloudy, based on the last five years. Whether or not this model is still relevant this year is anyone's guess, but it's a model nonetheless. As another example, a biologist can produce a model for predicting the number of goslings raised by a pair of geese per year, even though you have no idea what the pair in your backyard will do. Is his model correct? Here's your chance to find out.

In this chapter, you build models for the proportion of outcomes that fall into each category for a categorical variable. You then test these models by collecting data and comparing what you observe in your data to what you expect from the model. You do this evaluation through a goodness-of-fit test that's based on the Chi-square distribution. In a way, a goodness-of-fit test is likened to a reality check of a model for categorical data.

# Finding the Goodness-of-Fit Statistic

The general idea of a *goodness-of-fit* procedure involves determining what you expect to find and comparing it to what you actually observe in your own sample through the use of a test statistic. This test statistic is called the *goodness-of-fit test statistic* because it measures how well your model (what you expected) fits your actual data (what you observed).

In this section, you see how to figure out the numbers that you should expect in each category given your proposed model, and you also see how to put those expected values together with your observed values to form the goodness-of-fit test statistic.

## What's observed versus what's expected

For an example of something that can be observed versus what's expected, look no further than a bag of tasty M&M'S Milk Chocolate Candies. A ton of different kinds of M&M'S are out there, and each kind has its own variation of colors and tastes. For this study, any reference I give to M&M'S is to the original milk chocolate candy — my favorite.

The percentage of each color of M&M'S that appear in a bag is something Mars (the company that makes M&M'S) spends a lot of time thinking about. Mars wants specific percentages of each color in its M&M'S bags, which it determines through comprehensive marketing research based on what people like and want to see. Mars then posts its current percentages for each color of M&M'S on its Web site. Table 15-1 shows the percentage of M&M'S of each color in 2006.

| Table 15-1 | Expected Percentage of Each Color of M&M'S Milk Chocolate Candies (2006) |
|---|---|
| **Color** | **Percentage** |
| Brown | 13% |
| Yellow | 14% |
| Red | 13% |
| Blue | 24% |
| Orange | 20% |
| Green | 16% |

Now that you know what to expect from a bag of M&M'S, the next question is, how does Mars deliver? If you were to open a bag of M&M'S right now, would you get the percentages of each color that you're supposed to get? You know from your previous studies in statistics that sample results vary (for a quick review of this idea, see Chapter 3). So you can't expect each bag of M&M'S to have exactly the correct number of each color of M&M'S as listed in Table 15-1. However, in order to keep customers happy, Mars should get close to the expectations. How can you determine how close the company does get?

Table 15-1 tells you what percentages are expected to fall into each category in the entire population of all M&M'S (that means every single M&M'S Milk Chocolate Candy that's currently being made). This set of percentages is called the *expected model* for the data. You want to see whether the percentages in the expected model are actually occurring in the packages you buy. To start this process, you can take a sample of M&M'S (after all, you can't check every single one in the population) and make a table showing what percentage of each color you observe. Then you can compare this table of observed percentages to the expected model.

Some expected percentages are known, as they are for the M&M'S, or you can figure them out by using math techniques. For example, if you're examining a single die to determine whether or not it's a fair die, you know that if the die is fair, you should expect $\frac{1}{6}$ of the outcomes to fall into each category of 1, 2, 3, 4, 5, and 6.

As an example, I examined one 1.69-ounce bag of plain, milk-chocolate M&M'S (tough job, but someone had to do it), and you can see my results in Table 15-2, column two. (Think of this bag as a random sample of 56 M&M'S, even though it's not technically the same as reaching into a silo filled with M&M'S and pulling out a true random sample of 1.69 ounces. For the sake of argument, one bag is okay.)

**Table 15-2  Percentage of M&M'S Observed in One Bag (1.69 oz.) Versus Percentage Expected**

| Color | Percentage Observed | Percentage Expected |
|---|---|---|
| Brown | $\frac{4}{56} = 7.14$ | 13.00 |
| Yellow | $\frac{10}{56} = 17.86$ | 14.00 |
| Red | $\frac{4}{56} = 7.14$ | 13.00 |
| Blue | $\frac{10}{56} = 17.86$ | 24.00 |
| Orange | $\frac{15}{56} = 26.79$ | 20.00 |
| Green | $\frac{13}{56} = 23.21$ | 16.00 |
| **TOTAL** | 100.00 | 100.00 |

Compare what I observed in my sample (column two of Table 15-2) to what I expected to get (column three of Table 15-2). Notice that I observed a lower percentage of brown and red M&M'S than expected and a lower percentage of blues than expected. I also observed a higher percentage of yellow, orange, and green M&M'S than expected. Sample results vary by random chance, from sample to sample, and the difference I observed may just be due to this chance variation. But could the differences indicate that the expected percentages reported by Mars aren't being followed?

It stands to reason that if the differences between what you observed and what you expected are small, you should attribute that difference to chance and let the expected model stand. On the other hand, if the differences between what you observed and what you expected are large enough, you may have enough evidence to indicate that the expected model has some problems. How do you know which conclusion to make? The operative phrase is, "if the differences are large enough." You need to quantify this term *large enough*. Doing so takes a bit more machinery, which I cover in the next section.

## Calculating the goodness-of-fit statistic

The goodness-of-fit statistic is one number that puts together the total amount of difference between what you expect in each cell compared to the number you observe. The term *cell* is used to express each individual category within a table format. For example, with the M&M'S example, the first columns of Tables 15-1 and 15-2 contain six cells, one for each color of M&M. For any cell, the number of items you observe in that cell is called the *observed cell count*. The number of items you expect in that cell (under the given model) is called the *expected cell count*. You get the expected cell count by taking the expected cell percentage times the sample size.

The expected cell count is just a proportion of the total, so it doesn't have to be a whole number. For example, if you roll a fair die 200 times, you should expect to roll ones $\frac{1}{6}$, or 16.67 percent, of the time. In terms of the number of ones you expect, it should be 0.1667 * 200 = 33.33. Use the 33.33 in your calculations for goodness-of-fit; don't round to a whole number. Your final answer is more accurate that way.

The reason the goodness-of-fit statistic is based on the *number* in each cell rather than the *percentage* in each cell is because percents are a bit deceiving. If you know that 8 out of 10 people support a certain view, that's 80 percent. But 80 out of 100 is also 80 percent. Which one would you feel is a more-precise statistic? The 80 out of 100 percent because it uses more information. Using percents alone disregards the sample size. Using the counts (the number in each group) keeps track of the amount of precision you have.

For example, if you roll a fair die, you expect the percentage of ones to be $\frac{1}{6}$. If you roll that fair die 600 times, the expected number of ones will be $\frac{1}{6} * 600 = 100$. That number (100) is the expected cell count for the cell that represents the outcome of one. If you roll this die 600 times and get 95 ones, then 95 is the observed cell count for that cell.

The formula for the goodness-of-fit statistic is given by the following:

$\sum_{all\ cells} \frac{(O-E)^2}{E}$, where $E$ is the expected number in a cell and $O$ is the observed

number in a cell. The steps for this calculation are as follows:

1. **For the first cell, find the expected number for that cell (*E*) by taking the percentage expected in that cell times the sample size.**

2. **Take the observed value in the first cell (*O*) minus the number of items that are expected in that cell (*E*).**

3. **Square that difference.**

4. **Divide the answer by the number that's expected in that cell, (*E*).**

5. **Repeat steps one through four for each cell.**

6. **Add up the results to get the goodness-of-fit statistic.**

The reason you divide by the expected cell count in the goodness-of-fit statistic (step four) is to take into account the magnitude of any differences you find. For example, if you expected 100 items to fall in a certain cell and you got 95, the difference is 5. But in terms of a percentage, this difference is only $\frac{5}{100} = 5$ percent. However, if you expected 10 items to fall into that cell and you observed 5 items, the difference is still 5, but in terms of a percentage, it's $\frac{5}{10} = 50$ percent. This difference is much larger in terms of its impact. The goodness-of-fit statistic operates much like a percentage difference. The only added element is to square the difference to make it positive. (That's done because whether you expected 10 and got 15 or expected 10 and got 5 makes no difference to others; you're still off by 50 percent.)

Table 15-3 shows the step-by-step calculation of the goodness-of-fit statistic for the M&M'S example, where $O$ indicates observed cell counts and $E$ indicates expected cell counts. To get the expected cell counts, you take the expected percentages shown in Table 15-1 and multiply by 56 because 56 is the number of M&M'S I had in my sample. The observed cell counts are the ones found in my sample, shown in Table 15-2.

| Table 15-3 | | Goodness-of-Fit Statistic for M&M'S Example | | | |
|---|---|---|---|---|---|
| Color | O | E | O − E | (O − E)² | $\frac{(O-E)^2}{E}$ |
| Brown | 4 | 0.13 ∗ 56 = 7.28 | 4 − 7.28 = −3.28 | 10.76 | 1.48 |
| Yellow | 10 | 0.14 ∗ 56 = 7.84 | 10 − 7.84 = 2.16 | 4.67 | 0.60 |
| Red | 4 | 0.13 ∗ 56 = 7.28 | 4 − 7.28 = −3.28 | 10.76 | 1.48 |
| Blue | 10 | 0.24 ∗ 56 = 13.44 | 10 − 13.44 = −3.44 | 11.83 | 0.88 |
| Orange | 15 | 0.20 ∗ 56 = 11.20 | 15 − 11.20 = 3.80 | 14.44 | 1.29 |
| Green | 13 | 0.16 ∗ 56 = 8.96 | 13 − 8.96 = 4.04 | 16.32 | 1.82 |
| TOTAL | 56 | 56 | | | 7.55 |

The goodness-of-fit statistic for the M&M'S example turns out to be 7.55, the bolded number in the lower-right corner of Table 15-3. This number represents the total squared difference between what I expected and what I observed, adjusted for the magnitude of each expected cell count. The next question is how to interpret this value of 7.55. Is it large enough to indicate that colors of M&M'S in the bag aren't following the percentages posted by Mars? The next section addresses how to make sense of these results.

# Interpreting the Goodness-of-Fit Statistic Using a Chi-Square

After you get your goodness-of-fit statistic, your next job is to interpret it. To do this, you need to figure out the possible values you could have gotten and where your statistic fits in among them. You can accomplish this task with a Chi-square goodness-of-fit test.

The values of a goodness-of-fit statistic actually follow a Chi-square distribution with $k - 1$ degrees of freedom, where $k$ is the number of categories in your particular population (see Chapter 14 for the full details on the Chi-square). You use the Chi-square table (Table A-3 in the appendix) to find the $p$-value of your Chi-square test statistic.

If your Chi-square goodness-of-fit statistic is large enough, you conclude that the original model doesn't fit and you have to chuck it; there's too much of a difference between what you observed and what you expected under the model. However, if your goodness-of-fit statistic is relatively small, you don't reject the model. (What constitutes a large or small value of a Chi-square test statistic depends on the degrees of freedom.)

The goodness-of-fit statistic follows the main characteristics of the Chi-square distribution. The smallest-possible value of the goodness-of-fit statistic is zero. Continuing the example from the previous section, if the M&M'S in my sample followed the exact percentages found in Table 15-1, the goodness-of-fit statistic would be zero. That's because the observed counts and the expected counts would be the same, so the values of the observed cell count minus the expected cell count would all be zero.

The largest-possible value of Chi-square isn't specified, although some values are more likely to occur than others. Each Chi-square distribution has its own set of likely values, as you can see in Figure 15-1. This figure shows a simulated Chi-square distribution with 6 − 1 = 5 degrees of freedom (relevant to the M&M'S example). It basically gives a breakdown of all the possible values you could have for the goodness-of-fit statistic in this situation and how often they occur. You can see in Figure 15-1 that a Chi-square test statistic of 7.55 isn't unusually high, indicating that the model for M&M'S colors probably can't be rejected. However, more particulars are needed before you can formally make that conclusion.

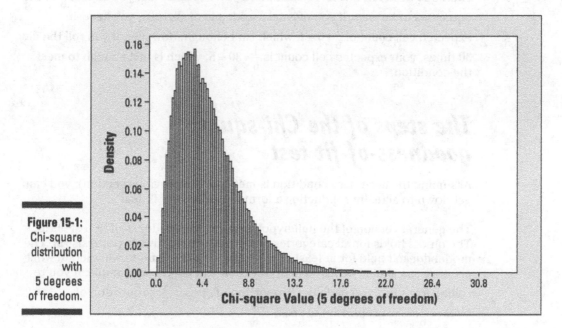

**Figure 15-1:**
Chi-square distribution with 5 degrees of freedom.

## Checking the conditions before you start

Every statistical technique seems to have a catch, and this case is no exception. In order to use the Chi-square distribution to interpret your goodness-of-fit statistic, you have to be sure you have enough information to work with in each cell. The stats gurus usually recommend that the expected count for each cell turns out to be greater than or equal to five. If it doesn't, one option is to combine categories to increase the numbers.

In the M&M'S example, the expected cell counts are all above seven (see Table 15-3), so the conditions are met. If this weren't the case, you should have taken a larger sample size, because you calculate the expected cell counts by taking the expected percentage in that cell times the sample size. If you increase the sample size, you increase the expected cell count. A higher sample size also increases your chances of detecting a real deviation from the model. This idea is related to the power of the test (see Chapter 3 for information on power).

After you collect your data, it's not right to go back and take a new and larger sample. It's best to set up the appropriate sample size ahead of time, and you can do this by determining what sample size you need to get the expected cell counts to be at least five. For example, if you roll a fair die, you expect $\frac{1}{6}$ of the outcomes to be ones. If you only take a sample of six rolls, you have an expected cell count of $\frac{1}{6} * 6 = 1$, which isn't enough. However, if you roll the die 30 times, your expected cell count is $\frac{1}{6} * 30 = 5$, which is just enough to meet the condition.

## The steps of the Chi-square goodness-of-fit test

Assuming the necessary condition is met (see the previous section), you can get down to actually conducting a formal goodness-of-fit test.

The general version of the null hypothesis for the goodness-of-fit test is Ho: The model holds for all categories; versus the alternative hypothesis Ha: The model doesn't hold for at least one category. Each situation will dictate what proportions should be listed in Ho for each category. For example, if you're rolling a fair die, you have Ho: Proportion of ones = $\frac{1}{6}$; proportion of twos = $\frac{1}{6}$; . . . ; proportion of sixes = $\frac{1}{6}$.

Following are the general steps for the Chi-square goodness-of-fit test, with the M&M'S example illustrating how you can carry out each step:

1. **Write down Ho using the percentages that you expect in your model for each category.**

   Using a subscript to indicate the proportion *(p)* of M&M'S you expect to fall into each category (see Table 15-1), your null hypothesis is Ho: $p_{brown} = 0.13$, $p_{yellow} = 0.14$, $p_{red} = 0.13$, $p_{blue} = 0.24$, $p_{orange} = 0.20$, and $p_{green} = 0.16$. All these proportions must hold in order for the model to be upheld.

2. **Write your Ha: This model doesn't hold for at least one of the percentages.**

   Your alternative hypothesis, Ha, in this case, would be: One (or more) of the probabilities given in Ho isn't correct. In other words, you conclude that at least one of the colors of M&M'S has a different proportion than what's stated in the model.

3. **Calculate the goodness-of-fit statistic using the steps in the earlier section "Calculating the goodness-of-fit statistic."**

   The goodness-of-fit statistic for M&M'S, from the earlier section, is 7.55. As a reminder, you take the observed number in each cell minus the expected number in that cell, square it, and divide by the expected number in that cell. Do that for every cell in the table and add up the results. For the M&M'S example, that total is equal to 7.55, the goodness-of-fit statistic.

4. **Look up the Chi-square distribution with $k - 1$ degrees of freedom, where $k$ is the number of categories you have.**

   You compare this statistic (7.55) to the Chi-square distribution with $6 - 1 = 5$ degrees of freedom (because you have $k = 6$ possible colors of M&M'S). (See Table A-3 in the appendix)

   Looking at Figure 15-1 you can see that the value of 7.55 is nowhere near the high end of this distribution, so you likely don't have enough evidence to reject the model provided by Mars for M&M'S colors.

5. **Find the *p*-value of your goodness-of-fit statistic.**

   You use a Chi-square table to find the *p*-value of your test statistic (see Table A-3 in the appendix). (For more info on the Chi-square distribution, refer to Chapter 14.)

   Because the Chi-square table can only list a certain number of results for each of the degrees of freedom, the exact *p*-value for your test statistic may fall between two *p*-values listed on the table.

To find the *p*-value for the test statistic in the M&M'S example (7.55), find the row for 5 degrees of freedom on the Chi-square table (Table A-3 in the appendix) and look at the numbers (the degrees of freedom is $k - 1 = 6 - 1 = 5$, where *k* is the number of categories). You see that the number 7.55 is less than the first value in the row (9.24), which has a *p*-value of 0.10. (Find the *p*-value by looking at the column heading above the number.) So the *p*-value for 7.55, which is the area to the right of 7.55 on Figure 15-1, must be greater than 0.10, because 7.55 is to the left of 9.24 on that Chi-square distribution.

Many computer programs exist (online or via a graphing calculator) that will find exact *p*-values for a Chi-square test, saving time and headaches when you have access to them (the technology, not the headaches). Using one such online *p*-value calculator, I found that the exact *p*-value for the goodness-of-fit test for the M&M'S example (test statistic 7.55 with 5 degrees of freedom for Chi-square) is 0.1828. To find online *p*-value calculators, simply type the name of the distribution and the word "p-value" in an Internet search engine. For this example, search "Chi-square p-value."

6. **If your *p*-value is less than your predetermined cutoff (α), reject Ho; the model doesn't hold. If your *p*-value is greater than α, you can't reject the model.**

   A typical value of α is 0.05. Some data analysts may use a higher value (up to 0.10), and others may go lower (for example, 0.010). See Chapter 3 for more information on choosing α and comparing your *p*-value to it.

   Going again to the M&M'S example, the *p*-value, 0.18, is greater than 0.05, so you fail to reject Ho. You can't say the model is wrong. So, Mars does appear to deliver on the percentages of M&M'S of each color as advertised. At least, you can't say it doesn't. (I'm sure Mars already knew that.)

Although some hypothesis tests are two-sided tests, the goodness-of-fit test is always a *right-tailed test*. You're only looking at the right tail of the Chi-square distribution when you're doing a goodness-of-fit test. That's because a small value of the goodness-of-fit statistic means that the observed data and the expected model don't differ much, so you stick with the model. If the value of the goodness-of-fit statistic is way out on the right tail of the Chi-square distribution, however, that's a different story. That situation means the difference between what you observed and what you expected is larger than what you should get by chance, and therefore, you have enough evidence to say the expected model is wrong.

You use the Chi-square goodness-of-fit test to check to see whether a specified model fits. A *specified model* is a model in which each possible value of the variable *x* is listed, along with its associated probability according to the model. For example, if you want to test whether three local hospitals take in the same percentage of emergency room patients, you test Ho: $p_1 = p_2 = p_3$, where each *p* represents the percentage of ER patients going to each hospital, respectively. In this case each *p* must equal 0.30 if the hospitals share the ER load equally.

# Part V

# Nonparametric Statistics: Rebels without a Distribution

The 5th Wave      By Rich Tennant

"Ted and I spent over 120 man-hours together analyzing the survey data, and here's what we discovered: Ted borrows pens and never returns them, he intentionally squeaks his chair to annoy me, and, evidently, I talk in my sleep."

## In this part . . .

**S**uppose you're driving home and one of the streets is blocked. What do you do? You back up and find another way to get home. Nonparametric statistics is that alternative route you take if the regular parametric statistical methods aren't allowed. What's more, this alternate route actually turns out to be better when the regular route isn't available. In this part, you find out just how much better nonparametric statistics is using the sign test, the signed rank test, and many more.

# Chapter 16

# Going Nonparametric

• • • • • • • • • • • • • • • • • • • • • • • • • • • • • • • • • • • • • • • • • • • • • •

### In This Chapter

▶ Understanding the need for nonparametric techniques

▶ Distinguishing regular methods from nonparametric methods

▶ Laying the groundwork: The basics of nonparametric statistics

• • • • • • • • • • • • • • • • • • • • • • • • • • • • • • • • • • • • • • • • • • • • • •

**M**any researchers do analyses involving hypothesis tests, confidence intervals, Chi-square tests, regression, and ANOVA. But nonparametric statistics doesn't seem to gain the same popularity as the other methods. It's more in the background — an unsung hero, if you will. However, nonparametric statistics is, in fact, a very important and very useful area of statistics because it gives you accurate results when other, more common methods fail.

In this chapter, you see the importance of nonparametric techniques and why they should have a prominent place in your data-analysis toolbox. You also discover some of the basic terms and techniques involved with nonparametric statistics.

## Arguing for Nonparametric Statistics

Nonparametric statistics plays an important role in the world of data analysis in that it can save the day when you can't use other methods. The problem is that researchers often disregard, or don't even know about, nonparametric techniques and don't use them when they should. In that case, you never know what kind of results you get; what you do know is they could very well be wrong.

In the following sections, you see the advantages and the flexibility of using a nonparametric procedure. You also find out just how minimal the downside is, which makes it a win-win situation most of the time.

# *No need to fret if conditions aren't met*

Many of the techniques that you typically use to analyze data, including many shown in this book, have one very strong condition on the data that must be met in order to use them: The populations from which your data are collected typically require a normal distribution. Methods requiring a certain type of distribution (such as a normal distribution) in order to use them are called *parametric* methods.

The following are ways to help you decide whether a population has a normal distribution, based on your sample:

✔ You can graph the data using a histogram, and see whether it appears to have a bell shape (a mound of data in the middle, trailing down on each side).

To make a histogram in Minitab, enter your data into a column. Go to Graph>Histogram, and click OK. Click on your variable in the left-hand box, and it appears in the Graph Variables box. Click OK, and check out your histogram.

✔ You can make a normal probability plot, which compares your data to that of a normal distribution, using an *x-y* graph (similar to the ones used when you graph a straight line). If the data do follow a normal distribution, your normal probability plot will show a straight line. If the data don't follow a normal distribution, the normal probability plot won't show a straight line; it may show a curve off to one side or the other, for example.

To make a normal probability plot in Minitab, enter your data in a column. Go to Graph>Probability Plot, and click OK. Click on your variable in the left-hand column, and it appears in the Graph Variables column. Click OK, and you see your normal probability plot.

When you find that the normal distribution condition is clearly not met, that's where nonparametric methods come in. *Nonparametric methods* are those data-analysis techniques that don't require the data to have a specific distribution. Nonparametric procedures may require one of the following two conditions (and these are only in certain situations):

✔ The data come from a symmetric distribution (which looks the same on each side when you cut it down the middle).

✔ The data from two populations come from the same type of distribution (they have the same general shape).

Note also that the normal distribution centers solely on the mean as its main statistic (for example, the $Z$-value for the hypothesis test for one population mean is calculated by taking the data value, subtracting the mean, and dividing by the standard deviation). So the condition that the population has a normal distribution automatically says you're working with the mean. However, many nonparametric procedures work with the *median,* which is a much more flexible statistic because it isn't affected by *outliers* (extreme values either above or below the mean) or *skewness* (a peak on one side and a long tail on the other side) as the mean is.

## The median's in the spotlight for a change

Many times a particular statistics question at hand revolves around the center of a population —that is, the number that represents a typical value, or a central value, in the population. One of those measures of center is the *mean.* The *population mean* is the average value over the entire population, which is typically not known (that's why you take a sample). Many data analysts focus heavily on the population mean; they want to estimate it, test it, compare the means of two or more populations, or predict the mean value of a *y* variable given an *x* variable. However, the mean isn't the only measure of the center of a population; you also have the good ol' median.

You may recall that the *median* of a data set is the value that represents the exact middle when you order the data from smallest to largest. For example, in the data set 1, 5, 4, 2, 3, you order the data to get 1, 2, 3, 4, 5 and find that the number in the middle is 3, the median. If the data set has an even number of values, for example, 2, 4, 6, 8, then you average the two middle numbers to get your median — (4 + 6) ÷ 2 = 5 in this case.

As you may recall from Stats I, you can find the mean and the median of a data set and compare them to each other. You first organize your data into a histogram, and you look at its shape.

- ✔ **If the data set is symmetric,** meaning it looks the same on either side when you draw a line down the middle, the mean and median are the same (or close). Figure 16-1a shows an example of this situation. In this case, the mean and median are both 5.

- ✔ **If the histogram is skewed to the right,** meaning that you have a lot of smaller values and a few larger values, the mean increases due to those few large values, but the median isn't affected. In this case, the mean is larger than the median. Figure 16-1b shows an example of this situation, in which the mean is 4.5 and the median is 4.0.

✔ **If the histogram is skewed to the left,** you have many larger values that pile up, but only a few smaller values. The mean goes down because of the few small values, but the median still isn't affected. In this case, the mean is lower than the median. Figure 16-1c illustrates this situation with a 6.5 mean and a 7.0 median.

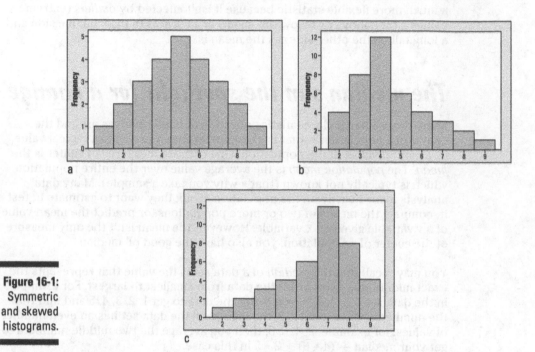

**Figure 16-1:**
Symmetric and skewed histograms.

REMEMBER

My point is that the median is important! It's a measure of the center of a population or a sample data set. The median competes with the mean and often wins. Researchers use nonparametric procedures when they want to estimate, test, or compare the median(s) of one or more populations. They also use the median in cases where their data are symmetric but don't necessarily follow a normal distribution, or when they want to focus on a measure of center that's not influenced by outliers or skewness.

For example, if you look at house prices in your neighborhood, you may find a large number of houses within a certain relatively small price range as well as a few homes that cost a great deal more. If a real estate agent wants to sell a house in your neighborhood and intends to justify a high price for it, she may report the mean price of homes in your neighborhood because the mean is affected by outliers. The mean is higher than the median in this case. But if the agent wants to help someone buy a house in your 'hood, she looks at the median of the house prices in the neighborhood because the median isn't affected by those few higher-priced homes and is lower than the mean.

Now suppose you want to come up with a number that describes the typical house price in your entire county. Should you use the mean or the median? You gathered techniques in Stats I for estimating the mean of a population (see Chapter 3 for a quick review), but you probably didn't hear about how to come up with a confidence interval for the median of a population. Oh sure, you can take a random sample and calculate the median of that sample. But you need a margin of error to go with it. And I'll tell you something — the formula for the margin of error for the mean doesn't work for the margin of error associated with the median. (Hang on, this book has you covered on that score.)

## So, what's the catch?

You may be wondering, what's the catch if I use a nonparametric technique? A downside must be lurking around here somewhere. Well, many researchers believe that nonparametric techniques water down statistical results; for example, suppose you find an actual difference between two population means, and the populations really do have a normal distribution. A parametric technique, the hypothesis test for two means, would likely detect this difference (if the sample size was large enough).

The question is, if you use a nonparametric technique (which doesn't need the populations to be normal), do you risk the chance of not finding the difference? The answer is maybe, but the risk isn't as big as you think. More often than not, nonparametric procedures are only slightly less efficient than parametric procedures (meaning they don't work quite as well at detecting a significant result or at estimating a value as parametric procedures) when the normality condition is met, but this difference in efficiency is small.

But the big payoff occurs when the normal distribution conditions aren't met. Parametric techniques can make the wrong conclusion, and corresponding nonparametric techniques can lead to a correct answer. Many researchers don't know this, so spread the word!

The bottom line: Always check for normality first. If you're very confident that the normality condition is met, go ahead and use parametric procedures because they're more precise. If you have any doubt about the normality condition, use nonparametric procedures. Even if the normality condition is met, nonparametric procedures are only a little less precise than parametric procedures. If the normality condition isn't met, nonparametrics provide appropriate and justifiable results where parametric procedures may not.

# Mastering the Basics of Nonparametric Statistics

Because you may not have run into nonparametric statistics during your Stats I class, your first step toward using these techniques is figuring out some of the basics. In this section, you get to know some of the terminology and major concepts involved in nonparametric statistics. These terms and concepts are commonly used in Chapters 17 through 20 of this book.

## Sign

The *sign* is a value of 0 or 1 that's assigned to each number in the data set. The sign for a value in the data set represents whether that data value is larger or smaller than some specified number. The value of +1 is given if the data value is greater than the specified number, and the value of 0 is given if the data value is less than or equal to the specified number. For example, suppose your data set is 10, 12, 13, 15, 20, and your specified number for comparison is 16. Because 10, 12, 13, and 15 are all less than 16, they each receive a sign of 0. Because 20 is greater than 16, it receives a sign of +1.

Several uses of the sign statistic appear in nonparametric statistics. You can use signs to test to see if the median of a population equals some specified value. Or you can use signs to analyze data from a matched-pairs experiment (where subjects are matched up according to some variable and a treatment is applied and compared). You also can use signs in combination with other nonparametric statistics. For example, you can combine signs with ranks to develop statistics for comparing the median of two populations. (Ranks are discussed in the next section and are used in a hypothesis test for two population medians in Chapter 18.)

In the following sections, you see exactly how to use the sign statistic to test the median of a population and analyze data in a matched-pairs experiment.

### Testing the median

You can use signs to test whether the median of a population is equal to some value $m$. You do this by conducting a hypothesis test based on signs. You have Ho: Median $= m$ versus Ha: Median $\neq m$ (or, you can use a > or < sign in Ha also). Your test statistic is the sum of the signs for all the data. If this sum is significantly greater or significantly smaller than what's expected if Ho were true, you reject Ho. Exactly how large or how small the sum of the signs must be to reject Ho is given by the sign test (refer to Chapter 17).

Suppose you're testing whether the median of a population is equal to 5. That is, you're testing Ho: Median = 5 versus Ha: Median ≠ 5. You collect the following data: 4, 4, 3, 3, 2, 6, 4, 3, 3, 5, 7, 5. Ordering the data, you get 2, 3, 3, 3, 3, 4, 4, 4, 5, 5, 6, 7. Now you find the sign for each value in the data set, determined by whether the value is greater than 5. The sign of the first data value, 2, is 0, because it's below 5. Each of the 3s receives a sign of 0, as do the 4s and 5s, for the same reason. Only the numbers 6 and 7 receive a sign of +1, being the only values in the data set that are greater than 5 (the number of interest for the median).

By summing the signs, you're in essence counting the number of values in the data set that are greater than the given quantity in Ho. For example, the total of all the signs of the ordered data values is

$$0 + 0 + 0 + 0 + 0 + 0 + 0 + 0 + 0 + 0 + 1 + 1 = 2$$

You can see that the total number of data values above 5 (the number of interest for the median) is 2. The fact that the total of the signs (2) is much less than half the sample size gives you some evidence that the median is probably not 5 here because the median represents the middle of the population. If the median were truly 5 in the population, your sample should yield about six values below it and six values above.

### Doing a matched-pairs experiment

You can use signs in a *matched-pairs experiment,* where you use the same subject twice or pair up subjects on some important variables. For example, you can use signs to test whether or not a certain treatment resulted in an improvement in patients, compared to a control. In the cases where the sign statistic is used, improvement is measured not by the mean of the differences in the responses for treatment versus control (as in a paired *t*-test), but rather by the median of the differences in the responses.

Suppose you're testing a new antihistamine for allergy patients. You take a sample of 100 patients and have each patient assess the severity of his or her allergy symptoms before and after taking the medication on a scale from 1 (best) to 10 (worst). (Of course, you do a controlled experiment in which some of the patients get a placebo to adjust for the fact that some people may perceive their symptoms to be going away just because they took something.)

In this study, you're not interested in what level the patients' symptoms are at, but rather in how many patients had a lower level of symptoms after taking the medicine. So you take the symptom level before the experiment minus the symptom level after the experiment for each subject.

✔ If that difference is positive, the medicine appears to have helped, and you give that person a sign of +1 (in other words, count them as a success).

✔ If the difference is zero, the medicine had no effect, and you give that person a sign of 0.

Remember, though, that the difference could be negative, indicating that the symptoms before were lower than the symptoms after; in other words, the medicine made their symptoms worse. This scenario results in a sign of 0 as well.

After you've found the sign for each value or pair in the data set, you're ready to analyze it by using the sign test or the signed rank test (see Chapter 17).

## Rank

*Ranks* are a nice way to use important information from a data set without using the actual values of the data themselves. Rank comes into play in nonparametric statistics when you're not interested in the values of the data, but rather where they stand compared to some supposed value for the median or compared to the ranks of values in another data set from another population. (You can see ranks in action in Chapter 18.)

The *rank* of a value in a data set is the number that represents its place in the ordering, from smallest to largest, within the data set. For example, if your data set is 1, 10, 4, 2, 1,000, you can assign the ranks in the following way: 1 gets the rank of 1 (because it's the smallest), 2 gets the rank of 2, 4 gets the rank of 3 (being the third-smallest number in the ordered data set), 10 gets the rank of 4, and 1,000 gets the rank of 5 (being the largest).

Now suppose your data set is 1, 2, 20, 20, 1,000. How do you assign the ranks? You know that 1 gets the rank of 1 (being the smallest), 2 gets the rank of 2, and 1,000 gets the rank of 5 (being the largest). But what about the two 20s in this data set? Should the first 20 get a rank of 3 and the second 20 get the rank of 4? That order doesn't seem to make sense, because you can't distinguish between the two 20s.

When two values in a data set are the same, it's called a tie. To assign ranks when there are ties, you take the average of the two ranks that the values need to fill and assign each tied value that average rank. If you have a tie between three numbers, you have three ranks, so take the sum of the ranks divided by three.

In this case, because both 20s are vying for the ranks of 3 and 4, assign each of them the rank of 3.5, the average of the two ranks they must share. I show the final ranking for the data set 1, 2, 20, 20, 1,000 in Table 16-1.

| Table 16-1 | Ranks of the Values in the Data Set 1, 2, 20, 20, 1,000 |
|---|---|
| **Data Value** | **Rank Assigned** |
| 1 | 1 |
| 2 | 2 |
| 20 | 3.5 |
| 20 | 3.5 |
| 1,000 | 5 |

The lowest a rank can be is 1, and the highest a rank can be is *n*, where *n* is the number of values in the data set. If you have a negative value in a data set, for example, if your data set is –1, –2, –3, you still assign the ranks 1 through 3 to those data values. Never assign negative ranks to negative data. (By the way, when you order the data set –1, –2, –3, you get –3, –2, –1, so –3 gets the rank of 1, –2 gets the rank of 2, and –1 gets the rank of 3.)

## Signed rank

A *signed rank* combines the idea of the sign and the rank of a value in a data set, with a small twist. The sign indicates whether that number is greater than, less than, or equal to a specified value. The rank indicates where that number falls in the ordering of the data set from smallest to largest.

To calculate the signed rank for each value in the data set, follow these steps:

1. **Assign a sign of +1 or 0 to each value in the data set, according to whether it's greater than some value specified in the problem.**

   If it's greater than the specified value, give it a sign of +1; if it's less than or equal to the specified value, give it a sign of 0.

2. **Rank the original data from smallest to largest, according to their absolute values.**

   Statisticians call these values the *absolute ranks.* The absolute value of any number is the positive version of that number. The notation for absolute value is | |, with the number between those lines. For example, |–2| = 2, |+2| = 2, and |0| = 0.

3. **Multiply the sign times the absolute rank to get the signed rank for each value in the data set.**

One scenario in which you can use signed ranks is an experiment in which you compare a response variable for a treatment group versus a control group. You can test for difference due to a treatment by collecting the data in pairs, either both from the same person (pretest versus post-test) or from two individuals who are matched up to be as similar as possible.

For example, suppose you compare four patients regarding their weight loss on a diet program. You're really wondering whether the overall change in weight is less than zero for the population. The following two factors are important:

✔ Whether or not the person lost weight

✔ How the person's weight change measures up, compared to everyone else in the data set

You measure the person's weight before the program (the pretest) as well as his weight after the program (the post-test). The change is the important facet of the data you're interested in, so you apply the signs to the changes in weight. You give the change a sign of +1 if the person lost weight (constituting a success for the program) and a sign of 0 if the person stayed the same or gained weight (thus not contributing to the success of the program). You convert all the changes in weight loss to their absolute values, and then you rank the absolute values (in other words, you've found the absolute ranks of the changes in weight). The signed rank is the product of the sign and the absolute rank. After determining the signed rank, you can really compare the effectiveness of the program: Large signed ranks indicate a big weight loss.

For example, weight changes of –20, –10, +1, and +5 have signs of +1, +1, 0, and 0. The absolute values of the weight changes are 20, 10, 1, and 5. Their absolute ranks, respectively, are 4, 3, 1, and 2. The signed ranks are 4 * 1 = 4, 3 * 1 = 3, 1 * 0 = 0, and 2 * 0 = 0.

## Rank sum

A *rank sum* is just what it sounds like: The sum of all the ranks. You typically use rank sums in situations when you're comparing two or more populations to see whether one has a central location that's higher than the other. (In other words, if you looked at the populations in terms of their histograms, one would be shifted to the right of the other on the number line.)

Here's a way in which researchers use rank sums: Suppose you're comparing quiz scores for two classes, and they don't have a normal distribution, hence you want to use nonparametric techniques to compare them. The total possible points on this quiz is 30. You collect random samples of five quiz scores from each of the classes. Suppose you collect the following sample data:

| Class Number One | Class Number Two |
|---|---|
| 22 | 23 |
| 23 | 30 |
| 20 | 27 |
| 25 | 28 |
| 26 | 25 |

The twist here is to combine all the data into one big data set, rank all the values, and sum the ranks for the first sample and then the second sample. Then compare the two rank sums. If one rank sum is higher, this outcome may indicate that a particular class did better on the quiz.

For the quiz example, the ordered data for the combined classes appear in the first line, with the respective ranks appearing in the second line. Circled scores came from the first class.

| Ordered Data | (20) | (22) | (23) | 23 | (25) | 25 | (26) | 27 | 28 | 30 |
|---|---|---|---|---|---|---|---|---|---|---|
| Respective Ranks | 1 | 2 | 3.5 | 3.5 | 5.5 | 5.5 | 7 | 8 | 9 | 10 |

The rank sum for the first class is $1 + 2 + 3.5 + 5.5 + 7 = 19$, which is quite a bit lower than the rank sum for the second class, $3.5 + 5.5 + 8 + 9 + 10 = 36$. This result tells you that the second class did better on the quiz than the first class, for this sample.

Chapter 18 shows you how to use a rank sum test to see whether the shapes of two population distributions are the same, meaning the values they take on and how often those values occur in each population. In Chapter 19, you can find even more uses for rank sums, including the Kruskal-Wallis test.

Note that taking the mean of each data set and comparing them by using a two-sample t-test would be wrong in the quiz example because the quiz scores admittedly don't have a normal distribution. Indeed if the quiz were easy, you'd get many high scores and few low ones, and the population would be skewed left. On the other hand, if the quiz were hard, you'd get many low scores and few high ones, and the population would be skewed right (don't think too much about that scenario). In either case, you need a nonparametric procedure. See Chapter 18 for more on the nonparametric equivalent of the t-test.

# Chapter 17

# All Signs Point to the Sign Test and Signed Rank Test

The hypothesis tests you see in Stats I use well-known distributions like the normal distribution or the *t*-distribution (see Chapter 3). Using these tests requires that certain conditions be met, such as the type of data you're using, the distribution of the population the data came from, and the size of your data set. Such procedures that involve such conditions are called *parametric procedures*. In general, parametric procedures are very powerful and precise, and statisticians use them as often as they can.

But, situations do arise in which your data don't meet the conditions for a parametric procedure. Perhaps you just don't have enough data (the biggest hurdle is whether the data come from a population with a normal distribution), or your data are just of a different type than quantitative data, such as ranks (where you don't collect numerical data, but instead just order the data from low to high or vice versa).

In these situations, your best bet is a *nonparametric procedure* (see Chapter 16 for background info). In general, nonparametric procedures aren't as powerful as parametric procedures, but they have very few assumptions tied to them. Moreover, nonparametric procedures are easy to carry out, and their formulas make sense. Most importantly, nonparametric procedures give accurate results compared to the use of parametric procedures when the conditions of parametric procedures aren't met or aren't appropriate.

In this chapter, you use the sign test and the Wilcoxon signed rank test to test or estimate the median of one population. These nonparametric procedures are the counterparts to the one-sample and matched pairs *t*-tests, which require data from a normal population.

# Reading the Signs: The Sign Test

You use the one-sample $t$-test from Stats I to test whether or not the population mean is equal to a certain value. It requires the data to have a normal distribution. When this condition isn't met, the *sign test* is a nonparametric alternative for the one-sample $t$-test. It tests whether or not the population median is equal to a certain value.

What makes the sign test so nice is that it's based on a very basic distribution, the binomial. You use the binomial distribution when you have a sequence of $n$ trials of an experiment, with only two possible outcomes each time (success or failure). The probability of success is denoted by $p$, and is the same for each trial. The variable is $x$, the number of success in the $n$ trials. (For more info on the binomial see your Stats I text.)

The only condition of the sign test is that the data are ordinal or quantitative — not categorical. However, this is no big deal because if you're interested in the median, you don't collect categorical data anyway.

Here are the steps for conducting the sign test. Note that Minitab can do steps four through eight for you; however, understanding what Minitab does behind the scenes is important, as always.

1. **Set up your null hypothesis: Ho: $m = m_o$.**

   The true value of the median is $m$, and $m_o$ is the claimed value of the median (the value you're testing).

2. **Set up your alternative hypothesis. Your choices are Ha: $m \neq m_o$; or Ha: $m > m_o$; or Ha: $m < m_o$.**

   Which Ha you choose depends on what conclusion you want to make in the case that Ho is rejected. For example, if you only want to know when the median is greater than some number $m$, use Ho: $m > m_o$. Chapter 3 tells you more about setting up alternative hypotheses.

3. **Collect a random sample of (ordinal or quantitative) data from the population.**

4. **Assign a plus or minus sign to each value in the data set.**

   If an observation is less than $m_o$, assign it a minus (–) sign. If the observation is greater than $m_o$, give it a plus (+) sign. If the observation equals $m_o$, disregard it and let the sample size decrease by one.

   In terms of the binomial distribution, you have $n$ values in the data set, and each one has one of two outcomes: It falls either below $m_o$ or above it. (This is akin to success and failure.)

5. **Count up all the plus signs. This sum is your test statistic, noted by *k*.**

In terms of the binomial, this sum represents the total number of successes, where a plus (+) sign is the designated success.

6. **Locate the test statistic *k* (from step five) on the binomial distribution (using Table A-2 in the appendix).**

You determine where your test statistic falls on the binomial distribution by looking it up in a binomial distribution table (check your textbook). To do this, you need to know *n*, *k*, and *p*.

Your sample size is *n*, your test statistic is *k* from step five, but what's your value of *p*, the probability of success? If the null hypothesis Ho is true, 50 percent of the data should lie below $m_o$ and 50 percent should lie above it. This corresponds to a success (+) having a probability of $p = 0.50$ on the binomial distribution.

7. **Find the *p*-value of your test statistic:**

   • If Ha has a < sign, add up all the probabilities on the binomial table for $x \leq k$.

   • If Ha has a > sign, add up all the probabilities on the binomial table for $x \geq k$.

   • If Ha has a ≠ sign, add up the probabilities on the binomial table of *x* being greater than or equal to *k* and double this value. This gives you the *p*-value of the test.

8. **Make your conclusion.**

If the *p*-value from step six is less than the predetermined value of α (typically 0.05), reject Ho and say the median is greater than, less than, or ≠ $m_o$, depending on Ha. Otherwise, you can't reject Ho.

To run a sign test in Minitab, enter your data in a single column. Go to Stat>Nonparametric>One-sample Sign. Click on your variable in the left-hand box, and click Select. The variable will appear in the Variables box. Then click OK, and you get the results of the sign test.

In the sections that follow, I show you two different ways in which you can use the sign test:

✔ To test or estimate the median of one population

✔ To test or estimate the median difference of data where the observations come in pairs, either from the same individual (pretest versus post-test) or individuals paired up according to relevant characteristics

## Testing the median

Situations arise in which you aren't interested in the mean, but rather the median of a population. (Chapter 16 has more on the median.) For example, perhaps the data don't have a normal, or even a symmetric, distribution. When you want to estimate or test the median of a population (call it $m$), the sign test is a great option.

Suppose you're a real estate agent selling homes in a particular neighborhood, and you hear from other agents that the median house price in that neighborhood is $110,000. You think the median is actually higher. Because you're interested in the median price of a home rather than the mean price, you decide to test the claim by using a sign test. Follow the steps of the sign test:

1. Set up your null hypothesis. Because the original claim is that the median price of a home is $110,000, you have Ho: $m$ = $110,000.

2. Set up the alternative hypothesis. Because you believe the median is higher than $110,000, your alternative hypothesis is Ha: $m$ > $110,000.

3. Take a random sample of ten homes in the neighborhood. You can see the data in Table 17-1; its histogram is shown in Figure 17-1.

   Now the question is, is the median selling price of all homes in the neighborhood equal to $110,000, or is it more than that (as you suspect)?

| Table 17-1 | Sample of House Prices in a Neighborhood | |
|---|---|---|
| House | Price | Sign (Compared to $110,000) |
| 1 | $132,000 | + |
| 2 | $107,000 | – |
| 3 | $111,000 | + |
| 4 | $105,000 | – |
| 5 | $100,000 | – |
| 6 | $113,000 | + |
| 7 | $135,000 | + |
| 8 | $120,000 | + |
| 9 | $125,000 | + |
| 10 | $126,000 | + |

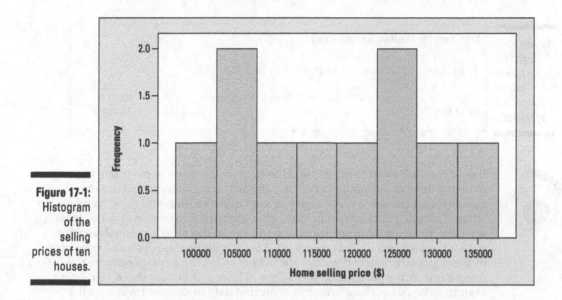

**Figure 17-1:**
Histogram of the selling prices of ten houses.

4. Assign a plus sign to any house price more than $110,000 and a minus sign to any house less than $110,000. (See column three of Table 17-1.)

5. Find your test statistic. Your test statistic is 7, the number of "+" signs in your data set (see Table 17-1), representing the number of houses in your sample whose prices are above $110,000.

6. Compare your test statistic to the binomial distribution (refer to a binomial distribution table) to find the *p*-value.

   For this case, look at the row in the binomial table where $n = 10$ (the sample size) and $k = 7$ (the test statistic) and the column where $p = 0.50$ (because if the population median equals $m_o$, 50 percent of the values in the population should be above it and 50 percent below it). According to the table, you find the probability that $x$ equals 7 is 0.117.

   Because you have a right-tailed test (meaning Ha has a > sign in it), you add up the probabilities of being at or beyond 7 to get the *p*-value. The *p*-value in this case is $0.117 + 0.044 + 0.010 + 0.001 = 0.172$.

7. To conclude, compare the *p*-value (0.172) to the predetermined $\alpha$ (I always use 0.05). Because the *p*-value is greater than 0.05, you can't reject Ho. You don't have enough evidence to say the median house selling price is more than $110,000.

Figure 17-2 shows these results as calculated by Minitab.

Figure 17-2:
Sign test
for house
prices
conducted
by Minitab.

```
Sign Test for Median: Selling Price

Sign test of median = 110000 versus > 110000

                       N    Below   Equal   Above        P    Median
Selling  Price        10        3       0       7   0.1719    116500
```

**WARNING!**

If your data are close to normal and the mean is the more appropriate measure of center for your situation, don't use the sign test. Instead, use the one-sample *t*-test (or *Z*-test). The sign test isn't quite as powerful (able to reject Ho when it should) as the *t*-test in situations where the conditions for the *t*-test are met. More importantly, though, don't run to the *t*-test to reanalyze your data if the sign test doesn't reject Ho. That would be improper and unethical. In general, statisticians consider the idea of following a nonparametric procedure with a parametric procedure in hopes of getting more significant results to be *data fishing,* which is analyzing data in different ways until a statistically significant result appears.

## Estimating the median

You can also use the sign test to find a confidence interval for one population median. This comes in handy when you're interested in estimating what the median value of a population is, such as the median income of a household in the United States or the median salary of people fresh out of an MBA program.

Following are the steps for conducting a confidence interval for the median by using the test statistic for the sign test, assuming your random sample of data has already been collected. Note that Minitab can calculate the confidence interval for you (steps two to five), but knowing how Minitab does the steps is important:

1. **Determine your level of confidence, 1 – α (that is, how confident you want to be that this process will correctly estimate *m* over the long term).**

   The typical confidence level data analysts use is 95 percent (see Chapter 3 for more information).

2. **On the binomial table (Table A-2 in the appendix), find the section for *n* equal to your sample size, and the column where *p* = 0.50 (because the median is the point where 50 percent of the data lies below and 50 percent lies above).**

You'll find probabilities for values of $x$ from 0 to $n$ in that section.

3. **Starting at each end ($x = 0$ and $x = n$) and moving one step at a time toward the middle of the $x$ values, add up the probabilities for those values of $x$ until you pass the total of $\alpha$ (which is one minus your confidence level).**

4. **Record the number of steps that you had to make just before you passed the value of $1 - \alpha$. Call this number $c$.**

5. **Order your data set from smallest to largest. Starting at each end, work your way to the middle until you reach the $c$th number from the bottom and the $c$th number from the top.**

6. **Use these numbers as the low end and the high end of an interval. This result is your confidence interval for the median.**

You can use these steps to find a confidence interval for the median in the house-price example from the preceding section. Here's how this example breaks down:

1. Let your confidence level be set at $1 - \alpha = 0.95$.

2. On the binomial table (Table A-2 in the appendix), look at the section where $n = 10$ (the sample size) and $p = 0.50$. These values are listed in Table 17-2.

**Table 17-2**      **Binomial Probabilities to Help Calculate a Confidence Interval for the Median ($n = 10$, $p = 0.50$)**

| x | p(x) |
|---|------|
| 0 | 0.001 |
| 1 | 0.010 |
| 2 | 0.044 |
| 3 | 0.117 |
| 4 | 0.205 |
| 5 | 0.246 |
| 6 | 0.205 |
| 7 | 0.117 |
| 8 | 0.044 |
| 9 | 0.010 |
| 10 | 0.001 |

3. Start with the outermost values of $x$ ($x = 0$ and $x = 10$) and sum those probabilities to get $0.001 + 0.001 = 0.002$. Because you haven't yet passed 0.05 (the value of $\alpha$), you go to the second-innermost values of $x$ ($x = 1$ and $x = 9$). Add their probabilities to what you have so far to get 0.002 (old total) + 0.010 + 0.010 = 0.022. You're still not past 0.05 ($\alpha$), so go one more step. Add the third-innermost probabilities for $x = 2$ and $x = 8$ to the grand total to get 0.022 (old total) + 0.044 + 0.044 = 0.110. You've now passed the value of $\alpha = 0.05$. The value of $c$ equals 2 because you passed 0.05 at the third-innermost values of $x$, and you back off one step from there to get your value of $c$.

4. Order your data (Table 17-1) from smallest to largest, giving you (in dollars): 100,000, 105,000, 107,000, 111,000, 113,000, 120,000, 125,000, 126,000, 132,000, and 135,000.

5. Work your way in from each end of the data set to take the second-innermost values (because $c = 2$): the numbers \$105,000 and \$132,000. Put these two numbers together to form an interval, and you conclude that a 95 percent confidence interval for the median selling price for a home in this neighborhood is between \$105,000 and \$132,000.

To find a $1 - \alpha$ percent confidence interval for the median using Minitab based on the sign test, enter your data into a single column. Go to Stat>Nonparametrics>One-sample Sign. Click on the variable in the left-hand column for which you want the confidence interval, and it appears in the Variables column. Click the circle that says Confidence Interval, and type in the value of $1 - \alpha$ you want for your confidence level. (The default is 95 percent, written as 95.) Click OK to get the confidence interval.

## Testing matched pairs

The most useful application of the sign test is in testing matched pairs of data — that is, data that come in pairs and represent two observations from the same person (pretests versus post-tests, for instance) or one set of data from each pair of people who are matched according to relevant characteristics. In this section, you see how you can compare data from a matched-pairs study to look for a treatment effect, using a sign test for the median.

The idea of using a sign test for the median difference with matched-pairs data is similar to using a $t$-test for the mean differences with matched-pairs data. (For details on matched-pairs data and the $t$-test, see your Stats I text.) You use a test of the median (rather than the mean) when the data don't necessarily have a normal distribution, or if you're only interested in the median difference rather than the mean difference.

First, you set up your hypothesis, Ho: The median is zero (indicating no difference between the pairs). Your alternative hypothesis is Ha: The median is ≠ 0, > 0, or < 0, depending on whether you want to know if the treatment made any difference, made a positive difference, or made a negative difference compared to the control. Then you collect your data (two observations per person or a pair of observations from each pair of people you've matched up). After that, you use Minitab to conduct steps four to seven of the sign test.

For example, suppose you wonder whether taking a test while chewing gum decreases test anxiety. You pair 20 students according to relevant factors such as GPA, score on previous exams, and so on. One member of each pair is randomly selected to chew gum during the exam, and the other member of the pair doesn't. You measure test anxiety of each person via a very short survey right after they turn in their exams. You measure the results on a scale of 1 (lowest anxiety level) to 10 (highest anxiety level). Table 17-3 shows the data based on a sample of ten pairs.

| Table 17-3 | Testing the Effectiveness of Chewing Gum in Lowering Test Anxiety | | | |
|---|---|---|---|---|
| Pair | Anxiety Level — Gum | Anxiety Level — No Gum | Difference (Gum/No Gum) | Sign |
| 1 | 9 | 10 | −1 | − |
| 2 | 6 | 8 | −2 | − |
| 3 | 3 | 1 | +2 | + |
| 4 | 3 | 5 | −2 | − |
| 5 | 4 | 4 | 0 | none |
| 6 | 2 | 7 | −5 | − |
| 7 | 2 | 6 | −4 | − |
| 8 | 8 | 10 | −2 | − |
| 9 | 6 | 8 | −2 | − |
| 10 | 1 | 3 | −2 | − |

The actual levels of test anxiety aren't important here; what matters is the difference between anxiety levels within each pair. So, instead of looking at all the individual anxiety levels, you can look at the difference in anxiety levels for each pair. This method gives you one data set, not two. (In this case, to calculate the differences in each pair, you can use the formula test anxiety without gum minus text anxiety with gum, and look for an overall difference that's positive.) Typically, in the case of matched-pairs data, you're testing whether the median difference equals zero. In other words, Ho: $m$ = 0; the same holds in the test anxiety example.

The differences in anxiety levels for each pair in your data set now become a single data set (see column four of Table 17-3). You can now use the regular sign test methods to analyze this data, using Ho: $m = 0$ (no median difference in test anxiety of gum versus no gum) versus Ha: $m < 0$ (chewing gum reduces test anxiety).

Assign each difference a plus or minus sign, depending on whether it's greater than zero (plus sign) or less than zero (minus sign). Your test statistic is the total number of plus signs, 1, and the relevant sample size is $10 - 1 = 9$. (You don't count the data that hit the median of zero right on the head.)

Now compare this test statistic to the binomial distribution with $p = 0.50$ and $n = 9$, using the binomial table (Table A-2 in the appendix). You have a test statistic of $k = 1$, and you want to find the probability that $x \leq 1$ (because you have a left-tailed test, see step six of the sign test from the earlier section "Reading the Signs: The Sign Test"). Under the column for $p = 0.50$ in the section for $n = 9$, you get the probability of 0.018 for $x = 1$ and 0.002 for $x = 0$. Add these values to get 0.020, your $p$-value. This result means that you reject Ho at the predetermined $\alpha$ level of 0.05. This tells you the anxiety levels for gum versus no gum are different. Now, how are they different? Based on this data, you conclude that chewing gum on an exam appears to decrease test anxiety because there are more negative differences than positive differences.

# Going a Step Further with the Signed Rank Test

The signed rank test is more powerful than the sign test at detecting real differences in the median. The most common use of the signed rank test is testing matched-pairs data for a median difference due to some treatment (like chewing-gum use during an exam and its effect on test anxiety). In this section, you find out what the signed rank test is and how it's carried out, and I walk you through an application involving the test of a weight-loss program.

## A limitation of the sign test

The sign test has the advantage of being very simple and easy to do by hand. However, because it only looks at whether a value is above or below the median, it doesn't take the magnitude of the difference into account.

Looking at Tables 17-1 and 17-3, you see that for each data value, the test statistic for the sign test only counts whether or not each data value is greater than or equal to the median in the null hypothesis, $m_o$. It doesn't count how great those differences are. For example, in Table 17-3, you can see that the sixth pair had a huge reduction in test anxiety when chewing gum (from 7 down to 2), but the first pair had a very small reduction in test anxiety (from 10 down to 9). Yet both of these differences received the same outcome (a minus sign) in the test statistic for the sign test.

Because it doesn't take into account how much the values in the data differ from the median, the sign test is less powerful (meaning less able to detect when Ho is false) than it could be. So if you want to test the median and you want to take the magnitude of the differences into account (and you're willing to jump through some math hoops to get there), you can conduct the *signed rank test*, also known as the *Wilcoxon signed rank test*. The next section walks you through it.

# Stepping through the signed rank test

Just like the sign test, the only condition of the signed rank test is that the data are ordinal or quantitative.

Following are the steps for carrying out the signed rank test on paired data:

1. **Set up your hypotheses.**

   The null hypothesis is Ho: $m = 0$. Your choices for an alternative hypothesis are Ha: $m \neq 0$; Ha: $m > 0$; or Ha: $m < 0$, depending on whether you want to detect any difference, a positive difference, or a negative difference in the pairs.

2. **Collect a random sample of paired data.**

3. **For each observation, calculate the difference for each pair of observations.**

4. **Calculate the absolute value of each of the differences.**

5. **Rank the absolute values from smallest to largest.**

   If two of the absolute values are tied, give each one the average rank of the two values. For example, if the fourth and fifth numbers in order are tied, give each one the rank of 4.5.

6. **Add up the ranks that correspond to those original differences from step three that are positive.**

   The sum of the positive differences is your signed rank test statistic, denoted by SR.

7. **Find the *p*-value.**

Look at all possible ways that the absolute differences could have appeared in a sample, with either plus or minus signs, assuming that Ho is true. Find all their test statistics (SR values) from all these possible arrangements by using steps four through six, and compare your SR value to those. The percentage of SR values that are at or beyond your test statistic is your *p*-value.

Minitab can do this step for you.

**8. Make your conclusion.**

If the *p*-value is less than the predetermined α (typically 0.05), reject Ho and conclude the median difference is not zero. Otherwise, you can't reject Ho.

To conduct the Wilcoxon signed rank test using Minitab, enter the differences from step three in a single column. Go to Stat>Nonparametrics>1-Sample Wilcoxon. Click on the name of the variable for your differences in the left-hand box, and it appears in the right-hand Variables box. Click on the circle that says Test Median, and indicate which Ha you want (> 0, < 0, or ≠). Click OK, and your test is done. (Note that Minitab calculates the test statistic for the signed rank test a little differently than what you'd get by hand, although the results are close. The reason for the slight calculation difference is beyond the scope of this book.)

How do you handle situations in which a piece of data is exactly equal to the median? Most of the time (including all data sets you will encounter) this occurrence is rare and can be handled by ignoring those data values and reducing the sample size by one for each time the matchup occurs.

## Losing weight with signed ranks

This section shows you the signed rank test in action. I first show you each step as if you were doing the process by hand. Then you see the results in Minitab.

Suppose you want to test whether or not a weight-loss plan is effective. You want to look at the median weight loss for people on the plan by using a matched-pairs experiment. You want the magnitude of weight loss to factor into the analysis, which means you use a signed rank test to analyze the data. Here are the steps you follow to conduct the test in this example:

1. Set up your hypotheses. Test Ho: $m = 0$, where $m$ represents the median weight loss (before the program versus after the program). Your alternative hypothesis is Ha: $m > 0$, indicating the median difference in weight loss is positive.

2. Take a random sample of, say, three people and measure them before and after an eight-week weight-loss program. You calculate the difference in weight of each person (weight before the program minus weight after the program). A positive difference means the person lost weight, and a negative difference means they gained weight.

   Table 17-4 shows the data and relevant statistics for the weight-loss signed rank test. (Note that I have only three people in this study; this is for illustrative purposes only.) You can see the differences in weight (before – after) in column four.

### Table 17-4      Data on Weight Loss Before and After Program

| Person | Before | After | Difference | \|Difference\| | Rank |
|--------|--------|-------|------------|----------------|------|
| 1 | 200 | 205 | −5 | 5 | 1 |
| 2 | 180 | 160 | +20 | 20 | 2* |
| 3 | 134 | 110 | +24 | 24 | 3* |

*Represents ranks associated with a positive difference in weight loss

3. Take the absolute values of the differences. You can see those in column five of Table 17-4.

4. Rank the absolute differences. Column six reflects the ranks of those absolute values, from 1 to 3.

5. Find your test statistic, which is the sum of the ranks corresponding to positive differences. (In other words, you only count ranks of people who lost weight.) For this data set, those ranks you can count are indicated by * in Table 17-4. The sum turns out to be 2 + 3 = 5. This number, 5, is your test statistic; you can call it SR to designate the signed rank test statistic.

6. Calculate the *p*-value. Now you need to compare that test statistic to some distribution to see where it stands. To do this, you determine all the possible ways that the three absolute differences (column five of Table 17-4) — 5, 20, and 24 — could have appeared in a sample, with their actual differences taking on plus signs or minus signs. (Assume Ho is true and the actual differences have a 50-percent chance of being positive or negative, like the flip of a coin.)

   Then you find all their test statistics (SR values) from all these possible arrangements, and compare your SR value, 5, to those. The percentage of the other SR values that are at or beyond your test statistic is your *p*-value.

For the weight-loss example, you have eight possible ways that you can have absolute differences of 5, 20, and 24 by including either plus or minus signs on each difference (two possible signs for each equals $2 * 2 * 2 = 8$). Those eight possibilities are listed in separate columns of Table 17-5. SR denotes the sum of the positive ranks in each case (these are the test statistics for each possible arrangement).

| Table 17-5 | | | Possible Samples with Absolute Differences of 5, 20, and 24 | | | | | |
|---|---|---|---|---|---|---|---|---|
| **1** | **2** | **3** | **4** | **5** | **6** | **7** | **8** | **Rank of \|Diff\|** |
| 5 | –5* | 5 | 5 | –5* | –5* | 5 | –5* | 1 |
| 20 | 20 | –20* | 20 | –20* | 20 | –20* | –20* | 2 |
| 24 | 24 | 24 | –24* | 24 | –24* | –24* | –24* | 3 |
| SR = 6 | SR = 5 | SR = 4 | SR = 3 | SR = 3 | SR = 2 | SR = 1 | SR = 0 | — |

*Denotes negative differences*

To make sense of Table 17-5, consider the following: The three absolute differences you have in your data set are 5, 20, and 24, which have ranks 1, 2, and 3, respectively (which you can see in Table 17-4). You can find the eight different combinations of 5, 20, and 24 that exist, where you can put either a minus or plus sign on any of those values. For each scenario, I found the signed rank statistic by summing the ranks for only those differences that are positive (the person lost weight). Those ranks are the column nine values in Table 17-5 for data values without an asterisk (*).

For example, column seven has two negative differences, –20 and –24, and one positive difference of 5 (whose rank among the absolute differences is 1 because it's the smallest; see column nine). Summing the positive ranks in column seven produces a signed rank statistic (SR) of 1 because 5 is the only positive number. (You can see in column two the data that you actually observed in the sample.)

Now compare the test statistic, 5 (from step five), to all the values of SR in the last row of Table 17-5. Because you're using Ha: $m > 0$, you can find the percentage of signed ranks (SR) that are at or above the value of 5. You have two of them out of eight, so your $p$-value (the percentage of possible teststatistics beyond or the same as yours if Ho were true), is $\frac{2}{8} = 0.25$, or 25 percent.

7. Because the *p*-value (0.25) is greater than the predetermined value of α (typically 0.05), you can't reject Ho, and you can't say there's positive weight loss via this program. (*Note:* With a sample size of only three, it's difficult to find any real difference, so the weight-loss program may actually be working and this small data set just couldn't determine that, and one person actually gained weight, which doesn't help.)

Figure 17-3 shows the Minitab output for this test, using the data from Table 17-4. The *p*-value turns out to be 0.211 due to a slight difference in the way that Minitab calculates the test statistic. Note the estimated median found in Figure 17-3 refers to a calculation made over all possible samples and the medians you would get from them.

**Figure 17-3:**
Computer
output for
signed
rank test of
weight-loss
data.

```
Wilcoxon Signed Rank Test: Wt loss

Test of median = 0.000000 versus median > 0.000000

                    N
                   for    Wilcoxon              Estimated
           N      Test    Statistic      P       Median
Wt loss    3                  5.0     0.211      14.75
```

You also can use the SR statistic to estimate the median of one population (or the median of the difference in a matched-pairs situation). To find a $1 - α$ percent confidence interval for the median using Minitab based on the signed rank test, enter your data into a single column. (If your data represents differences from a matched-pairs data set, enter those differences as one column.) Go to Stat>Nonparametrics>1-Sample Wilcoxon. Click on the name of the variable in the left-hand column, and it appears in the Variables column on the right-hand side. Click the circle that says Confidence Interval, and type in the value of $1 - α$, your confidence level. Click OK.

Because the positive 20 is greater than the predetermined value of frequency of reason that i > 0 (i.e. H0), and you don't say there's positive weight loss (weight) problem. (Note: With a stop at zero (zero effect), it picked up to be any real difference for its weight-loss problem may actually be weight-loss that the good score that you can't really determine that and one person actually gained weight? It's probably not being.)

Figure 17-3 shows the Minitab output for the test, using the data from Table 17-1. The p-value turns out to be 0.34, showing a slight difference is there on the Minitab, calculate the test statistic. Note the estimated upper limb number in Figure 17-3 where a calculation made over-impressive numbers and the median you would get from there.

| | **Wilcoxon Signed Rank Test: Wtloss** |
| --- | --- |
| | Test of median = 0 versus median not = 0 |

You also can use the test statistic to estimate the median of one population (or the median of the difference in a matched-pairs situation also). To find a 95% confidence interval for the median, using Minitab, read on the signed-rank test, enter your data into a single column. If your data represent a difference from two measured-pairs data (or repeated measures data as explained in Chapter 16), put the difference values in. Click on the main at the top left and choose Nonparametric, and then Wilcoxon. Put the variable column on the right-hand side. Click the Confidence Interval and type in the value that you want your confidence level. Click OK.

# Chapter 18

# Pulling Rank with the Rank Sum Test

· · · · · · · · · · · · · · · · · · · · · · · · · · · · · · · · · · · · · · · · · · · · · · · · · · · · · ·

## In This Chapter

▶ Comparing two populations by using medians, not means

▶ Conducting the rank sum test

· · · · · · · · · · · · · · · · · · · · · · · · · · · · · · · · · · · · · · · · · · · · · · · · · · · · · ·

*I*n Stats I, when you want to compare two populations, you conduct a hypothesis test for two population means. The most common tool for comparing population means is a *t*-test (see Chapter 3). However, a *t*-test has the condition that the data come from a normal distribution. When conditions for parametric procedures (ones involving normal distributions) aren't met, a nonparametric alternative is always there to save the day.

In this chapter, you work with a nonparametric test that compares the centers of two populations — the *rank sum test*. This test focuses on the *median*, which is the measure of center that's most appropriate in situations where the data isn't symmetric.

# Conducting the Rank Sum Test

This section addresses the conditions for the rank sum test and walks you through the steps for conducting the test. You can put your understanding and skill to the test (pun intended) in the section "Performing a Rank Sum Test: Which Real Estate Agent Sells Homes Faster?" later in this chapter.

## Checking the conditions

Before you can think about conducting the rank sum test to compare the medians of two populations, you have to make sure your data sets meet the conditions for the test. The conditions for the rank sum test are the following:

✔ **The two random samples, one taken from each population, are independent of each other.**

You take care of the first condition in the way you collect your data. Just make sure you aren't using matched pairs, for example, using data from the same person in a pretest and post-test manner. Then the two sets of data would be dependent.

✔ **The two populations have the same distribution — that is, their histograms have the same shape.**

You can check this condition by making histograms to compare the shapes of the sample data from the two populations. (See your Stats I textbook or my book *Statistics For Dummies* (Wiley) for help making histograms.)

✔ **The two populations have the same variance, meaning that the amount of spread in the values is the same.**

You can check this condition by finding the variances or standard deviations of the two samples. They should be close. (A hypothesis test for two variances actually exists, but that's outside the scope of this book.)

Notice that the centers of the two populations need not be equal; that's what the test is going to decide.

More sophisticated methods for checking the second and third conditions listed here fall outside the scope of this book. However, checking the conditions as I describe them allows you to find and stay clear of any major problems.

## Stepping through the test

The *rank sum test* is a test for the equality of the two population medians — call them $\eta_1$ and $\eta_2$. After you've checked the conditions for using the rank sum test (see the preceding section), you conduct the test by following these steps. (*Note:* Minitab can run this test for you, but you should still know what it's doing behind the scenes.)

1. Set up Ho: $\eta_1 = \eta_2$ versus Ha: $\eta_1 > \eta_2$ (a one-sided test); Ha: $\eta_1 < \eta_2$ (a one-sided test); or Ha: $\eta_1 \neq \eta_2$ (a two-sided test), depending on whether you're looking for a positive difference, a negative difference, or any difference between the two population medians.

2. **Think of the data as one combined group and assign overall ranks to the values from lowest (rank = 1) to highest.**

   In the case of ties, give both values the average of the ranks they normally would have been given. For example, suppose the third and fourth numbers (in order) are the same. If the two numbers were different, they would get ranks of 3 and 4, respectively. But because they're the same, you give them the same rank, 3.5, which is the average of 3 and 4. Note that the next number (in order) is the fifth number, which receives rank 5.

3. **Sum the ranks assigned to the sample that has the smallest sample size; call this statistic *T*.**

   You use the smallest sample in this step according to convention — statisticians like to be consistent. If the sample sizes are equal, sum the ranks for the first sample to get *T*. If the value of *T* is small (relative to the total sum of all the ranks from both data sets), the numbers from the first sample tend to be smaller than the second sample, hence the median of the first population may be smaller than the median of the second one.

4. **Look at Table A-4(a) and (b) in the appendix, the rank sum tables. For the chosen table, find the column and row for the sample sizes of group one and two, respectively.**

   You see two critical values, $T_L$ (the lower critical value) and $T_U$ (the upper critical value). These critical values are the boundaries between rejecting Ho and not rejecting Ho.

5. **Compare your test statistic, *T*, to the critical values in Table A-4 in the appendix to conclude whether you can reject Ho — that the population medians are different.**

   The method you use to compare these values depends on the type of test you're conducting:

   - **One-sided test (Ha has a > or < sign in it):** Table A-4 in the appendix shows the critical values for $\alpha$ level 0.05. For a right-sided test (that means you have Ha: $\eta_1 > \eta_2$), reject Ho if $T \geq T_U$. For a left-sided test (that means where Ha: $\eta_1 < \eta_2$), reject Ho if $T \leq T_L$. If you reject Ho, conclude that the population medians are different and that one of them is greater than the other depending on Ha. (Otherwise you can't conclude that there's a difference in their medians.)

   - **Two-sided test:** Table A-4 in the appendix shows the critical values for $\alpha$ level 0.025. Reject Ho if *T* falls outside of the interval $(T_L, T_U)$; that is, reject Ho if $T \leq T_L$ or $\geq T_U$. Conclude that the population medians are not equal. (Otherwise you can't conclude that there's a difference in their medians.)

To conduct a rank sum test in Minitab, enter your data from the first sample in Column 1 and your data from the second sample in Column 2. Go to Stat>Nonparametrics>Mann-Whitney. Click on the name of your Column 1 variable; it appears in the First Sample box. Click on the name of your Column 2 variable; it appears in the Second Sample box. Under Alternative, there's a pull-down menu to select whether your Ha is not equal, greater than, or less than (as indicated by your particular problem). Click OK, and the test is done.

## Stepping up the sample size

After the sample sizes reach a certain point, the table values run out. Table A-4 in the appendix (which shows the critical values for rejecting Ho in the rank sum test) only shows the critical values for sample sizes between three and ten. If both sample sizes are larger than ten, you use a two-sample Z-test to get an approximation for your answer. That's because for large sample sizes the test statistic T for the rank sum test resembles a normal distribution. (So why not use it? As long as you can leave the proof to the professionals!) The larger the two sample sizes are, the better the approximation will be.

So if both sample sizes are more than ten, you conduct steps one through three of the rank sum test as before. Then, instead of looking up the value of T on Table A-4 in the appendix in step four of the rank sum test, you change it to a Z-value (a value on the standard normal distribution) by subtracting its mean and dividing by its standard error.

The formula you use to get this Z-value for the test statistic is

$$Z = \frac{T - \frac{n_1(n_1 + n_2 + 1)}{2}}{\sqrt{\frac{n_1 n_2 (n_1 + n_2 + 1)}{12}}}$$, where T is given by step three in the previous section,

$n_1$ is the sample size for the first data set (taken from the first population), and $n_2$ is the sample size for the second data set (taken from the second population). After you have the Z-value, follow the same procedures that you do for any test involving a Z-value, such as the test for two population means. That is, find the p-value by looking up the Z-value on a Z-table (see your Stats I text) or look on the bottom row of the t-table (which you can find in the appendix), and finding the area beyond it. (If the test is a two-sided test, double the p-value.) If your p-value is less than α, reject Ho. Otherwise fail to reject Ho.

In the case where n is large and you use a Z-value for the test statistic, you can still use Minitab (in fact, that's recommended in order to save you the tedium of working through a big example by hand). Refer to the earlier section "Stepping through the test" for the Minitab directions.

# Performing a Rank Sum Test: Which Real Estate Agent Sells Homes Faster?

Suppose you want to choose a real estate agent to sell your house, and two agents are in your area. Your most important criteria is to get the house sold fast, so you decide to find out whether one agent sells homes faster. You choose a random sample of eight homes each agent sold in the last year, and for each home, you record the number of days it was on the market before being sold. You can see the data in Table 18-1.

| Table 18-1 | Market Time for Homes Sold by Two Real Estate Agents | |
|---|---|---|
| | *Agent Suzy Sellfast* | *Agent Tommy Nowait* |
| House 1 | 48 days | 109 days |
| House 2 | 97 days | 145 days |
| House 3 | 103 days | 160 days |
| House 4 | 117 days | 165 days |
| House 5 | 145 days | 185 days |
| House 6 | 151 days | 250 days |
| House 7 | 220 days | 251 days |
| House 8 | 300 days | 350 days |

Check out the data summarized in *boxplots* (a graph summarizing the data by showing its minimum, first quartile, median, third quartile, and maximum values) in Figure 18-1a and the descriptive statistics in Figure 18-1b. In the following sections, you use this data to see the rank sum test in action. Prepare to be amazed.

To make two boxplots side by side in Minitab, go to Graph>Boxplots>Simple Multiple Y's. Click on each of your two variables in the left-hand box; they'll appear in the right-hand Variables box. Click OK.

## Checking the conditions for this test

In checking the conditions, you know that the data from the two samples are independent, assuming that Suzy and Tommy are competitors.

The boxplots in Figure 18-1a show the same basic shape and amount of variability for each data set. (You don't have enough data to make histograms to check this further.) So based on this data, it isn't unreasonable to assume that the two population distributions of days on the market are the same for the two agents. In Figure 18-1b, the sample standard deviations are close: 79.2 days for Suzy and 77.4 days for Tommy. Because the data meet the conditions for the rank sum test, you can go ahead and apply it to analyze your data.

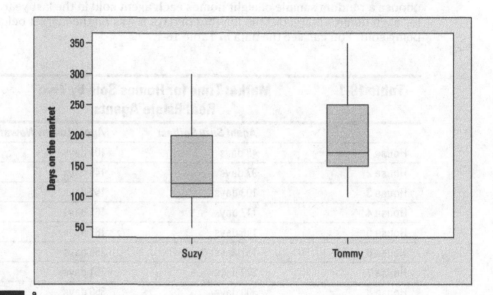

a

**Figure 18-1:**
Boxplots and descriptive statistics for real estate agent data.

**Descriptive Statistics: Suzy, Tommy**

| Variable | Total Count | Mean | StDev | Minimum | Median | Maximum |
|---|---|---|---|---|---|---|
| Suzy | 8 | 147.6 | 79.2 | 48.0 | 131.0 | 300.0 |
| Tommy | 8 | 201.9 | 77.4 | 109.0 | 175.0 | 350.0 |

b

To find descriptive statistics (such as the standard deviation) in Minitab, go to Stat>Basic Statistics>Display Descriptive Statistics. Click on Options. Click on the box for each statistic you want to calculate. If a box is checked for a statistic you don't want, click on it again and the check mark disappears.

Figure 18-1b shows that the median for Suzy (131 days on the market) is less than the median for Tommy (175 days). So, on the surface you may say that Suzy sells houses faster, and that's that. But the median doesn't tell the whole story. Looking at Figure 18-1, you see a portion of the two boxplots overlap each other. This means some of the selling times for Suzy and Tommy were close. There's also a great deal of variability in the selling times for each agent, as you can see from the range of values in each boxplot. This tells you that the evidence isn't entirely clear cut. While Suzy may actually be the fastest agent, you can't tell for sure by comparing the boxplots from two samples. You need a hypothesis test to make that final determination.

## Testing the hypotheses

The null hypothesis for the real estate agent test is Ho: $\eta_1 = \eta_2$, where $\eta_1$ = the median number of days on the market for the population of all Suzy's homes sold in the last year, and $\eta_2$ = the median number of days on the market for the population of all Tommy's homes sold in the last year. The alternative hypothesis is Ha: $\eta_1 \neq \eta_2$.

Suppose you looked at the data and developed a hunch that if one of the agents sold homes faster, it was Suzy. However, before you saw the data, you had no preconceived notion as to who was faster. You must base your Ho and Ha on what your thoughts were *before* you looked at the data, not after. Setting up your hypotheses after you collect the data is unfair and unethical.

After you determine your Ho and Ha, the time comes to test your data.

### Combining and ranking

The first step in the data analysis is to combine all the data and rank the days on the market from lowest (rank = 1) to highest. You can see the overall ranks for the combined data in Table 18-2.

In the case of ties, you give both values the average of the ranks they normally would have received. You can see in Table 18-2 that two values of 145 are in the data set. Because they represent the sixth and seventh numbers in the ordered data set, you give each of them the same rank of $(6 + 7) + 2 = 6.5$.

## Table 18-2  Ranks of Combined Data from the Real Estate Example

| Agent Suzy Sellfast | Overall Rank | Agent Tommy Nowait | Overall Rank |
|---|---|---|---|
| 48 days | 1 | 109 days | 4 |
| 97 days | 2 | 145 days | 6.5 |
| 103 days | 3 | 160 days | 9 |
| 117 days | 5 | 165 days | 10 |
| 145 days | 6.5 | 185 days | 11 |
| 151 days | 8 | 250 days | 13 |
| 220 days | 12 | 251 days | 14 |
| 300 days | 15 | 350 days | 16 |

### Finding the test statistic

After you've ranked your data, you determine which group is group one so you can find your test statistic, *T*. Because the sample sizes are equal, let group one be Suzy because her data is given first. Now sum the ranks from Suzy's data set. The sum of Suzy's ranks is $1 + 2 + 3 + 5 + 6.5 + 8 + 12 + 15 = 52.5$; this value of *T* is your rank sum test statistic.

### Determining whether you can reject Ho

Suppose you want to use an overall $\alpha$ level of 0.05 for this test; using this cutoff means that you use Table A-4(a) in the appendix, because you have a two-sided test at level $\alpha = 0.05$ with 0.025 on each side. Go to the column for $n_1 = 8$ and the row for $n_2 = 8$. You see $T_L = 49$ and $T_U = 87$. You reject Ho if *T* is outside this range; in other words, reject Ho if $T \le T_L = 49$ or if $T \ge T_U = 87$. Your statistic $T = 52.5$ doesn't fall outside this range; you don't have enough evidence to reject Ho at the $\alpha = 0.05$ level. So you can't say there's a statistical difference in the median number of days on the market for Suzy and Tommy.

These results may seem very strange given the fact that the medians for the two data sets were so different: 131 days on the market for Suzy compared to 175 days on the market for Tommy. However you have two strikes against you in terms of being able to find a real difference here:

- ✔ **The sample sizes are quite small (only eight in each group).** A small sample size makes it very hard to get enough evidence to reject Ho.

- ✔ **The standard deviations are both in the high 70s, which is quite large compared to the medians.**

Both of these problems make it hard for the test to actually find anything through all the variability the data show. In other words, it has low power (see Chapter 3).

To conduct the rank sum test by using Minitab, click on Stat>Nonparametric> Mann-Whitney. Select your two samples and choose your alternate Ha as >, <, or ≠. The Confidence Level is equal to one minus your value of $\alpha$. After you make all these settings, click OK.

Figure 18-2 shows the Minitab output when you conduct the rank sum test, or Mann-Whitney test, on the real estate data. To interpret the results in Figure 18-2, you must note that Minitab writes ETA rather than $\eta$ for the medians. The results at the bottom of the output say that the test for equal (versus non-equal) medians is significant at the level 0.1149, when adjusting for ties. This is your *p*-value adjusted for ties. (If no ties are present in your data, you use the results just above that line. That gives you the *p*-value not adjusted for ties.)

To make your final conclusion, compare your *p*-value to your predetermined level of $\alpha$ (typically 0.05.) If your *p*-value is at or below 0.05, you reject Ho;

otherwise you can't. In this case, because 0.1149 is greater than 0.05, you can't reject Ho. That means you don't have enough evidence to say the population medians for days on the market for Suzy's versus Tommy's houses are different based on this data. These results confirm your conclusions from the previous section.

**Figure 18-2:**
Using the
rank sum
test to figure
out who
sells homes
faster.

```
Mann-Whitney Test and CI: Suzy, Tommy
          N   Median
Suzy      8    131.0
Tommy     8    175.0


Point estimate for ETA1-ETA2 is -49.0
95.9 Percent CI for ETA1-ETA2 is (-137.0, 36.0)
W = 52.5
Test of ETA1 = ETA2 vs ETA1 not = ETA2 is significant at 0.1152
The test is significant at 0.1149 (adjusted for ties)
```

The Minitab output in Figure 18-2 also provides a confidence interval for the difference in the medians between the two populations, based on the data from these two samples. The difference in the sample medians (Suzy – Tommy) is 131.0 – 175.0 = –44.0. Adding and subtracting the margin of error (these calculations are beyond the scope of this book), Minitab finds the confidence interval for the difference in medians (Suzy – Tommy) is –137.0, +36.0; the difference in the population medians could be anywhere from –137.0 to 36.0. Because 0, the value in Ho, is in this interval, you can't reject Ho in this case. So again, you can't say that the medians are different, based on this (limited) data set.

# Using a rank sum test to compare judges' scoring practices

You can use rank sum tests to compare two groups of judges of a competition to see whether there's a difference in their scores. For example, in figure-skating competitions, the gender of the judges is sometimes suspected to play a role in the scores they give to certain skaters. Suppose you have a men's figure-skating competition with ten judges: five males and five females. You want to know whether male and female judges score the competitors in the same way, so you do a rank sum test to compare their median scores. Your hypotheses

*(continued)*

*(continued)*

are Ho: Male and female judges have the same median score versus Ha: They have different median scores. For your sample, you let each judge score every individual. You rank their scores in order from lowest to highest and label M for a male judge and F for a female judge. Your results are the following: F, M, M, M, M, F, F, F, F, M. The value of the test statistic $T$ is the sum of the ranks for group one (the males),

which gives you $T = 2 + 3 + 4 + 5 + 10 = 24$. Now compare that to the critical values in Table A-4 in the appendix, where both sample sizes equal five, and you get $T_L = 18$ and $T_U = 37$. Because your test statistic, $T = 24$, is inside this interval, you fail to reject Ho: Judging is the same for male and female judges. In this situation you don't have enough evidence to say that they differ.

# Chapter 19

# Do the Kruskal-Wallis and Rank the Sums with the Wilcoxon

---

*In This Chapter*

▶ Comparing more than two population medians with the Kruskal-Wallis test

▶ Determining which populations are different by using the Wilcoxon rank sum test

---

$S$tatisticians who are in the nonparametrics business make it their job to always find a nonparametric equivalent to a parametric procedure (one that doesn't depend on the normal distribution). And in the case of comparing more than two populations, these stats superheroes don't let us down. In this chapter, you see how the Kruskal-Wallis test works to compare more than two populations as a nonparametric procedure versus its parametric counterpart, ANOVA (see Chapter 9). If Kruskal-Wallis tells you at least two populations differ, this chapter also helps you figure out how to use the Wilcoxon rank sum test to determine which population is different in the same way multiple comparison procedures follow ANOVA (see Chapter 10).

## Doing the Kruskal-Wallis Test to Compare More than Two Populations

The Kruskal-Wallis test compares the medians of several (more than two) populations to see whether or not they're different. The basic idea of Kruskal-Wallis is to collect a sample from each population, rank all the combined data from smallest to largest, and then look for a pattern in how those ranks are distributed among the various samples. For example, if one sample gets all the low ranks and another sample gets all the high ranks, perhaps their population medians are different. Or if all the samples have an equal mix of all the ranks, perhaps the medians of the populations are all deemed to be the same. In this section, you see exactly how to conduct the Kruskal-Wallis test using ranks and sums and all that good stuff, and you see it applied to an example comparing airline ratings.

Suppose your boss flies a lot, and she wants you to determine which of three airlines gets the best ratings from customers. You know that ratings involve data that's just not normal (pun intended), so you opt to use the Kruskal-Wallis test. You take three random samples of nine people each from three different airlines. You ask each person to rate his satisfaction with their one airline. Each person uses a scale from 1 (the worst) to 4 (the best). You can see the data from your samples in Table 19-1.

You may be thinking of using ANOVA, the test that compares the means of several populations (see Chapter 9), to analyze this data. But the data from each airline consist of ratings from 1 to 4, which blows the strongest condition of ANOVA — the data from each population must follow a normal distribution. (A *normal distribution* is continuous, meaning it takes on all real numbers in a certain range. Data that are whole numbers like 1, 2, 3, and 4 don't fall under this category.) But don't sweat; a nonparametric alternative fits the bill. The Kruskal-Wallis test compares the medians of several (more than two) populations to see whether they're all the same or not. In other words, it's like ANOVA except that it's done with medians not means.

| Table 19-1 | Customer Ratings of Three Airlines | |
|---|---|---|
| **Airline A Rating** | **Airline B Rating** | **Airline C Rating** |
| 4 | 2 | 2 |
| 3 | 3 | 3 |
| 4 | 3 | 3 |
| 4 | 3 | 2 |
| 3 | 4 | 2 |
| 3 | 4 | 1 |
| 2 | 3 | 3 |
| 3 | 4 | 2 |
| 4 | 3 | 2 |

In looking at the data in Table 19-1, it appears that airlines A and B have better ratings than airline C. However, the data have a lot of variability, so you have to conduct a hypothesis test before you can make any general conclusions beyond this data set.

In this section, you discover how to check the conditions of the Kruskal-Wallis test, set it up, and carry it out step by step.

## Checking the conditions

The following conditions must be met in order to conduct the Kruskal-Wallis test:

- ✔ The random samples taken from each population are independent. (This means matched-pairs data like in Chapter 17 are out of this picture.)

- ✔ All the populations have the same distribution, meaning their shapes are the same as seen on a histogram. (Note they don't specify what that distribution is.)

- ✔ The variances of the populations are the same. The amount of spread in the population values is the same from one population to the next.

Note that these conditions mention shape and spread, but not the center of the distributions. The test is trying to determine whether the populations are centered at the same place.

In nonparametrics, you often see the word *location* used in reference to a population distribution rather than the word *center,* although the two words mean about the same thing. Location indicates where the distribution is sitting on the number line. If you have two bell-shaped curves with the same variance and one has mean 10 and the other has mean 15, the second distribution is located 5 units to the right of the first. In other words, its location is a 5-unit shift to the right of the first distribution. In nonparametrics, where you don't have bell-shaped distributions, you typically use the median as a measure of location of a distribution. So throughout this discussion, you could use the word *median* instead of location (although location leaves it a bit more open).

Regarding the airline survey, you know that the samples are independent, because you didn't use the same person to rate more than one airline. The other two conditions have to do with the distributions the samples came from; each population must have the same shape and the same spread. You can examine both conditions by looking at boxplots of the data (see Figure 19-1) and descriptive statistics, such as the median, standard deviation, and the rest of the summary statistics making up the boxplots (see Figure 19-2).

The boxplots in Figure 19-1 all have the same shape, and their standard deviations, shown in Figure 19-2, are very close. All this evidence taken together allows you to go ahead with the Kruskal-Wallis test. (Looking at the overlap in the boxplots for airlines A and B in Figure 19-1, you also can make an early prediction that airlines A and B have similar ratings. Whether C is different enough from A and B is impossible to say without running the hypothesis test.)

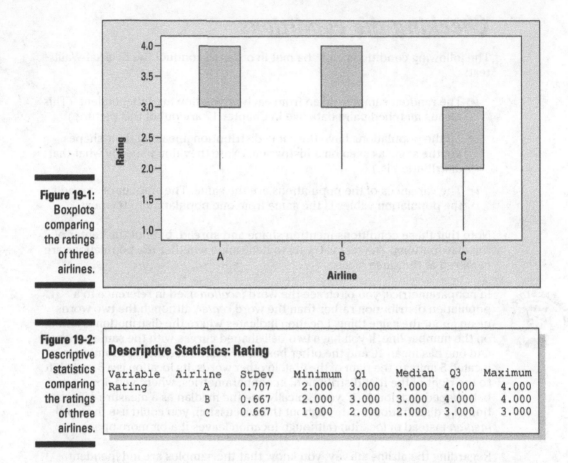

Figure 19-1:
Boxplots
comparing
the ratings
of three
airlines.

Figure 19-2:
Descriptive
statistics
comparing
the ratings
of three
airlines.

**Descriptive Statistics: Rating**

| Variable | Airline | StDev | Minimum | Q1 | Median | Q3 | Maximum |
|----------|---------|-------|---------|-------|--------|-------|---------|
| Rating | A | 0.707 | 2.000 | 3.000 | 3.000 | 4.000 | 4.000 |
| | B | 0.667 | 2.000 | 3.000 | 3.000 | 4.000 | 4.000 |
| | C | 0.667 | 1.000 | 2.000 | 2.000 | 3.000 | 3.000 |

Either a boxplot or a histogram can tell you about the shape and spread of a distribution (as well as the center). The *boxplot* is a common type of graph to use for nonparametric procedures because it displays the median (the non-parametric statistic of choice) rather than the mean. At its best, a *histogram* shows the shape of the data; it doesn't directly tell where the center is — you just have to eyeball it.

Note that the boxplots in Figure 19-1 don't have lines through them, as you may expect. That's because in each case the median happens to equal Q1, the first quartile (see Figure 19-2). This situation can happen with rank data when many ranks take on the same value.

To make boxplots of each sample of data show up side by side on one graph (called *side-by-side boxplots,* cleverly) in Minitab, click on Graph>Box Plots and select the Multiple Y's Simple version. In the left-hand box, click on each of the column names for your data sets. They each appear in the Graph Variables window on the right. Click OK and you get a set of boxplots that are side by side, all on the same graph using the same scale (slick, huh?).

## Setting up the test

The Kruskal-Wallis test assesses Ho: All $k$ populations have the same location versus Ha: The locations of at least two of the $k$ populations are different. Here, $k$ is the number of populations you're comparing.

In Ho, you see that all the populations have the same location (which means they all sit on top of each other on the number line and are in essence the same population). Ha is looking for the opposite situation in this case. However, the opposite of "the locations are all equal" isn't "the locations are all different." The opposite is that at least two of them are different. Failure to recognize this difference will lead you to believe all the populations differ when, in reality, there may only be two that differ, and the rest are all the same. That's why you see Ha stated the way it is in the Kruskal-Wallis test. (The same idea holds for comparing means using ANOVA; see Chapter 9.)

For the airline satisfaction example (see Table 19-1), your setup looks like this: Ho: The satisfaction ratings of all three airlines have the same median versus Ha: The median satisfaction ratings of at least two airlines are different.

## Conducting the test step by step

After you've determined your hypotheses and checked the conditions, you can carry out the test. Here are the steps for conducting the Kruskal-Wallis test using the airline example to show how each step works:

1. **Rank all the numbers in the entire data set from smallest to largest (using all samples combined); in the case of ties, use the average of the ranks that the values would have normally been given.**

   Figure 19-3 shows the results for ranking and summing the data in the airline example; you can see how to rank the ties. For example, you have only one 1, which is given rank 1. Then you have seven 2s, which normally would have gotten ranks 2, 3, 4, 5, 6, 7, and 8. Because the 2s are all equal, you give each of them the average of all these ranks, which is $\frac{(2+3+4+5+6+7+8)}{7} = 5$. Similarly, you see twelve 3s, whose ranks would be 9 through 20. Because they're all equal, give them each a rank equal to $\frac{(9+10+\ldots+20)}{12} = 14.5$. Finally, you see seven 4s, each with rank 24, which is the average of their would-be ranks, ranging from 21 to 27.

| Airline A | | Airline B | | Airline C | |
|---|---|---|---|---|---|
| Rating | Rank | Rating | Rank | Rating | Rank |
| 4 | 24 | 2 | 5 | 2 | 5 |
| 3 | 14.5 | 3 | 14.5 | 3 | 14.5 |
| 4 | 24 | 3 | 14.5 | 3 | 14.5 |
| 4 | 24 | 3 | 14.5 | 2 | 5 |
| 3 | 14.5 | 4 | 24 | 2 | 5 |
| 3 | 14.5 | 4 | 24 | 1 | 1 |
| 2 | 5 | 3 | 14.5 | 3 | 14.5 |
| 3 | 14.5 | 4 | 24 | 2 | 5 |
| 4 | 24 | 3 | 14.5 | 2 | 5 |
| | $T_1 = 159$ | | $T_2 = 149.5$ | | $T_3 = 69.5$ |

**Figure 19-3:**
Rankings and rank sum for the airline example.

2. **Total the ranks for each of the samples; call those totals $T_1$, $T_2$, ..., $T_k$, where $k$ is the number of populations.**

   The totals of the ranks in each column of Figure 19-3 are $T_1 = 159$ (the total ranks for airline A), $T_2 = 149.5$, and $T_3 = 69.5$. In the steps that follow, you use these rank totals in the Kruskal-Wallis test statistic (denoted KW). (Note $T_1$ and $T_2$ are close to equal, but $T_3$ is much lower, giving the impression that airline C may be the odd man out.)

3. **Calculate the Kruskal-Wallis test statistic, $KW = \dfrac{12}{n(n+1)} \sum \dfrac{T_j^2}{n_j} - 3(n+1)$, where $n$ is the total number of observations (all sample sizes combined).**

   For the airline example, the Kruskal-Wallis test statistic is

   $KW = \dfrac{12}{27(27+1)}\left( \dfrac{159^2}{9} + \dfrac{149.5^2}{9} + \dfrac{69.5^2}{9} \right) - 3(27+1)$, which equals

   $0.0159 * 5,829.056 - 3(28) = 8.52$.

4. **Find the $p$-value for your KW test statistic by comparing it to a Chi-square distribution with $k - 1$ degrees of freedom (see Table A-3 in the appendix).**

   For the airline example, you look at the Chi-square table (Table A-3 in the appendix) and find the row with $3 - 1 = 2$ degrees of freedom. Then look at where your test statistic (8.52) falls in that row. Because 8.52 lies between 7.38 and 9.21 (shown on the table in row two), the $p$-value for 8.52 lies between 0.025 and 0.010 (shown in their respective column headings.)

5. **Make your conclusion about whether you can reject Ho by examining the $p$-value.**

You can reject Ho: All populations have the same location, in favor of Ha: At least two populations have differing locations, if the $p$-value associated with KW is $< \alpha$, where $\alpha$ is 0.05 (or your prespecified $\alpha$ level). Otherwise, you fail to reject Ho.

Following the airline example, because the $p$-value is between 0.010 and 0.025, which are both less than $\alpha$ = 0.05, you can reject Ho. You conclude that the ratings of at least two of the three airlines are different.

To conduct the Kruskal-Wallis test by using Minitab, enter your data in two columns, the first column representing the actual data values and the second column representing which population the data came from (for example, 1, 2, 3). Then click on Stat>Nonparametrics>Kruskal-Wallis. In the left-hand box, click on column one; it appears on the right side as your *response variable*. Then click on column two in the left-hand box. This column appears on the right side as the *factor variable*. Click OK, and the KW test is done. The main results of the KW test are shown in the last two lines of the Minitab output.

The results of the Minitab data analysis of the airline data are shown in Figure 19-4. On the second-to-last line of Figure 19-4, you can see the KW test statistic for the airline example is 8.52, which matches the one you found by hand (whew!). The exact $p$-value from Minitab is 0.014.

**Figure 19-4:**
Comparing
ratings
of three
airlines by
using the
Kruskal-
Wallis test.

```
Kruskal-Wallis Test: Rating versus Airline

Kruskal-Wallis Test on Rating

Airline        N    Median    Ave Rank       Z
A              9     3.000        17.7     1.70
B              9     3.000        16.6     1.21
C              9     2.000         7.7    -2.91
Overall       27                  14.0

H = 8.52   DF = 2   P = 0.014
H = 9.70   DF = 2   P = 0.008 (adjusted for ties)
```

However, this data set has quite a few ties, and the formulas have to adjust a bit for that (in ways that go outside the scope of this book). Taking those ties into account, the computer gives you KW = 9.70 with a $p$-value of 0.008. The total evidence here says the same result loud and clear — reject Ho: The ratings for the three airlines have the same location. You conclude that the ratings of at least two of the airlines are different. (But which ones? The answer comes in the next section.)

# Pinpointing the Differences: The Wilcoxon Rank Sum Test

Suppose you reject Ho in the Kruskal-Wallis test, meaning you have enough evidence to conclude that at least two of the populations have different medians. But you don't know which ones are different. When you find that a set of populations don't all share the same median, the next question is very likely to be, "Well then, which ones are different?" To find out which populations are different after the Kruskal-Wallis test has rejected Ho, you can use the *Wilcoxon rank sum test* (also known as the *Mann-Whitney test*).

You can't go looking for differences in specific pairs of populations until you've first established that at least two populations differ (that is, Ho is rejected in the Kruskal-Wallis test). If you don't make this check first, you can encounter a ton of problems, not the least of which being a much-increased chance of making the wrong decision.

In the following sections, you see how to conduct pairwise comparisons and interpret them in order to find out where the differences lie among the *k* population medians you're studying.

## Pairing off with pairwise comparisons

The rank sum test is a nonparametric test that compares two population locations (for example, their medians). When you have more than two populations, you conduct the rank sum test on every pair of populations in order to see whether differences exist. This procedure is called conducting *pairwise comparisons* or *multiple comparisons*. (See Chapter 10 for info on the parametric version of multiple comparisons.) For example, because you're comparing three airlines in the airline satisfaction example (see Table 19-1), you have to run the rank sum test three times to compare airlines A and B, A and C, and B and C. So you need three pairwise comparisons to figure out which populations are different.

To determine how many pairs of comparisons you need if you're given *k* populations, you use the formula $\frac{k(k-1)}{2}$. You have *k* populations to choose from first and then *k* − 1 populations left to compare them with. Finally, you don't care what the order is among the populations (as long as you keep track of them); so you divide by two because you have two ways to order any pair (for example, comparing A and B gives you the same results as comparing B and A). In the airlines example, you have *k* = 3 populations, so you should have $\frac{k(k-1)}{2} = \frac{3(3-1)}{2} = 3$ pairs of populations to compare, which matches what was determined previously. (For more information and examples on how to count the number of ways to choose or order a group of items by using permutations and combinations, see another book I authored, *Probability For Dummies* [Wiley].)

## Carrying out comparison tests to see who's different

The Wilcoxon rank sum test assesses Ho: The two populations have the same location versus Ha: The two populations have different locations. Here are the general steps for using the Wilcoxon rank sum test for making comparisons:

1. **Check the conditions for the test by using descriptive statistics and histograms for the last two and proper sampling procedures for the first one:**

   • The two samples must be from independent populations.

   • The populations must have the same distribution (shape).

   • The populations must have the same variance.

2. **Set up your Ho: The two medians are equal versus Ha: The two medians aren't equal.**

3. **Combine all the data and rank the values from smallest to largest.**

4. **Add up all the ranks from the first sample (or the smallest sample if the sample sizes are not equal).**

   This result is your test statistic, $T$.

5. **Compare $T$ to the critical values in Table A-4 in the appendix, in the row and column corresponding to the two sample sizes (denoted $T_L$ and $T_U$).**

   If $T$ is at or beyond the critical values (less than or equal to the lower one [$T_L$] or greater than or equal to the upper one [$T_U$]), reject Ho and conclude the two population medians are different. Otherwise, you can't reject Ho.

6. **Repeat steps one through five on every pair of samples in the data set and draw conclusions.**

   Sort through all the results to see the overall picture of which pairs of populations have the same median and which ones don't.

To conduct the Wilcoxon rank sum test for pairwise comparisons in Minitab, refer to Chapter 18. Note that Minitab calls this test by its other name, the Mann-Whitney test.

You can see the Minitab results of the three Wilcoxon rank sum tests comparing airlines A and B, A and C, and B and C in Figures 19-5a, 19-5b, and 19-5c, respectively.

Figure 19-5a compares the ratings of airlines A and B. The *p*-value (adjusted for ties) is 0.7325, which is much higher than the 0.05 you need to reject Ho. So you can't conclude that airlines A and B have satisfaction ratings with different medians. Figure 19-5b shows that the *p*-value for comparing airlines A and C is 0.0078. Because this *p*-value is a lot smaller than the typical α level of 0.05, it's very convincing evidence that airlines A and C don't have the same median ratings. Figure 19-5c also has a small *p*-value (0.0107), which gives evidence that airlines B and C have significantly different ratings.

**Mann-Whitney Test and CI: Airline A, Airline B**

```
    N   Median
A   9   3.000
B   9   3.000


Point estimate for ETA1-ETA2 is -0.000
95.8 Percent CI for ETA1-ETA2 is (-1.000,1.000)
W = 89.5
Test of ETA1 = ETA2 vs ETA1 not = ETA2 is significant at 0.7573
The test is significant at 0.7325   (adjusted for ties)
```

a

**Mann-Whitney Test and CI: Airline A, Airline C**

```
    N   Median
A   9   3.000
C   9   2.000


Point estimate for ETA1-ETA2 is 1.000
95.8 Percent CI for ETA1-ETA2 is (0.000,2.000)
W = 114.5
Test of ETA1 = ETA2 vs ETA1 not = ETA2 is significant at 0.0118
The test is significant at 0.0078 (adjusted for ties)
```

b

**Figure 19-5:**
Wilcoxon
rank sum
tests
comparing
ratings of
two airlines
at a time.

c

**Mann-Whitney Test and CI: Airline B, Airline C**

```
    N   Median
B   9   3.000
C   9   2.000


Point estimate for ETA1-ETA2 is 1.000
95.8 Percent CI for ETA1-ETA2 is (0.000,2.000)
W = 113.0
Test of ETA1 = ETA2 vs ETA1 not = ETA2 is significant at 0.0171
The test is significant at 0.0107 (adjusted for ties)
```

## Examining the medians to see how they're different

Rejecting Ho for a multiple comparison means you conclude the two populations you examined have different medians. There are two ways to proceed from here to see how the medians differ for all pairwise comparisions:

- ✔ You can look at side-by-side boxplots of all the samples and compare their medians (located at the line in the middle of each box).

- ✔ You can calculate the median of each sample and see which ones are higher and which ones are lower from the populations you have concluded are statistically different.

From the previous section, you see that the pairwise comparisons for the airline data conducted by Wilcoxon rank sum tests conclude that the ratings of airlines A and B aren't found to be different, but both of them are found to be different from airline C.

But you can say even more; you can say how the differing airline compares to the others. Going back to Figure 19-2, you see the medians of both airlines A and B are 3.0, while the median of airline C is only 2.0. That difference means airlines A and B have similar ratings, but airline C has lower ratings than A and B. The boxplots in Figure 19-1 confirm these results.

# Chapter 20

# Pointing Out Correlations with Spearman's Rank

. . . . . . . . . . . . . . . . . . . . . . . . . . . . . . . . . . . . . . . . . . . . . . . . . .

### In This Chapter

▶ Understanding correlation from a nonparametric point of view

▶ Finding and interpreting Spearman's rank correlation

. . . . . . . . . . . . . . . . . . . . . . . . . . . . . . . . . . . . . . . . . . . . . . . . . .

**D**ata analysts commonly look for and try to quantify relationships between two variables, $x$ and $y$. Depending on the type of data you're dealing with in $x$ and $y$, you use different procedures for quantifying their relationship.

When $x$ and $y$ variables are *quantitative* (that is, their possible outcomes are measurements or counts), the correlation coefficient (also known as the *Pearson's correlation coefficient*) measures the strength and direction of their linear relationship. (See Chapter 4 for all the info on Pearson's correlation coefficient, denoted by $r$.) If $x$ and $y$ are both *categorical* variables (meaning their possible outcomes are categories that have no numerical meaning, such as male and female), you use Chi-square procedures and conditional probabilities to look for and describe their relationship. (I lay out all that machinery in Chapters 13 and 14.)

Then there's a third type of variable, called *ordinal* variables; their values fall into categories, but the possible values can be placed into an order and given a numerical value that has some meaning, for example, grades on a scale of A = 4, B = 3, C = 2, D = 1, and E = 0 or a student's evaluation of a teacher on a scale from best = 5 to worst = 1. To look for a relationship between two ordinal variables like these, statisticians use *Spearman's rank correlation,* the nonparametric counterpart to Pearson's correlation coefficient (covered in Chapter 4). In this chapter, you see why ordinal variables don't meet Pearson's conditions, and you find out how to use and interpret Spearman's rank correlation to correctly quantify and interpret relationships involving ordinal variables.

# Pickin' On Pearson and His Precious Conditions

Pearson's correlation coefficient is the most common correlation measure out there, and many data analysts think it's the *only* one out there. Trouble is, Pearson's correlation has certain conditions that must be met before using it. If those conditions aren't met, Spearman's correlation is waiting in the wings.

The Pearson correlation coefficient $r$ (the correlation) is a number that measures the direction and strength of the linear relationships between two variables $x$ and $y$. (For more info on the correlation, see Chapter 4.)

Several conditions have to be met for ol' Pearson:

- **Both variables $x$ and $y$ must be numerical (or quantitative).** They must represent measurements with no restriction on their level of precision. For example, numbers with many places after the decimal point (such as 12.322 or 0.219) must be possible.

- **The variables $x$ and $y$ must have a linear relationship (as shown on a scatterplot; see Chapter 4).**

- **The $y$ values must have a normal distribution for each $x$, with the same variance at each $x$.**

Typically with ordinal variables, you won't see many different categories offered or compared for reasons of simplicity. This means there won't be enough numerical values to try to build a linear regression model involving ordinal variables like you can with two quantitative variables.

In another scenario, if you have a gender variable with categories male and female, you can assign the numbers 1 and 2 to each gender, but those numbers have no numerical meaning. Gender isn't an ordinal variable; rather it's a *categorical variable* (a variable that places individuals into categories only). Categorical variables also don't lend themselves to linear relationships, so they don't meet Pearson's conditions either. (To explore relationships between categorical variables, see Chapter 14.)

## Who are these guys? A look at the people behind the statistics

Some people are lucky enough to have a statistic actually named after them. Typically, the person who came up with the statistic in the first place, recognizing a need for it and coming up with a solution, gets the honor. If the new statistic gets picked up and used by others, it eventually takes on the name of its inventor.

Spearman's rank correlation is named after its inventor, Charles Edward Spearman (1863–1945). He was an English psychologist who studied experimental psychology, worked in the area of human intelligence, and was a professor for many years at the University College London.

Spearman followed closely the work of Francis Galton, who originally developed the concept of correlation. Spearman developed his rank correlation in 1904.

Pearson's correlation coefficient was developed several years prior, in 1893 by Karl Pearson, one of Spearman's colleagues at University College London and another follower of Galton. Pearson and Spearman didn't get along; Pearson had an especially strong and volatile personality and had problems getting along with quite a few people, in fact.

# Scoring with Spearman's Rank Correlation

Spearman's rank correlation doesn't require the relationship between the variables $x$ and $y$ to be linear, nor does it require the variables to be numerical. Rather than examining a linear relationship between $x$ and $y$, Spearman's rank correlation tests whether two ordinal and/or quantitative variables are independent (in other words, not related to each other). *Note:* Spearman's rank applies to ordinal data only. To test to see if two categorical (and nonordinal) variables are independent, you use a Chi-square test; see Chapter 14.

Spearman's rank correlation is the same as Pearson's correlation except that it's calculated based on the *ranks* of the $x$ and $y$ variables (that is, where they stand in the ordering; see Chapter 16) rather than their actual values. You interpret the value of Spearman's rank correlation, $r_s$, the same way you interpret Pearson's correlation, $r$ (see Chapter 4). The values of $r_s$ can go between –1 and +1. The higher the magnitude of $r_s$ (in the positive or negative directions), the stronger the relationship between $x$ and $y$. If $r_s$ is zero, $x$ and $y$ are independent. And as with $r$, if the correlation between $x$ and $y$ is not zero, you can't say whether or not they're independent.

In this section, you see how to calculate and interpret Spearman's rank correlation and apply it to an example.

## Figuring Spearman's rank correlation

The notation for Spearman's rank correlation is $r_s$, where $s$ stands for Spearman. To find $r_s$, you do the steps listed in this section. Minitab does the work for you in steps two through six, although some professors may ask you to do the work by hand (not me of course).

1. **Collect the data in the form of pairs of values $x$ and $y$.**

2. **Rank the data from the $x$ variable where 1 = lowest to $n$ = highest, where $n$ is the number of pairs of data in the data set.**

   This step gives you a new set of data for the $x$ variable called the *ranks of the x values*. If any of the values appear more than once, Minitab assigns each tied value the average of the ranks they would normally be given if they weren't tied.

3. **Complete step two with the data from the $y$ variable.**

   This step gives you a new data set called the *ranks of the y-values*.

4. **Find the standard deviation of the ranks of the $x$-values, using the**

   usual formula for standard deviation, $\sqrt{\dfrac{\sum(x-\bar{x})^2}{n-1}}$; call it $s_{xx}$. In

   a similar manner find the standard deviation of the ranks of the

   $y$-values using $\sqrt{\dfrac{\sum(y-\bar{y})^2}{n-1}}$; call it $s_{yy}$.

   Note that $n$ is the sample size, $\bar{x}$ is the mean of the ranks of the $x$ values, and $\bar{y}$ is the mean of the ranks of the $y$ values.

5. **Find the *covariance* of the $x$-$y$ values, using the formula**

   $\text{Cov}(x,y) = \dfrac{\sum\sum(x-\bar{x})(y-\bar{y})}{n-1}$; call it $s_{xy}$.

   The covariance of $x$ and $y$ is a measure of the total deviation of the $x$ and $y$ values from the point $(\bar{x},\bar{y})$.

6. **Calculate the value of Spearman's rank correlation by using the**

   formula $r_s = \dfrac{s_{xy}}{s_{xx}s_{yy}}$.

The formula for Spearman's rank correlation is just the same as the formula for Pearson's correlation coefficient, except the data Spearman uses is the ranks of $x$ and $y$ rather than the original $x$- and $y$-values as used by Pearson. So Spearman just cares about the order of the values of the $x$'s and the $y$'s, not their actual values.

To calculate Spearman's rank correlation straightaway by using Minitab, rank the *x*-values, rank the *y*-values, and then find the correlation of the ranks. That is, go to Data>Rank and click on the *x* variable to get *x* ranks. Then do the same thing to get the *y* ranks. Go to Stat>Basic Statistics>Correlation, click on the two columns representing ranks, and click OK.

## Watching Spearman at work: Relating aptitude to performance

Knowing the process of how to calculate Spearman's rank correlation is one thing, but if you can apply it to real-world situations, you'll be the golden child of the statistics world (or at least your statistics class). So, try putting yourself in this section's scenario to get the full effect of Spearman's rank correlation.

You're a statistics professor, and you give exams every now and then (it's a dirty job, but someone's got to do it). After looking at students' final grades over the years (yes, you're an old professor, or at least in your mid-40s), you notice that students who do well in your class tend to have a better aptitude (background ability) for math and statistics. You want to check out this theory, so you give students a math and statistics aptitude test on the first day of the course; you want to compare students' aptitude test scores with their final grades at the end of the course.

Now for the specifics. Your variables are *x* = aptitude test score (using a 100-point pretest on the first day of the course) and *y* = final grade on a scale from 1 to 5 where 1 = F (failed the course), 2 = D (passed), 3 = C (average), 4 = B (above average), and 5 = A (excellent). The *y* variable, final grade, is an ordinal variable, and the *x* variable, aptitude, is a numerical variable. You want to find out whether there's a relationship between *x* and *y*. You collect data on a random sample of 20 students; the data are shown in Table 20-1. This is step one of the process of calculating Spearman's rank correlation (from the steps listed in the previous section).

### Table 20-1    Aptitude Test Scores and Final Grades in Statistics

| Student | Aptitude | Final Grade |
|---|---|---|
| 1 | 59 | 3 |
| 2 | 47 | 2 |
| 3 | 58 | 4 |
| 4 | 66 | 3 |
| 5 | 77 | 2 |

*(continued)*

**Table 20-1** *(continued)*

| Student | Aptitude | Final Grade |
|---------|----------|-------------|
| 6 | 57 | 4 |
| 7 | 62 | 3 |
| 8 | 68 | 3 |
| 9 | 69 | 5 |
| 10 | 36 | 1 |
| 11 | 48 | 3 |
| 12 | 65 | 3 |
| 13 | 51 | 2 |
| 14 | 61 | 3 |
| 15 | 40 | 3 |
| 16 | 67 | 4 |
| 17 | 60 | 2 |
| 18 | 56 | 3 |
| 19 | 76 | 3 |
| 20 | 71 | 5 |

Using Minitab, you get a Spearman's rank correlation of 0.379. The following discussion walks you through steps two through six as you do this correlation yourself. This is likely what you may be asked to do on an exam.

Steps two and three of finding Spearman's rank correlation are to rank the aptitude test scores *(x)* from lowest (1) to highest; then rank the final grades *(y)* from lowest (1) to highest. Note that the final exam grades have several ties, so you use average ranks. For example, in column three of Table 20-1 you see a single 1 for Student 10, which gets rank 1. Then you see four 2s for Students 2, 5, 13, and 17. Those ranks, had they not been tied, would have been 2, 3, 4, and 5. The average of these four ranks is $\frac{2+3+4+5}{4} = \frac{14}{4} = 3.5$. Each of the 2s in column three, therefore, receives rank 3.5.

Table 20-2 shows the original data, the ranks of the aptitude scores *(x)*, and the ranks of the final grades *(y)* as calculated by Minitab.

### Table 20-2 Aptitude Test Scores, Final Exam Grades, and Rank

| Student | Aptitude | Rank of Aptitude | Final Grade | Rank of Final Grade |
|---------|----------|------------------|-------------|---------------------|
| 1 | 59 | 9 | 3 | 10.5 |
| 2 | 47 | 3 | 2 | 3.5 |
| 3 | 58 | 8 | 4 | 17.0 |
| 4 | 66 | 14 | 3 | 10.5 |
| 5 | 77 | 20 | 2 | 3.5 |
| 6 | 57 | 7 | 4 | 17.0 |
| 7 | 62 | 12 | 3 | 10.5 |
| 8 | 68 | 16 | 3 | 10.5 |
| 9 | 69 | 17 | 5 | 19.5 |
| 10 | 36 | 1 | 1 | 1.0 |
| 11 | 48 | 4 | 3 | 10.5 |
| 12 | 65 | 13 | 3 | 10.5 |
| 13 | 51 | 5 | 2 | 3.5 |
| 14 | 61 | 11 | 3 | 10.5 |
| 15 | 40 | 2 | 3 | 10.5 |
| 16 | 67 | 15 | 4 | 17.0 |
| 17 | 60 | 10 | 2 | 3.5 |
| 18 | 56 | 6 | 3 | 10.5 |
| 19 | 76 | 19 | 3 | 10.5 |
| 20 | 71 | 18 | 5 | 19.5 |

For step four of finding Spearman's rank correlation, you have Minitab calculate the standard deviation of the aptitude test score ranks (located in column two of Table 20-2) and the standard deviation of the final grades (located in column four of Table 20-2). In step five, you have Minitab calculate the covariance of the ranks of aptitude test scores and final grade ranks. These statistics are shown in Figure 20-1.

**Figure 20-1:**
Standard
deviations
and
covariance
of ranks of
aptitude *(x)*
and final
grade *(y)*.

**Descriptive Statistics: Ranks of X, Ranks of Y**

| Variable | StDev |
|---|---|
| Ranks of X | 5.92 |
| Ranks of Y | 5.50 |

**Covariances: Ranks of X, Ranks of Y**

| | Ranks of X | Ranks of Y |
|---|---|---|
| Ranks of X | 35.0000 | |
| Ranks of Y | 12.3421 | 30.2632 |

For the sixth and final step of finding Spearman's rank correlation, calculate $r_s$ by taking the covariance of the ranks of $x$ and $y$ (found in the lower left corner) divided by the standard deviation of the ranks of $x$ (called $s_{xx}$) times the standard deviation of the ranks of

$y$ (called $s_{yy}$). You get $\frac{12.34}{5.92 * 5.50} = 0.379$. This matches the value for Spearman's correlation that Minitab found straightaway.

This Spearman's rank correlation of 0.379 is fairly low, indicating a weak relationship between aptitude scores before the course and final grades at the end of the course. The moral of the story? If a student isn't the sharpest tack in the bunch, he can still hope, and if he comes in on top, he may not go out the same way. Although, there's still something to be said about working hard during the course (and buying *Statistics II For Dummies* certainly doesn't hurt!).

# Part VI

# The Part of Tens

The 5th Wave · By Rich Tennant

"What do you mean I don't fit your desired sample population at this time?"

## In this part . . .

This part is a staple in *For Dummies* books, and for good reason. In this part, you get useful insight about what statistical conclusions to be wary of, along with the many ways that statistics is used in the workplace. I also tell you why you shouldn't try to escape statistics, but rather plunge right in, because knowing statistics means job security in practically anything you do.

# Chapter 21

# Ten Common Errors in Statistical Conclusions

• • • • • • • • • • • • • • • • • • • • • • • • • • • • • • • • • • • • • • • • • •

## In This Chapter

▶ Recognizing and avoiding mistakes when interpreting statistical results

▶ Deciding whether or not someone's conclusions are credible

• • • • • • • • • • • • • • • • • • • • • • • • • • • • • • • • • • • • • • • • • •

**S**tats II is all about building models and doing data analysis. It focuses on looking at data and figuring out the story behind it. It's about making sure that the story is told correctly, fairly, and comprehensively. In this chapter, I discuss some of the most common errors I've seen as a teacher and statistical consultant for many moons. You can use this "not to do" list to pull ideas together for homework and reports or as a quick review before a quiz or exam. Trust me — your professor will love you for it!

# Claiming These Statistics Prove . . .

Be skeptical of anyone who uses *these statistics* and *prove* in the same sentence. The word *prove* is a definitive, end-all-be-all, case-closed, lead-pipe-lock sort of concept, and statistics by nature isn't definitive. Instead, statistics gives you evidence for or against someone's theory, model, or claim based on the data you collected; then it leaves you to your own conclusions. Because the evidence is based on data that changes from sample to sample, the results can change as well — that's the challenge, the beauty, and sometimes the frustration of statistics. The best you can say is that your statistics suggest, lead you to believe, or give you sufficient evidence to conclude — but never go as far as to say that your statistics prove anything.

# It's Not Technically Statistically Significant, But . . .

After you set up your model and test it with your data, you have to stand by the conclusions no matter how much you believe they're wrong. Statistics must lend objectivity to every process.

Suppose Barb, a researcher, has just collected and analyzed the heck out of her data, and she still can't find anything. However, she knows in her heart that her theory holds true, even if her data can't confirm it. Barb's theory is that dogs have ESP — in other words, a "sixth sense." She bases this theory on the fact that her dog seems to know when she's leaving the house, when he's going to the vet, and when a bath is imminent because he gets sad and finds a corner to hide in.

Barb tests her ESP theory by studying ten dogs, placing a piece of dog food under one of two bowls and asking each dog to find the food by pushing on a bowl. (Assume the bowl is thick enough that the dogs can't cheat by smelling the food.) She repeats this process ten times with each dog and records the number of correct responses. If the dogs don't have ESP, you would expect that they would be right 50 percent of the time because each dog has two bowls to choose from and each bowl has an equal chance of being selected.

As it turns out, the dogs in Barb's study were right 55 percent of the time. Now, this percentage is technically higher than the long-term expected value of 50 percent, but it's not enough (especially with so few dogs and so few trials) to warrant statistical significance. In other words, Barb doesn't have enough evidence for the ESP theory. But when Barb presents her results at the next conference she attends, she puts a spin on her results by saying, "The dogs were correct 55 percent of the time, which is more than 50 percent. These results are *technically* not enough to be statistically significant, but I believe they do show some evidence that dogs have ESP" (causing every statistician in the room to scream "NOT!").

Some researchers use this kind of conclusion all the time — skating around the statistics when they don't go their way. This game is very dangerous because the next time someone tries to replicate Barb's results (and believe me, someone always does), they find out what you knew from the beginning (through ESP?): When Barb starts packing to leave the house, her dog senses trouble coming and hides. That's all.

# Concluding That x Causes y

Do you see the word that makes statisticians nervous? The first two seem pretty tame, and *x* and *y* are just letters of the alphabet, it's got to be that word *cause*. Of all the words used too loosely in statistics, *cause* tops the list.

Here's an example of what I mean. For your final report in stats class, you study which factors are related to a student's final exam score. You collect data on 500 statistics students, asking each one a variety of questions, such as "What was your grade on the midterm?"; "How much sleep did you get the night before the final?"; and "What's your GPA?" You conduct a multiple linear regression analysis (using techniques from Chapter 5) and conclude that study time and the amount of sleep the night before the test are the most-important factors in determining exam scores. You write up all your analyses in a paper, and at the very end you say, "These results demonstrate that more study time and a good night of sleep the night before cause a student's exam grade to increase."

I was with you until you said the word *cause*. You can't say that more sleep or more study time causes an increase in exam score. The data you collected shows that people who get a lot of sleep and study a lot do get good grades, and those who don't do those things don't get the good grades. But that result doesn't mean a flunky can just sleep and study more and all will be okay. This theory is like saying that because an increase in height is related to an increase in weight, you can get taller by gaining weight.

The problem is that you didn't take the same person, change his sleep time and study habits, and see what happened in terms of his exam performance (using two different exams of the same difficulty). That study requires a *designed experiment*. When you conduct a *survey*, you have no way of controlling other related factors going on, which can muddy the waters, like quality of studying, class attendance, grades on homework, and so on.

The only way to control for other factors is to do a randomized experiment (complete with a treatment group, a control group, and controls for other factors that may ordinarily affect the outcome). Claiming causation without conducting a randomized experiment is a very common error some researchers make when they draw conclusions.

# Assuming the Data Was Normal

The operative word here is *assuming*. To break it down simply, an assumption is something you believe without checking. Assumptions can lead to wrong analyses and incorrect results — all without the person doing the assuming even knowing it.

Many analyses have certain requirements. For example, data should come from a normal distribution (the classic distribution that has a bell shape to it). If someone says, "I assumed the data was normal," she just assumed that the data came from a normal distribution. But is having a normal distribution an assumption you just make and then move on, or is more work involved? You guessed it — more work.

For example, in order to conduct a one-sample *t*-test (see Chapter 3), your data must come from a normal distribution unless your sample size is large, in which case you get an approximate normal distribution anyway by the Central Limit Theorem (remember those three words from Stats I?). Here, you aren't making an assumption, but examining a *condition* (something you check before proceeding). You plot the data, see if they meet the condition, and if they do, you proceed. If not, you can use nonparametric methods instead (discussed in Chapter 16).

Nearly every statistical technique for analyzing data has at least some conditions on the data in order for you to use it. Always find out what those conditions are, and check to see whether your data meet them (and if not, consider using nonparametric statistics; see Chapter 16). Be aware that many statistics textbooks wrongly use the word *assumption* when they actually mean *condition*. It's a subtle but very important difference.

# Only Reporting "Important" Results

As a data analyst, you must not only avoid the pitfall of reporting only the significant, exciting, and meaningful results but also be able to detect when someone else is doing so. Some number crunchers examine every possible option and look at their data in every possible way before settling on the analysis that gets them the desired result.

You can probably see the problem with that approach. Every technique carries the chance for error. If you're doing a *t*-test, for example, and the $\alpha$ level is 0.05, over the long term 5 out every 100 *t*-tests you conduct will result in a false alarm (meaning you declare a statistically significant result when it wasn't really there) just by chance. So, if an eager researcher conducts 20 hypothesis tests on the same data set, odds are that at least one of those tests could result in a false alarm just by chance, on average. As this researcher conducts more and more tests, he's unfairly increasing his odds of finding something that occurred just by chance and running the risk of a wrong conclusion in the process.

It's not all the eager researcher's fault, though. He's pressured by a results-driven system. It's a sad state of affairs when the only results that get broadcast on the news and appear in journal articles are the ones that show a statistically significant result (when Ho is rejected). Perhaps it was a bad move when statisticians came up with the term *significance* to denote rejecting Ho — as if to say that rejecting Ho is the only important conclusion you can come to. What about all the times when Ho couldn't be rejected? For example, when doctors failed to conclude that drinking diet cola causes weight gain, or when pollsters didn't find that people were unhappy with the president? The public would be better served if researchers and the media were encouraged to report the statistically insignificant but still important results along with the statistically significant ones.

The bottom line is this: In order to find out whether a statistical conclusion is correct, you can't just look at the analysis the researcher is showing you. You also have to find out about the analyses and results they're *not* showing you and ask questions. Avoid the urge to rush to reject Ho.

# Assuming a Bigger Sample Is Always Better

Bigger is better in some things, but not always with sample sizes. On one hand, the bigger your sample is, the more precise the results are (if no bias is present). A bigger sample also increases the ability of your data analysis to detect differences from a model or to deny some claim about a population (in other words, to reject Ho when you're supposed to). This ability to detect true differences from Ho is called the *power* of a test (see Chapter 3). However, some researchers can (and often do) take the idea of power too far. They increase the sample size to the point that even the tiniest difference from Ho sends them screaming to press that all-important reject Ho button.

Sample sizes should be large enough to provide precision and repeatability of your results, but there's such a thing as being too large, believe it or not. You can always take sample sizes big enough to reject any null hypothesis, even when the actual deviation from it is embarrassingly small. What can you do about this? When you read or hear that a result was deemed statistically significant, ask what the sample mean actually was (before it was put into the *t*-formula) and judge its significance to you from a practical standpoint. Beware of someone who says, "These results are statistically significant, and the large sample size of 100,000 gives even stronger evidence for that."

Suppose research claims that the typical in-house dog watches an average of ten hours of TV per week. Bob thinks the true average is more, based on the fact that his dog Fido watches at least ten hours of cooking shows alone each week. Bob sets up the following hypothesis test: Ho: $\mu = 10$ versus Ha: $\mu > 10$. He takes a random sample of 100 dogs and has their owners record how much TV their dogs watch per week. The result turns out a sample mean of 10.1 hours, and the sample standard deviation is 0.8 hours. This result isn't what Bob hoped for because 10.1 is so close to 10. He calculates the

test statistic for this test using the formula $t = \dfrac{(\bar{x} - \mu)}{\frac{s}{\sqrt{n}}}$ and comes up with a

value of $t = \dfrac{(10.1 - 10.0)}{\frac{0.8}{\sqrt{100}}} = \dfrac{0.1}{0.08}$, which equals 1.25 for $t$. Because the test is a

right-tailed test (> in Ha), Bob can reject Ho at $\alpha$ if $t$ is beyond 1.645, and his $t$-value of 1.25 is far short of that value. Note that because $n = 100$ here, you find the value of 1.645 by looking at the very last row of the $t$-distribution table (visit Table A-1 in the appendix). The row is marked with the infinity sign to indicate a large sample. So Bob can't reject Ho. To add insult to injury, Bob's friend Joe conducts the same study and gets the same sample mean and standard deviation as Bob did, but Joe uses a random sample of 500 dogs rather

than 100. Consequently, Joe's $t$-value is $t = \dfrac{(10.1 - 10.0)}{\frac{0.8}{\sqrt{500}}} = \dfrac{0.1}{0.036}$, which equals

2.78. Because 2.78 is greater than 1.645, Joe gets to reject Ho (to Bob's dismay).

Why did Joe's test find a result that Bob's didn't? The only difference was the sample size. Joe's sample was bigger, and a bigger sample size always makes the standard error smaller (see Chapter 3). The standard error sits in the denominator of the $t$-formula, so as it gets smaller, the $t$-value gets larger. A larger $t$-value makes it easier to reject Ho. (See Chapter 3 for more on precisions and margin of error.)

Now, Joe could technically give a big press conference or write an article on his results (his mom would be so proud), but you know better. You know that Joe's results are technically *statistically* significant, but not *practically* significant — they don't mean squat to any person or dog. After all, who cares that he was able to show evidence that dogs watch just a tiny bit more than ten hours of TV per week versus exactly ten hours per week? This news isn't exactly earth-shattering.

# *It's Not Technically Random, But . . .*

When you take a sample on which to build statistical results, the operative word is *random*. You want the sample to be randomly selected from the population. A problem is that people often collect a sample that they think is mostly random, sort of random, or random enough — and that doesn't cut it. A plan for taking a sample is either random or it isn't.

One day I gave each of the 50 students in my class a number from 1 to 50, and I drew two numbers randomly from a hat. The two students I picked were sitting in the first row, and not only that, they were right next to each other. My students immediately cried foul!

After this seemingly odd result, I took the opportunity to talk to my class about truly random samples. A *random sample* is chosen in such a way that every member of the original population has an equal chance of being selected. Sometimes people who sit next to each other are chosen. In fact, if these seemingly strange results never happen, you may worry about the process; in a truly random process, you're going to get results that may seem odd, weird, or even fixed. That's part of the game.

In my consulting experiences, I always ask how my clients chose or plan to choose their samples. They always say they'll make sure it's random. But when I ask them how they'll do this, I sometimes get less-than-stellar answers. For example, someone needed to get a random sample from a population of 500 free-range chickens in a farmyard. He needed five chickens and said that he'd select them randomly by choosing the five that came up to him first. The problem is, animals that come up to you may be friendlier, more docile, older, or perhaps more tame. These characteristics aren't present in every chicken in the yard, so choosing a sample this way isn't random. The results are likely biased in this case.

Always ask the researcher how he or she selected a sample, and when you select your own samples, stay true to the definition of random. Don't use your own judgment to choose a random sample; use a computer to do it for you!

# Assuming That 1,000 Responses Is 1,000 Responses

A newspaper article on the latest survey says that 50 percent of the respondents said blah blah blah. The fine print says the results are based on a survey of 1,000 adults in the United States. But wait — is 1,000 the actual number of people selected for the sample, or is it the final number of respondents? You may need to take a second look; those two numbers hardly ever match.

For example, Jenny wants to know the percentage of Americans who have ever knowingly cheated on their taxes. In her statistics class, she found out that if she gets a sample of 1,000 people, the margin of error for her survey is only ±3 percent, which she thinks is groovy. So she sets out to achieve the goal of 1,000 responses to her survey. She knows that in these days it's hard to get people to respond to a survey, and she's worried that she may lose a great deal of her sample that way, so she has an idea. Why not send out more surveys than she needs, so that she gets 1,000 surveys back?

Jenny looks at several survey results in the newspapers, magazines, and on the Internet, and she finds that the response rate (the percentage of people who actually responded to the surveys) is typically around 25 percent. (In terms of the real world, I'm being generous with this number, believe it or not. But think about it: How many surveys have you thrown away lately? Don't worry, I'm guilty of it too.) So, Jenny does the math and figures that if she sends out 4,000 surveys and gets 25 percent of them back, she has the 1,000 surveys she needs to do her analysis, answer her question, and have that small margin of error of ±3 percent.

Jenny conducts her survey, and just like clockwork, out of the 4,000 surveys she sends out, 1,000 come back. She goes ahead with her analysis and finds that 400 of those people reported cheating on their taxes (40 percent). She adds her margin of error, and reports, "Based on my survey data, 40 percent of Americans cheat on their taxes, ±3 percentage points."

Now hold the phone, Jenny. She only knows what those 1,000 people who returned the survey said. She has no idea what the other 3,000 people said. And here's the kicker: Whether or not someone responds to a survey is often related to the reason the survey is being done. It's not a random thing. Those nonrespondents (people who don't respond to a survey) carry a lot of weight in terms of what they're not taking time to tell you.

For the sake of argument, suppose that 2,000 of the people who originally received the survey were uncomfortable with the question because they *do* cheat on their taxes; they just didn't want anyone to know about it, so they threw the survey in the trash. Suppose that the other 1,000 people don't cheat on their taxes, so they didn't think it was an issue and didn't return the survey. If these two scenarios were true, the results would look like this:

Cheaters = 400 (respondents) + 2,000 (nonrespondents) = 2,400

These results raise the total percentage of cheaters to 2,400 divided by 4,000 = 60 percent. That's a huge difference!

You could go completely the other way with the 3,000 nonrespondents. You can suppose that none of them cheat, but they just didn't take time to say so. If you knew this info, you would get 600 (respondents) + 3,000 (nonrespondents) = 3,600 noncheaters. Out of 4,000 surveyed, this would mean 90 percent didn't cheat, and only 10 percent did. The truth is likely to be somewhere between the two examples I just gave you, but nonrespondents make it too hard to tell.

And the worst part is that the formulas Jenny uses for margin of error don't know that the information she put into them is based on biased data, so her reported 3 percent margin of error is wrong. The formulas happily crank out results no matter what. It's up to you to make sure that you put good, clean info into the formulas.

Getting 1,000 results when you send out 4,000 surveys is nowhere near as good as getting 1,000 results when sending out 1,000 surveys (or even 100 results from 100 surveys). Plan your survey based on how much follow-up you can do with people to get the job done, and if it takes a smaller sample size, so be it. At least the results have a better chance of being statistically on target.

# Of Course the Results Apply to the General Population

Making conclusions about a much broader population than your sample actually represents is one of the biggest no-no's in statistics. This kind of problem is called *generalization*, and it occurs more often than you may think. People want their results instantly; they don't want to wait for them, so well-planned surveys and experiments take a back seat to instant Web surveys and convenience samples.

For example, a researcher wants to know how cable news channels have influenced the way Americans get their news. He also happens to be a statistics professor at a large research institution and has 1,000 students in his class. He decides that instead of taking a random sample of Americans, which would be difficult, time-consuming, and expensive, he'll just put a question on his final exam to get his students' answers. His data analysis shows him that only 5 percent of his students read the newspaper and/or watch network news programs; the rest watch cable news. For his class, the ratio of students who exclusively watch cable news compared to those students who don't is 20 to 1. The professor reports this and sends out a press release about it. The cable news channels pick up on it and the next day are reporting, "Americans choose cable news channels over newspapers and network news by a 20 to 1 margin!"

Do you see what's wrong with this picture? The professor's conclusions go way beyond his study, which is wrong. He used the students in his statistics class to obtain the data that serves as the basis for his entire report and the resulting headline. Yet the professor reports that these results are true for all Americans. I think it's safe to say that a sample of 1,000 college students taking a statistics class at the same time at the same college doesn't represent a cross section of America.

If the professor wants to make conclusions in the end about America, he has to select a random sample of Americans to take his survey. If he uses 1,000 students from his class, his conclusions can only be made about that class and no one else.

To avoid or detect generalization, identify the population that you're intending to make conclusions about and make sure the sample you selected represents that population. If the sample represents a smaller group within that population, you have to downsize the scope of your conclusions also.

## Deciding Just to Leave It Out

It seems easier sometimes to just leave out information. I see this all too often when I read articles and reports based on statistics. But, this error isn't the fault of only one person or group. The guilty parties can include

✔ **The producers:** Some researchers may leave statistical details out of their reports for a variety of reasons, including time and space constraints. After all, you can't write about every element of the experiment from beginning to end. However, other items they leave out may be indicative of a bigger problem. For example, reports often say very little about how they collected the data or chose the sample. Or they may discuss the results of a survey but not show the actual questions they asked. Ten out of 100 people may have dropped out of an experiment, and the researchers don't tell you why. All these items are important to know before making a decision about the credibility of someone's results.

Another way in which some data analysts leave information out is by removing data that doesn't fit the intended model (in other words, "fudging" the data). Suppose a researcher records the amount of time spent surfing the Internet and relates it to age. He fits a nice line to his data indicating that younger people surf the Internet much more than older people and that surf time decreases as age increases. All is good except for Claude the outlier who's 80 years old and surfs the Internet day and night, leading his own bingo chat rooms and everything. What to do with Claude? If not for him, the relationship looks beautiful on the graph; what harm would it do to remove him? After all, he's only one person, right?

No way. Everything is wrong with this idea. Removing undesired data points from a data set is not only very wrong but also very risky. The only time it's okay to remove an observation from a data set is if you're certain beyond doubt that the observation is just plain wrong. For example, someone writes on a survey that she spends 30 hours a day surfing the Internet or that her IQ is 2,200.

✔ **The communicators:** When reporting statistical results, the media leave out important information all the time, which is often due to space limitations and tight deadlines. However, part of it is a result of the current, fast-paced society that feeds itself on sound bites. The best example is survey results in which the margin of error isn't communicated. You can't judge the precision of the results without it.

✔ **The consumers:** The general public also plays a role in the leave-things-out mindset. People hear a news story and instantly believe it to be true, ignoring any chance for error or bias in the results. For example, you need to make a decision about what car to buy, and you ask your neighbors and friends rather than examine the research and the resulting meticulous, comprehensive ratings. At one time or another everyone neglects to ask questions as much as they should, which indirectly feeds the entire problem.

In the chain of statistical information, the producers (researchers) need to be comprehensive and forthcoming about the process they conducted and the results they got. The communicators of that information (the media) need to critically evaluate the accuracy of the information they're getting and report it fairly. The consumers of statistical information (the rest of us) need to stop taking results for granted and to rely on credible sources of statistical studies and analyses to help make important life decisions.

In the end, if a data set looks too good, it probably is. If the model fits too perfectly, be suspicious. If it fits exactly right, run and don't look back! Sometimes what's left out speaks much louder than what's put in.

# Chapter 22

# Ten Ways to Get Ahead by Knowing Statistics

. . . . . . . . . . . . . . . . . . . . . . . . . . . . . . . . . . . . . . . . . . . . . . . . . . . . . . . . . . . .

*In This Chapter*

▶ Knowing what information to look for

▶ Being skeptical and confident

▶ Piecing together the statistics puzzle and checking your answers

▶ Knowing how best to present your findings

. . . . . . . . . . . . . . . . . . . . . . . . . . . . . . . . . . . . . . . . . . . . . . . . . . . . . . . . . . . .

*O*ne of my personal goals of teaching statistics is to help people get very good at being able to say "Wait a minute!" and stop a wrong analysis or a misleading graph in its tracks. I also want to help them become the stats gurus in their workplaces — those people who aren't afraid to work with statistics and do so correctly and confidently (and to also know when to consult a professional statistician). This chapter arms you with ten ways of trusting your statistics instincts and increasing your professional value through your understanding of the critical world of stats.

## Asking the Right Questions

Every study, every experiment, and every survey is done because someone had a question they wanted answered. For example "How long should this warranty last?"; "What's the chance of me developing complications during surgery?"; "What does the American public think about banning public smoking?" Only after a clear question has been defined can proper data collection begin.

Suppose a restaurant owner tells me that he wants to conduct a survey to learn more about the clientele at his restaurant. We talk about various variables to look at, including the number of people in the party, how often they've been there before, the type of food ordered, the amount they pay, how long they stay, and so on. After we collect some data and go over the results, he suddenly has a major realization: What he really wants to do is compare the clientele of his lunch crowd to his dinner crowd. Does the dinner crowd spend more money? Are they older? Do they stay longer? But sadly, he can't answer any of those questions because he didn't mention collecting data on whether the customers were there for lunch or dinner.

What happened here is a common mistake. The restaurant owner said he "just wanted to study" his clientele; he never mentioned comparisons because he hadn't thought that far ahead. If he had thought about it, he would have realized the real question was, "How does my lunch clientele compare with my dinner clientele?" Then including a question on whether diners were there for lunch or dinner would have been a no-brainer. Always ask the right questions to get the answers you need.

Testing the waters a bit before plunging into a full-blown study can be very helpful. One way to do this is to conduct what researchers call a pilot study. A *pilot study* is a small exploratory study that you use as a testing ground for the real thing. For example, you design a survey and try it out on a small group to see if they find any confusing questions, redundancies, spelling errors, and so on. Pilot studies are a quick and inexpensive way to help ensure that all goes well when the actual study takes place.

# *Being Skeptical*

Being statistically skeptical is a good thing (within reason). Some folks have given up on statistics, thinking that people can say anything they want if they manipulate the data enough. So those who have a healthy degree of skepticism can get ahead of the game.

Colorful charts and graphs can catch your eye, especially if they have neat little captions, and long and detailed professional reports may show you more information than you want to know, all laid out in neat tables, page after page. What's most important, however, is not how nice-looking the information is, or how professionally sound or scientific it looks. What's most important is what's happened behind the scenes, statistically speaking, in order to produce results that are correct, fair, and clear.

Many folks know only enough statistics to be dangerous. And many reported results are incorrect, either by mistake or by design (unfortunately). It's better to be skeptical than sorry!

Here's how to put your skepticism to good use:

- ✔ Get a copy of survey questions asked. If the questions are misleading, the survey results aren't credible.

- ✔ Find out about the data-collection process. When was the survey conducted? Who was selected to participate? How was the information collected? Surveys conducted on the Internet and those based on call-in polls are almost always biased, and their results should be thrown out the window.

✔ Find out about the response rate of the survey. How many people were initially contacted? How many responded? If many were contacted and few responded, the results are almost certainly biased because survey respondents typically have stronger feelings than those who don't respond.

# Collecting and Analyzing Data Correctly

On one hand it's very important to think very critically and even be skeptical at times about statistical results that you come across in everyday life and in the workplace. You should always ask questions before you deem the results to be credible.

On the other hand, it's also very important to remember that others are thinking critically about your results also, and you need to avoid the skepticism that you see others receiving. To avoid potential potshots that may be taken at your results, you need to make sure you've done everything right.

Because you're reading this book, by now you should have many tools to help you do data collection and analysis correctly. In each chapter you hear the same theme song: Using the wrong analysis or too many analyses isn't good. For each type of analysis I present, you see how to check to make sure that particular analysis is okay to use with the data you have. Chapters 1 and 2 serve as a reference to which techniques are needed, and where to find them in the book.

Ninety percent of the work involved in a statistical analysis happens before the data even goes into the computer. Here's a basic to-do list of what to check for:

✔ Design your survey, your experiment, or your study to avoid bias and ensure precision.

✔ Make sure you conduct the study at the right time and select a truly random sample of individuals to participate.

✔ Follow through with those participants to make sure your final results have a high response rate.

This to-do list can be challenging, but in the end, you'll be safe in knowing that your results will stand up to criticism because you did everything right.

# Calling for Help

One of the toughest things for nonstatisticians to get is that they don't have to do all the statistics themselves. In fact, it's not a good way to go in many instances. The six most important words for any nonstatistician are "Know when to consult a statistician." Know when to ask for help. And the best time to ask for help is *before* you collect any data.

So how can you tell when you're in a bit over your head and you need someone to throw you a statistical lifeline? Here are some examples to help give you an idea of when to call:

✔ If your boss wants no less than a 100-page marketing results report on her desk by Monday and you haven't collected data point #1, CALL.

✔ If you're reading *Cosmopolitan* on your lunch break and you want to analyze how you and your friends came out on the "Who's the Gossip Queen in Your Workplace?" quiz, DON'T CALL.

✔ If the list of questions on your survey becomes longer than you are tall, CALL.

✔ If you want to make a bar graph of how many of your Facebook friends are fans of the 1970s, 80s or 90s, DON'T CALL.

✔ If a scatterplot of your data looks like it should be in a Rorschach inkblot test, CALL (and fast!).

✔ If you want to know the odds that someone you haven't seen since high school is on the same plane to Africa as you are, DON'T CALL.

✔ If you have an important job to do that involves statistics and you are unsure of how to begin or how you'll analyze your data once you get it, CALL. The sooner you call, the more the professionals can do to help you look good!

# Retracing Someone Else's Steps

At some point in your work life, you'll take out a report, read it, and you'll have a question about it. You'll go to find the data, and after much searching, you'll bring up a spreadsheet with rows upon rows and columns upon columns of numbers and characters. Your eyes will glaze over; you'll have no idea what you're looking at. You'll tell yourself not to panic and just to find the person who entered all the data and find out what's going on.

But then comes the bad news. Someone named Bob collected the data and entered it a couple of years ago, and Bob doesn't work for the company anymore. Now what do you do? More than likely, you'll have to ditch the data and the report, start all over again from scratch, and lose valuable time and money in the process.

How could this disaster have been prevented? All the following issues should have been addressed before Bob passed on his report:

- ✔ The report should include a couple of paragraphs telling how and when the data were collected, the names of the variables in the data set, where they're located in the spreadsheet, and what their labels are.

- ✔ The report should include a note about missing data. Missing data are sometimes left blank, but they also can be written as a negative sign (–) or a decimal point. (Using zeroes for missing data is a special no-no because they will be confused with actual data values that equal zero.)

- ✔ The rows of the data set should be defined. For example, does each row represent one person? Do they have ID numbers?

Unfortunately, many people create statistical reports and then disappear without a trace, leaving behind a data mess that often can't be fixed. It's common courtesy to take steps to avoid leaving other people in the lurch, the way Bob did. Always leave a trail for the next person to pick up where you left off. And on the flipside, always ask for the explanation and background of a data set before using it.

# Putting the Pieces Together

You should never jump right into an analysis expecting to get a one-number answer and then walk away. Statistics requires much more work! You should view every statistical problem as a puzzle whose pieces need to be put together before you can see the big picture of what's really going on.

For example, suppose a coffee vendor wants to predict how much coffee she should have ready for an upcoming football game in Buffalo, New York. Her first step is to think about what variables may be related to coffee sales. Variables may be cost of the coffee, ease of carrying it, seat location (who wants to walk a mile for a cup of coffee?), and age of fans. The vendor also suspects that temperature at the game may affect coffee sales, with low temps translating into higher sales.

The vendor collects data on all these variables and explores the relationships. She finds that coffee sales and temperature are somewhat related. But is there more to this story than temperature?

To find out, the vendor compares coffee sales for two games with the same temperature and notices a big difference. Looking deeper, she notices one game was on a Sunday and one was on a Monday. Attendance was higher on Monday, and that game had more adults in attendance. By analyzing the data, the vendor found that temperature is related to coffee sales, but so is attendance, day of the week that the game was played, and age of the fans. Knowing this information, the vendor was able to predict coffee sales more accurately with a lower chance of running out of coffee or wasting it. This example illustrates that putting the pieces together to keep an eye on the big picture can really pay off.

# Checking Your Answers

After your data have been analyzed and you get your results, you need to take one more step before running giddily to your boss saying, "Look at this!" You have to be sure that you have the right answers.

By right answers, I don't mean that you need to have the results that your boss wants to hear (although that would be great, of course). Rather, you need to make sure your data analysis and calculations are correct and don't leave you high and dry when the questions start to come. Follow these basic steps:

1. **Double-check that you entered the data correctly, and weed out numbers that obviously make no sense (such as someone saying that he's 200 years old, or that he sold 500 billion light bulbs at his store last year).**

   Mistakes influence the data and the results, so catch them before it's too late.

2. **Make sure that your numbers add up when they're supposed to.**

   For example, if you collected data on number of employees for 100 companies and you don't list enough number groups to cover them all, you're in trouble! Also be on the lookout for data on an individual that have been entered twice. This error shows up if you sort the data by rows.

3. **If you intend to make conclusions, make sure you're using the right numbers to do so.**

   If you want to talk about how crime has increased in your area over the last five years, showing the number of crimes on a graph is incorrect. The number of crimes can increase simply because the population size increases. For correct statistical conclusions about crime, you need to report the crime rate, which is the number of crimes per person (per capita), or the number of crimes per 100,000 people. Just take the number of crimes divided by the population size, or divided by 100,000, respectively. This approach takes population size out of it.

# Explaining the Output

Computers certainly play a major role in the process of collecting and analyzing data. Many different statistical software packages exist, including MS Excel,

Minitab, SAS, SPSS, and a host of others. Each type has its own style of printing out results. Understanding how to read, interpret, and explain computer output is an art and a science that not everyone possesses. With your statistical knowledge, though, you can be that person!

Computer output is the raw form of the results of doing any statistical summary or analysis. It can be graphs, charts, scatterplots, tables, regression analysis results, an analysis of variance table, or a set of descriptive statistics. Often the analysis is labeled by the computer; for example, ANOVA indicates an Analysis of Variance has been conducted (see Chapter 9). Graphs, charts, and tables, however, require the user to tell the computer what labels, titles, or legends (if any) to include so that the audience can quickly understand what's what.

Interpreting computer output involves sifting through what can seem like an intimidating amount of information. The trick is knowing exactly what results you want and where the computer places them on the output. For example, in the output from a regression analysis, you find the equation of the regression line by looking in the COEF column of the output (see Chapter 4).

Most of the time there's information on a computer output that you don't need; sometimes there's also information that you don't understand. Before skipping everything, you may want to consult a statistician to make sure you aren't missing an important step, such as examining the correlation coefficient before doing a regression analysis (see Chapter 4).

Explaining what you found on computer output involves sizing up the knowledge of the person you're talking to, too. If you're writing an executive report, you don't need to explain every little thing you did and why; just use the parts of the output that tell the bottom line, and explain how it affects the company. If you're helping a colleague understand results, give him some reference information about the analyses (you can use this book). For example, you may discuss what a histogram does in general before you talk about the results of your histogram.

Most importantly, make sure the analysis is correct before explaining it to anyone. Sometimes it's easy when analyzing data to click on the wrong variable or to highlight the wrong column of data, which makes the analysis totally wrong.

# Making Convincing Recommendations

As one moves up the corporate ladder, the less time she has to read reports and carefully examine statistics. The best data analysis in the world won't mean squat if you can't communicate your results to someone who doesn't have the time or interest to get into the nitty-gritty. In this data-driven world, statistics can play a major role in good decision-making. The ability to use statistics to make an effective argument, make a strong case, or give solid recommendations is critical.

Put yourself in the following situation. You've done the work, you've collected marketing and sales data, and you've done the analyses and processed the results. Based on your study of product placement for your Sugar Surge Pop, you determined that the best strategy for placing this product on grocery store shelves is to put it in the checkout aisle at eye level so children can see it. (You never see nail clippers or hand sanitizers on the kid's eye-level shelves in the impulse aisle, do you?) Word is that your boss favors putting this product in the candy aisle of the store. (Of course she has no data to support this, just her own experience people-watching in the candy aisle.) How do you convince her to follow your recommendation?

Probably the worst thing you can do is go into her office with a 100-page report loaded with everything from soup to nuts. Loads of complex information may impress your mom, but it won't impress your boss. Save that report in case she asks for it (or in case you need a doorstop). What you need is a short, succinct, and straightforward report that makes the point. Here's how to craft it:

1. **Start out with a statement of the problem.**

   "We want to determine which location has the most sales of the Sugar Surge Pop."

2. **Briefly outline your data-collection process.**

   "We chose 50 stores at random and placed the product in the checkout aisle at 25 stores and in the candy aisle in the other 25. We controlled for other factors such as number of products placed."

3. **Describe what data you collected.**

   "We tracked the sales of the product over a six-month period, calculating weekly sales totals for each store." At this point, show your charts and graphs of the sales over time for the two groups.

4. **Tell briefly how you analyzed the data, but spend the most time on your findings.**

   Don't show the output — your boss doesn't need to see that. You know the expression, "Never let them see you sweat"? That's important here. What you do want to say is, "We did a statistical analysis comparing average sales at these locations, and we found sales in the checkout aisle to be significantly higher than sales in the candy aisle." You can quantify the difference with percentages.

Follow up with your recommendation for product placement at kids' eye-level in the checkout aisle, being sure to answer the original question you started with in Step 1. Then the most important tip is to let your boss think the optimal placement was all her idea!

# Establishing Yourself as the Statistics Go-To Guy or Gal

Nothing is more valuable than someone in the workplace who isn't afraid to do statistics. Every office has one person with the courage to calculate, the confidence to make confidence intervals, the willingness to wrestle with the output, and the gumption to graph. This person is eventually everyone's friend and the first person to get to know when starting a new job.

What are the perks of being the statistics go-to guy or gal? It's the glory of knowing that you're saving the day, taking one for the team, and standing tall in the face of disaster. Your colleagues will say, "I owe you one," and you can take them up on it.

But seriously, the statistics go-to guy has a more secure job because his boss knows that statistics is a staple of the workplace, and having someone to jump in when needed is invaluable.

Statistics and statistical analyses can be intimidating, yet they're critical for the workplace. In most any career these days, you need to know how to select samples, write surveys, set up a process for collecting the data, and analyze it.

# Chapter 23

# Ten Cool Jobs That Use Statistics

## In This Chapter

▶ Using statistics in a wide range of jobs (yours might be next!)

▶ Tracking data from birds to sports to crime

▶ Taking stats into the professional world of medicine, law, and the stock market

This book is meant to be a guide for folks who need to know statistics for their everyday life (which is all of us) as well as in their workplaces (which is most of us). If I think about it long enough, I can come up with some use of statistics in almost every job out there (except maybe a psychic advisor).

This chapter features a cross section of ten careers that all involve statistics in some way, shape, or form. You may be surprised at how often statistics turn up in the workplace! So don't burn this book when your stats course is over; you may find it to be useful in your job hunt or your job. (My accountant has a copy of this book on his shelf — what does that say? As long as he doesn't have a copy of *Accounting For Dummies* alongside it, I guess we'll be okay.)

One of my personal goals as a teacher of statistics is to help my students become the go-to folks in the workplace. You know, that person with a background in statistics who knows what she's doing, and when it's crunch time, she can do the statistics needed correctly and confidently. With experience and help from this book, you too can become that person. You'll become a hero, and your job will be all the more secure for it.

## Pollster

Pollsters collect information on people from populations they're interested in. Some of the big names in professional polling include the Gallup Organization, the Associated Press (AP), Zogby International, Harris

Interactive, and the Pew Research Center. Major news organizations such as NBC, CBS, and CNN also conduct polls, as do many other agencies and organizations.

The purposes of polls vary from the medical field trying to determine what's causing obesity, to political pollsters who want to keep up with the daily pulse of American opinion, to surveys that provide feedback and ideas to corporations.

Knowledge of statistics is considered golden in the polling industry, because jobs can include designing surveys; selecting a proper sample of participants; carrying out a survey to collect data; and then recording, analyzing, and presenting the results.

All these tasks are part of statistics — the art and science of collecting and making sense of data. But don't just take my word for it; here's a quote from a job posting for the Gallup Organization for a Research Analyst. I have to say it totally screams STATISTICS!

> If you have a strong academic record in the social sciences or economics, a familiarity with quantitative and categorical research and statistical tools in Market Research/Survey Research or Consulting, enjoy pulling together research data and abstract concepts to tell a meaningful story, while continually learning — this is the place to manage processes and projects that deliver perfect completion of client engagements.

And here's something you don't see every day. I found a job posting for a polling analyst with roughly the same requirements but a very different work setting. The job was for a company that provides security and intelligence for the United States government. For this job you need federal security clearance. You never know where your statistical background is going to take you!

Other positions related to polling that I've seen listed are quantitative research specialist and public polling research analyst.

A great Web site for finding out more about what pollsters do and what their work looks like is, appropriately, www.pollster.com.

# Ornithologist (Bird Watcher)

Everyone watches birds on occasion. I gladly admit to being a semi-serious birdwatcher, always trekking to Magee Marsh on Lake Erie in May for International Migratory Bird Day. But have you ever thought about getting paid to watch birds and other wildlife? Today's ever-increasing awareness of the environment includes a great deal of focus on identifying, studying, and protecting wildlife of all kinds.

Ornithology is the science of bird study. Ornithologists are always collecting data and finding and studying statistics on birds — often on a certain type of bird and its behavior. Some examples of common bird statistics include:

✔ Bird counts (number of birds per square unit of space on a particular day)

✔ Nest locations and territory maps

✔ Number of eggs laid and hatched

✔ Food preferences and foraging techniques

✔ Behaviors caught on tape and quantified

You can tap into a Web site totally dedicated to jobs that need birdwatchers and wildlife watchers and the use of their statistical skills. Here's one of the job postings supplied by the Ornithological Society:

> FLAMMULATED OWL SURVEY TECHNICIANS (2) needed for Idaho Bird Observatory study of Flammulated Owls and other forest birds in Idaho (approx. 2.5 months). Duties will consist mainly of standardized surveys and data entry. Qualifications of applicants should include: 1) good eyesight and hearing, 2) proficiency with standardized survey procedures, 3) ability to identify Western birds by sight and sound, and 4) willingness to give your all. Candidates must be physically fit and undaunted by the prospects of heat, humidity, bugs, and mud. (Indeed!)

With more experience and knowledge, you can eventually become a research wildlife biologist for the U.S. Department of the Interior U.S. Geological Survey. A job description for this position just found today actually requires 15 credit hours of statistics, proving that the government is onto the whole statistics thing.

# Sportscaster or Sportswriter

Every good sportscaster or sportswriter knows that you're nothing without good juicy statistics that no one else knows. You do your homework by studying training camps and pouring over printouts, spreadsheets, and historical data. You read newspapers, look at record books, and watch film. There's no shortage of data out there, and your audience can't get enough.

Sports fans are statistics addicts! (Being an Ohio State Buckeye, I'm as rabid as the rest of 'em.) Here's just a sampling of the statistics recorded and presented in my favorite sport of college football:

- Points scored
- Points against
- Rushing yards
- Receiving yards
- Passing yards
- Interceptions
- Fumbles
- Punt and kick returns
- Number and distance of field goals attempted and made
- Kicker's career longest
- Number of first downs
- Third down conversions
- Fourth down conversions
- Penalties
- All-purpose yards
- Total offense
- Total defense
- Number of sacks

- Rushing defense
- Passing defense
- Turnover margins
- Passing efficiency
- Scoring offense
- Scoring defense
- Scoring by special teams
- Coaches' Poll standings
- AP Poll standings
- BCS standings (that's another book for another day!)
- The most 12+ win seasons
- Coaching records
- Single game high scores
- Game attendance
- Toughness of schedule
- Winning and losing streaks
- Coin toss winners

It's obvious that we need a new saying: "Those who play sports play. Those who watch sports do statistics."

# Journalist

Journalists of all types at some point or another have to work with data. They have a hard row to hoe, because the data comes to them from an infinite number of possible avenues and channels on an infinite number of topics, and they need to make sense of it, pick out what they feel is most important, boil it down, present the results, and write a story around it, all under a very strict deadline (sometimes only a few hours). That's a big job!

As a consumer of the media, I see many good uses of statistics that are clear, correct, and make interesting and important points. However, I also see many incorrect and misleading statistics in the media, and I cringe every time.

Some of the most common problems include making simple math errors, reporting percentages above 100 percent, assuming cause and effect relationships that aren't proven, using misleading graphs, and leaving out information (such as the number of people surveyed, the rate of nonresponse, and the margin of error). But for me the biggest problem is reading a headline that sounds catchy, eye-opening, perhaps even shocking, only to find that it's not corroborated by the statistics in the article.

Having a couple of solid statistics courses under your belt puts you way ahead in that job interview for a journalist position. The statisticians around the world are counting on you to get out there on your white horse and do things right! (Don't forget to take this book with you in your saddle bag!)

To recognize the importance and appreciation of the difficult task that journalists have in using and reporting with statistics, the Royal Statistical Society has established an Award for Statistical Excellence in Journalism. Following is the description for the award, and I couldn't agree with it more!

> The Royal Statistical Society wishes to encourage excellence in journalists' use of statistics to question, analyze, and investigate the issues that affect society at large. Journalistic excellence in statistics helps to hold decision makers in all sectors to account — through accessible communication of complex information, highlighting of success, and exposure of important missing information.

# Crime Fighter

Crime statistics help the nation's crime fighters, such as police officers, determine which kinds of crimes occur where, how often, to whom, and by whom. Crime statistics for the entire nation are compiled and analyzed by the U.S. Department of Justice. National Crime Victimization Surveys are also conducted to help understand trends in crimes of various types.

Police officers record every incident they're involved in, forming large databases that city, county, and state officials can use to determine the number of police officers needed and which areas to focus most heavily on, and also to make changes in the policies and procedures of their police departments. The FBI can also use these huge databases to track criminals, look for patterns in crime types and occurrences, and keep track of overall trends in the number of crimes as well as the type of crimes that occur over time.

People looking for a new home or a new school can consult freely available information on crime statistics, and politicians use it to show that crime is going up or down, that money should or should not be spent on more police officers, and how safe their city or state has become with them in office.

Here's an overview of how the U.S. Department of Justice Web site uses data and statistics to help fight crime:

> All states have established a criminal record repository which maintains criminal records and identification data and responds to law enforcement inquiries and inquiries for other purposes such as background checks and national security. Criminal records include data provided by all components of the criminal justice system: law enforcement, prosecution, courts, and corrections. . . . Records developed for statistical purposes describe and classify each criminal incident and include data on offender characteristics, relationships between the offender and the victim, and offense impact. Statistical data are extracted from operational records using uniform criteria for classification and collection. Detailed statistical data permit localities to identify problem areas and to allocate manpower and limited financial resources in an efficient and effective manner.

# Medical Professional

People who work in the medical field depend on statistics to do research and find new cures, therapies, medicines, and procedures to increase the health and well-being of all people. Medical researchers conduct clinical trials to measure every conceivable side effect of every drug that goes through the approval process. Comparative studies are done all the time to determine what factors influence weight, height, intelligence level, and ability to survive a certain disease. Statistics are a lifeline to being more confident that what works for a sample of individuals will also work for the population for which it was meant.

In the medical profession, the use of statistics starts as soon as a patient's name is called in the waiting room. Suppose you're a nurse. The first thing you ask the patient to do is to "go ahead and step on the scale please" (eight dreaded words heard in doctors' offices round the world). From there, you check his vital signs (also known as vital statistics): temperature, blood pressure, pulse rate, and sometimes, respiration (breathing) rate. You record his numbers in your computer and compare them to what has been determined to be the normal range. Setting the normal ranges involves statistics as well, through analyzing historical data and medical research.

Like other kinds of statistics, the way in which a person's vital statistics are collected can greatly affect the results. For example, different types of blood pressure instruments exist, and some are better than others. And we all know the scale in the doctor's office is always at least 10 pounds over!

# Marketing Executive

Marketing is critical to any product's success. That's why companies spend millions of dollars for 30-second commercials during the Super Bowl. Researching who will buy your product, where, when, and for how much is a job that includes lots of statistics.

Some data is what statisticians call *quantitative,* such as surveys of existing, past, and potential customers, sales information and trends, economic and demographic information, and data regarding competitors. Other data is *categorical,* including in-depth one-on-one interviews and focus groups to get a general picture of what consumers think, how they feel, what ideas they have, and what additional information they need about your product. (See Chapter 2 or your Stats I textbook for a review of quantitative and categorical data.)

Consider the example of Mars, Incorporated, which makes M&M'S candy. How has the company's product become a national icon for kids of all ages? The secrets have to be Mars' innate ability to change with the times and knowledge of what its customers want.

Statistics plays a huge role in this success through collecting data on sales, but most importantly, through getting direct feedback from customers using interviews, focus groups, and surveys. (I can't imagine the strain of trying out M&M'S and talking about what I think. "Oh wait, I need another sample before I can give you a good answer.") By analyzing this data, Mars is able to determine some of the most important and intricate details that spell success and longevity of any company.

For example, in 1995 Mars conducted a nationwide survey asking customers to choose the newest M&M'S color. That's when blue came on the scene. In 2002, the survey went global and purple became the new addition to the M&M'S palette. The Mars company uses statistics to find ways to be innovative in making new colors, flavors, styles, and even allowing for customized M&M'S, yet it still retains the classic essence of the M&M'S that started it all.

# Lawyer

You've no doubt heard the phrase "beyond a reasonable doubt." It's the code that jurors use to make a decision of guilty or not guilty. The field of statistics plays a major role in determining whether laws are being followed or broken, whether a defendant is guilty or innocent, and whether laws need to be created or changed. Statistical information is very powerful evidence.

Lawsuits are often settled on the basis of statistical evidence collected in multiple situations over the course of years. Statistics also allow lawmakers to break down information in order to propose new laws. For example, using statistics to show that the first two hours are the most critical in terms of finding a missing child led to the Amber Alerts broadcast on TV and radio and posted on highways when children go missing.

Prosecutors and defense attorneys often use probability and statistics to help make their cases, too. They also have to make these statistics understandable to a jury. (Maybe copies of this book should be a requirement for sequestered juries!) Statisticians are often brought onto the legal team to help attorneys gather information, decipher the results, and use the data in a jury trial situation.

Attorneys may use correlation to show that certain variables have a linear relationship, such as skid distance and amount of a certain type of concrete in pavement, or the strength of a bridge beam related to the weight placed upon it.

Statistics also can help test claims. For example, suppose Shipping Company A claims its packages are delivered on average two hours faster than Company B. If a random sample of packages takes longer than two days to arrive and the difference is large enough to have strong evidence against Company A's claim, it could get in trouble for false advertising. Of course, any decision based on statistics can be wrong, just by chance. In this case, if the random sample of packages just happened to take longer than usual and doesn't represent the typical average delivery time, Shipping Company A can fight back, saying they were unjustly accused of false advertising. It's a tight line to walk, and statisticians try their best to set up procedures to help the real truth come to light.

# Stock Broker

Professional gamblers are folks who make a living by gambling. They know the ropes, they've been all over town, and they're good at what they do, which is basically play games that they feel they can win with skill as well as luck. There are different styles of professional gamblers. One type goes for casino games, such as poker and blackjack.

Another type of gambler is the stock broker who dresses up in a suit, has power lunches, and constantly changes his decisions throughout the day depending on the current state of the system. Stock brokers make predictions and decisions on buying and selling stocks on a minute-to-minute basis.

Statistics plays a critical role in a successful stock broker's decision making. To gain an edge, brokers use sophisticated data collection and analysis software, financial models, and numbers from all over the world, and they have to be able to analyze the information, interpret it, and act on it quickly.

However, one has to also keep an eye out for stock brokers who either ignore the real statistics or come up with their own statistics to give their clients an unrealistic view of what's going on with their money. There have even been instances where brokers have stolen their clients' money right out from under their noses to the tune of millions and even billions of dollars. These situations aren't common, but they make a huge impact on the confidence of investors and ultimately even on the state of the economy. Collecting and using verifiable data, and analyzing that data with legitimate techniques, should be the goal of all good stock brokers (and for everyone who uses statistics in the workplace).

# Appendix

# Reference Tables

• • • • • • • • • • • • • • • • • • • • • • • • • • • • • • • • • • • • • • • • •

*T*his appendix includes commonly used tables for five important distributions for Stats II: the *t*-distribution, the binomial distribution, the Chi-square distribution, the distribution for the rank sum test statistic, and the *F*-distribution.

## *t-Table*

Table A-1 shows right-tail probabilities for the *t*-distribution (refer to Chapter 3). To use Table A-1, you need four pieces of information from the problem you're working on:

- ✔ The sample size, *n*
- ✔ The mean of *x*, denoted μ
- ✔ The standard deviation of your data, *s*
- ✔ The value of *x* for which you want the right-tail probability

After you have this information, transform your value of *x* to a *t*-statistic (or *t*-value) by taking your value of *x*, subtracting the mean, and dividing by the standard error (see Chapter 3) by using the formula $t_{n-1} = \dfrac{\bar{x} - \mu}{\frac{s}{\sqrt{n}}}$.

Then look up this value of *t* on Table A-1 by finding the row corresponding to the degrees of freedom for the *t*-statistic (*n* – 1). Go across that row until you find two values between which your *t*-statistic falls. Then go to the top of those columns and find the probabilities there. The probability that *t* is beyond your value of *x* (the right-tail probability) is somewhere between these two probabilities. Note that the last row of the *t*-table shows df = ∞, which represents the values of the *Z*-distribution, because for large sample sizes *t* and *Z* are close.

---

| Table A-1 | | | | The t-Table | | | |

t-distribution showing area to the right

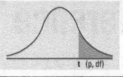

| df/p | 0.40 | 0.25 | 0.10 | 0.05 | 0.025 | 0.01 | 0.005 | 0.0005 |
|---|---|---|---|---|---|---|---|---|
| 1 | 0.324920 | 1.000000 | 3.077684 | 6.313752 | 12.70620 | 31.82052 | 63.65674 | 636.6192 |
| 2 | 0.288675 | 0.816497 | 1.885618 | 2.919986 | 4.30265 | 6.96456 | 9.92484 | 31.5991 |
| 3 | 0.276671 | 0.764892 | 1.637744 | 2.353363 | 3.18245 | 4.54070 | 5.84091 | 12.9240 |
| 4 | 0270722 | 0.740697 | 1.533206 | 2.131847 | 2.77645 | 3.74695 | 4.60409 | 8.6103 |
| 5 | 0.267181 | 0.726687 | 1.475884 | 2.015048 | 2.57058 | 3.36493 | 4.03214 | 6.8688 |
| 6 | 0.264835 | 0.717558 | 1.439756 | 1.943180 | 2.44691 | 3.14267 | 3.70743 | 5.9588 |
| 7 | 0.263167 | 0.711142 | 1.414924 | 1.894579 | 2.36462 | 2.99795 | 3.49948 | 5.4079 |
| 8 | 0.261921 | 0.706387 | 1.396815 | 1.859548 | 2.30600 | 2.89646 | 3.35539 | 5.0413 |
| 9 | 0.260955 | 0.702722 | 1.383029 | 1.833113 | 2.26216 | 2.82144 | 3.24984 | 4.7809 |
| 10 | 0260185 | 0.699812 | 1.372184 | 1.812461 | 2.22814 | 2.76377 | 3.16927 | 4.5869 |
| 11 | 0259556 | 0.697445 | 1.363430 | 1.795885 | 2.20099 | 2.71808 | 3.10581 | 4.4370 |
| 12 | 0259033 | 0.695483 | 1.356217 | 1.782288 | 2.17881 | 2.68100 | 3.05454 | 43178 |
| 13 | 0.258591 | 0.693829 | 1.350171 | 1.770933 | 2.16037 | 2.65031 | 3.01228 | 4.2208 |
| 14 | 0.258213 | 0.692417 | 1.345030 | 1.761310 | 2.14479 | 2.62449 | 2.97684 | 4.1405 |
| 15 | 0.257885 | 0.691197 | 1.340606 | 1.753050 | 2.13145 | 2.60248 | 2.94671 | 4.0728 |
| 16 | 0257599 | 0.690132 | 1.336757 | 1.745884 | 2.11991 | 2.58349 | 2.92078 | 4.0150 |
| 17 | 0.257347 | 0.689195 | 1.333379 | 1.739607 | 2.10982 | 2.56693 | 2.89823 | 3.9651 |
| 18 | 0.257123 | 0.688364 | 1.330391 | 1.734064 | 2.10092 | 2.55238 | 2.87844 | 3.9216 |
| 19 | 0.256923 | 0.687621 | 1.327728 | 1.729133 | 2.09302 | 2.53948 | 2.86093 | 3.8834 |
| 20 | 0.256743 | 0.686954 | 1.325341 | 1.724718 | 2.08596 | 2.52798 | 2.84534 | 3.8495 |
| 21 | 0.256580 | 0.686352 | 1.323188 | 1.720743 | 2.07961 | 2.51765 | 2.83136 | 3.8193 |
| 22 | 0256432 | 0.685805 | 1.321237 | 1.717144 | 2.07387 | 2.50832 | 2.81876 | 3.7921 |
| 23 | 0256297 | 0.685306 | 1.319460 | 1.713872 | 2.06866 | 2.49987 | 2.80734 | 3.7676 |
| 24 | 0.256173 | 0.684850 | 1.317836 | 1.710882 | 2.06390 | 2.49216 | 2.79694 | 3.7454 |
| 25 | 0.256060 | 0.684430 | 1.316345 | 1.708141 | 2.05954 | 2.48511 | 2.78744 | 3.7251 |
| 26 | 0.255955 | 0.684043 | 1.314972 | 1.705618 | 2.05553 | 2.47863 | 2.77871 | 3.7066 |
| 27 | 0.255858 | 0.683685 | 1.313703 | 1.703288 | 2.05183 | 2.47266 | 2.77068 | 3.6896 |
| 28 | 0.255768 | 0.683353 | 1.312527 | 1.701131 | 2.04841 | 2.46714 | 2.76326 | 3.6739 |
| 29 | 0.255684 | 0.683044 | 1.311434 | 1.699127 | 2.04523 | 2.46202 | 2.75639 | 3.6594 |
| 30 | 0.255605 | 0.682756 | 1.310415 | 1.697261 | 2.04227 | 2.45726 | 2.75000 | 3.6460 |
| ∞ | 0.253347 | 0.674490 | 1.281552 | 1.644854 | 1.95996 | 2.32635 | 2.57583 | 3.2905 |

# Binomial Table

Table A-2 shows probabilities for the binomial distribution (refer to Chapter 17). To use Table A-2, you need three pieces of information from the particular problem you're working on:

- The sample size, $n$
- The probability of success, $p$
- The value of $x$ for which you want the cumulative probability

Find the portion of Table A-2 that's devoted to your $n$, and look at the row for your $x$ and the column for your $p$. Intersect that row and column, and you can see the probability for $x$. To get the probability of being strictly less than, greater than, greater than or equal to, or between two values of $x$, you sum the appropriate values of Table A-2, using the steps found in Chapter 16.

## Table A-2          The Binomial Table

Numbers in the table represent the probabilities for values of x from 0 to $n$.

Binomial probabilities:

$$\binom{n}{x} p^x (1-p)^{n-x}$$

| n | x | 0.1 | 0.2 | 0.25 | 0.3 | 0.4 | 0.5 | 0.6 | 0.7 | 0.75 | 0.8 | 0.9 |
|---|---|-----|-----|------|-----|-----|-----|-----|-----|------|-----|-----|
| 1 | 0 | 0.900 | 0.800 | 0.750 | 0.700 | 0.600 | 0.500 | 0.400 | 0.300 | 0.250 | 0.200 | 0.100 |
|   | 1 | 0.100 | 0.200 | 0.250 | 0.300 | 0.400 | 0.500 | 0.600 | 0.700 | 0.750 | 0.800 | 0.900 |
| 2 | 0 | 0.810 | 0.640 | 0.563 | 0.490 | 0.360 | 0.250 | 0.160 | 0.090 | 0.063 | 0.040 | 0.010 |
|   | 1 | 0.180 | 0.320 | 0.375 | 0.420 | 0.480 | 0.500 | 0.480 | 0.420 | 0.375 | 0.320 | 0.180 |
|   | 2 | 0.010 | 0.040 | 0.063 | 0.090 | 0.160 | 0.250 | 0.360 | 0.490 | 0.563 | 0.640 | 0.810 |
| 3 | 0 | 0.729 | 0.512 | 0.422 | 0.343 | 0.216 | 0.125 | 0.064 | 0.027 | 0.016 | 0.008 | 0.001 |
|   | 1 | 0.243 | 0.384 | 0.422 | 0.441 | 0.432 | 0.375 | 0.288 | 0.189 | 0.141 | 0.096 | 0.027 |
|   | 2 | 0.027 | 0.096 | 0.141 | 0.189 | 0.288 | 0.375 | 0.432 | 0.441 | 0.422 | 0.384 | 0.243 |
|   | 3 | 0.001 | 0.008 | 0.016 | 0.027 | 0.064 | 0.125 | 0.216 | 0.343 | 0.422 | 0.512 | 0.729 |
| 4 | 0 | 0.656 | 0.410 | 0.316 | 0.240 | 0.130 | 0.063 | 0.026 | 0.008 | 0.004 | 0.002 | 0.000 |
|   | 1 | 0.292 | 0.410 | 0.422 | 0.412 | 0.346 | 0.250 | 0.154 | 0.076 | 0.047 | 0.026 | 0.004 |
|   | 2 | 0.049 | 0.154 | 0.211 | 0.265 | 0.346 | 0.375 | 0.346 | 0.265 | 0.211 | 0.154 | 0.049 |
|   | 3 | 0.004 | 0.026 | 0.047 | 0.076 | 0.154 | 0.250 | 0.346 | 0.412 | 0.422 | 0.410 | 0.292 |
|   | 4 | 0.000 | 0.002 | 0.004 | 0.008 | 0.026 | 0.063 | 0.130 | 0.240 | 0.316 | 0.410 | 0.656 |
| 5 | 0 | 0.590 | 0.328 | 0.237 | 0.168 | 0.078 | 0.031 | 0.010 | 0.002 | 0.001 | 0.000 | 0.000 |
|   | 1 | 0.328 | 0.410 | 0.396 | 0.360 | 0.259 | 0.156 | 0.077 | 0.028 | 0.015 | 0.006 | 0.000 |
|   | 2 | 0.073 | 0.205 | 0.264 | 0.309 | 0.346 | 0.312 | 0.230 | 0.132 | 0.088 | 0.051 | 0.008 |
|   | 3 | 0.008 | 0.051 | 0.088 | 0.132 | 0.230 | 0.312 | 0.346 | 0.309 | 0.264 | 0.205 | 0.073 |
|   | 4 | 0.000 | 0.006 | 0.015 | 0.028 | 0.077 | 0.156 | 0.259 | 0.360 | 0.396 | 0.410 | 0.328 |
|   | 5 | 0.000 | 0.000 | 0.001 | 0.002 | 0.010 | 0.031 | 0.078 | 0.168 | 0.237 | 0.328 | 0.590 |
| 6 | 0 | 0.531 | 0.262 | 0.178 | 0.118 | 0.047 | 0.016 | 0.004 | 0.001 | 0.000 | 0.000 | 0.000 |
|   | 1 | 0.354 | 0.393 | 0.356 | 0.303 | 0.187 | 0.094 | 0.037 | 0.010 | 0.004 | 0.002 | 0.000 |
|   | 2 | 0.098 | 0.246 | 0.297 | 0.324 | 0.311 | 0.234 | 0.138 | 0.060 | 0.033 | 0.015 | 0.001 |
|   | 3 | 0.015 | 0.082 | 0.132 | 0.185 | 0.276 | 0.313 | 0.276 | 0.185 | 0.132 | 0.082 | 0.015 |
|   | 4 | 0.001 | 0.015 | 0.033 | 0.060 | 0.138 | 0.234 | 0.311 | 0.324 | 0.297 | 0.246 | 0.098 |
|   | 5 | 0.000 | 0.002 | 0.004 | 0.010 | 0.037 | 0.094 | 0.187 | 0.303 | 0.356 | 0.393 | 0.354 |
|   | 6 | 0.000 | 0.000 | 0.000 | 0.001 | 0.004 | 0.016 | 0.047 | 0.118 | 0.178 | 0.262 | 0.531 |
| 7 | 0 | 0.478 | 0.210 | 0.133 | 0.082 | 0.028 | 0.008 | 0.002 | 0.000 | 0.000 | 0.000 | 0.000 |
|   | 1 | 0.372 | 0.367 | 0.311 | 0.247 | 0.131 | 0.055 | 0.017 | 0.004 | 0.001 | 0.000 | 0.000 |
|   | 2 | 0.124 | 0.275 | 0.311 | 0.318 | 0.261 | 0.164 | 0.077 | 0.025 | 0.012 | 0.004 | 0.000 |
|   | 3 | 0.023 | 0.115 | 0.173 | 0.227 | 0.290 | 0.273 | 0.194 | 0.097 | 0.058 | 0.029 | 0.003 |
|   | 4 | 0.003 | 0.029 | 0.058 | 0.097 | 0.194 | 0.273 | 0.290 | 0.227 | 0.173 | 0.115 | 0.023 |
|   | 5 | 0.000 | 0.004 | 0.012 | 0.025 | 0.077 | 0.164 | 0.261 | 0.318 | 0.311 | 0.275 | 0.124 |
|   | 6 | 0.000 | 0.000 | 0.001 | 0.004 | 0.017 | 0.055 | 0.131 | 0.247 | 0.311 | 0.367 | 0.372 |
|   | 7 | 0.000 | 0.000 | 0.000 | 0.000 | 0.002 | 0.008 | 0.028 | 0.082 | 0.133 | 0.210 | 0.478 |

*(continued)*

## Table A-2 *(continued)*

Binomial probabilities:

$$\binom{n}{x} p^x (1-p)^{n-x}$$

| | | | | | | | p | | | | | |
|---|---|---|---|---|---|---|---|---|---|---|---|---|
| n | x | 0.1 | 0.2 | 0.25 | 0.3 | 0.4 | 0.5 | 0.6 | 0.7 | 0.75 | 0.8 | 0.9 |
| 8 | 0 | 0.430 | 0.168 | 0.100 | 0.058 | 0.017 | 0.004 | 0.001 | 0.000 | 0.000 | 0.000 | 0.000 |
| | 1 | 0.383 | 0.336 | 0.267 | 0.198 | 0.090 | 0.031 | 0.008 | 0.001 | 0.000 | 0.000 | 0.000 |
| | 2 | 0.149 | 0.294 | 0.311 | 0.296 | 0.209 | 0.109 | 0.041 | 0.010 | 0.004 | 0.001 | 0.000 |
| | 3 | 0.033 | 0.147 | 0.208 | 0.254 | 0.279 | 0.219 | 0.124 | 0.047 | 0.023 | 0.009 | 0.000 |
| | 4 | 0.005 | 0.046 | 0.087 | 0.136 | 0.232 | 0.273 | 0.232 | 0.136 | 0.087 | 0.046 | 0.005 |
| | 5 | 0.000 | 0.009 | 0.023 | 0.047 | 0.124 | 0.219 | 0.279 | 0.254 | 0.208 | 0.147 | 0.033 |
| | 6 | 0.000 | 0.001 | 0.004 | 0.010 | 0.041 | 0.109 | 0.209 | 0.296 | 0.311 | 0.294 | 0.149 |
| | 7 | 0.000 | 0.000 | 0.000 | 0.001 | 0.008 | 0.031 | 0.090 | 0.198 | 0.267 | 0.336 | 0.383 |
| | 8 | 0.000 | 0.000 | 0.000 | 0.000 | 0.001 | 0.004 | 0.017 | 0.058 | 0.100 | 0.168 | 0.430 |
| 9 | 0 | 0.387 | 0.134 | 0.075 | 0.040 | 0.010 | 0.002 | 0.000 | 0.000 | 0.000 | 0.000 | 0.000 |
| | 1 | 0.387 | 0.302 | 0.225 | 0.156 | 0.060 | 0.018 | 0.004 | 0.000 | 0.000 | 0.000 | 0.000 |
| | 2 | 0.172 | 0.302 | 0.300 | 0.267 | 0.161 | 0.070 | 0.021 | 0.004 | 0.001 | 0.000 | 0.000 |
| | 3 | 0.045 | 0.176 | 0.234 | 0.267 | 0.251 | 0.164 | 0.074 | 0.021 | 0.009 | 0.003 | 0.000 |
| | 4 | 0.007 | 0.066 | 0.117 | 0.172 | 0.251 | 0.246 | 0.167 | 0.074 | 0.039 | 0.017 | 0.001 |
| | 5 | 0.001 | 0.017 | 0.039 | 0.074 | 0.167 | 0.246 | 0.251 | 0.172 | 0.117 | 0.066 | 0.007 |
| | 6 | 0.000 | 0.003 | 0.009 | 0.021 | 0.074 | 0.164 | 0.251 | 0.267 | 0.234 | 0.176 | 0.045 |
| | 7 | 0.000 | 0.000 | 0.001 | 0.004 | 0.021 | 0.070 | 0.161 | 0.267 | 0.300 | 0.302 | 0.172 |
| | 8 | 0.000 | 0.000 | 0.000 | 0.000 | 0.004 | 0.018 | 0.060 | 0.156 | 0.225 | 0.302 | 0.387 |
| | 9 | 0.000 | 0.000 | 0.000 | 0.000 | 0.000 | 0.002 | 0.010 | 0.040 | 0.075 | 0.134 | 0.387 |
| 10 | 0 | 0.349 | 0.107 | 0.056 | 0.028 | 0.006 | 0.001 | 0.000 | 0.000 | 0.000 | 0.000 | 0.000 |
| | 1 | 0.387 | 0.268 | 0.188 | 0.121 | 0.040 | 0.010 | 0.002 | 0.000 | 0.000 | 0.000 | 0.000 |
| | 2 | 0.194 | 0.302 | 0.282 | 0.233 | 0.121 | 0.044 | 0.011 | 0.001 | 0.000 | 0.000 | 0.000 |
| | 3 | 0.057 | 0.201 | 0.250 | 0.267 | 0.215 | 0.117 | 0.042 | 0.009 | 0.003 | 0.001 | 0.000 |
| | 4 | 0.011 | 0.088 | 0.146 | 0.200 | 0.251 | 0.205 | 0.111 | 0.037 | 0.016 | 0.006 | 0.000 |
| | 5 | 0.001 | 0.026 | 0.058 | 0.103 | 0.201 | 0.246 | 0.201 | 0.103 | 0.058 | 0.026 | 0.001 |
| | 6 | 0.000 | 0.006 | 0.016 | 0.037 | 0.111 | 0.205 | 0.251 | 0.200 | 0.146 | 0.088 | 0.011 |
| | 7 | 0.000 | 0.001 | 0.003 | 0.009 | 0.042 | 0.117 | 0.215 | 0.267 | 0.250 | 0.201 | 0.057 |
| | 8 | 0.000 | 0.000 | 0.000 | 0.001 | 0.011 | 0.044 | 0.121 | 0.233 | 0.282 | 0.302 | 0.194 |
| | 9 | 0.000 | 0.000 | 0.000 | 0.000 | 0.002 | 0.010 | 0.040 | 0.121 | 0.188 | 0.268 | 0.387 |
| | 10 | 0.000 | 0.000 | 0.000 | 0.000 | 0.000 | 0.001 | 0.006 | 0.028 | 0.056 | 0.107 | 0.349 |
| 11 | 0 | 0.314 | 0.086 | 0.042 | 0.020 | 0.004 | 0.000 | 0.000 | 0.000 | 0.000 | 0.000 | 0.000 |
| | 1 | 0.384 | 0.236 | 0.155 | 0.093 | 0.027 | 0.005 | 0.001 | 0.000 | 0.000 | 0.000 | 0.000 |
| | 2 | 0.213 | 0.295 | 0.258 | 0.200 | 0.089 | 0.027 | 0.005 | 0.001 | 0.000 | 0.000 | 0.000 |
| | 3 | 0.071 | 0.221 | 0.258 | 0.257 | 0.177 | 0.081 | 0.023 | 0.004 | 0.001 | 0.000 | 0.000 |
| | 4 | 0.016 | 0.111 | 0.172 | 0.220 | 0.236 | 0.161 | 0.070 | 0.017 | 0.006 | 0.002 | 0.000 |
| | 5 | 0.002 | 0.039 | 0.080 | 0.132 | 0.221 | 0.226 | 0.147 | 0.057 | 0.027 | 0.010 | 0.000 |
| | 6 | 0.000 | 0.010 | 0.027 | 0.057 | 0.147 | 0.226 | 0.221 | 0.132 | 0.080 | 0.039 | 0.002 |
| | 7 | 0.000 | 0.002 | 0.006 | 0.017 | 0.070 | 0.161 | 0.236 | 0.220 | 0.172 | 0.111 | 0.016 |
| | 8 | 0.000 | 0.000 | 0.001 | 0.004 | 0.023 | 0.081 | 0.177 | 0.257 | 0.258 | 0.221 | 0.071 |
| | 9 | 0.000 | 0.000 | 0.000 | 0.001 | 0.005 | 0.027 | 0.089 | 0.200 | 0.258 | 0.295 | 0.213 |
| | 10 | 0.000 | 0.000 | 0.000 | 0.000 | 0.001 | 0.005 | 0.027 | 0.093 | 0.155 | 0.236 | 0.384 |
| | 11 | 0.000 | 0.000 | 0.000 | 0.000 | 0.000 | 0.000 | 0.004 | 0.020 | 0.042 | 0.086 | 0.314 |

*(continued)*

## Table A-2 *(continued)*

Binomial probabilities:

$$\binom{n}{x} p^{x}(1-p)^{n-x}$$

|     |     | p |     |     |     |     |     |     |     |     |     |     |
|-----|-----|-------|-------|-------|-------|-------|-------|-------|-------|-------|-------|-------|
| n   | x   | 0.1   | 0.2   | 0.25  | 0.3   | 0.4   | 0.5   | 0.6   | 0.7   | 0.75  | 0.8   | 0.9   |
| 12  | 0   | 0.282 | 0.069 | 0.032 | 0.014 | 0.002 | 0.000 | 0.000 | 0.000 | 0.000 | 0.000 | 0.000 |
|     | 1   | 0.377 | 0.206 | 0.127 | 0.071 | 0.017 | 0.003 | 0.000 | 0.000 | 0.000 | 0.000 | 0.000 |
|     | 2   | 0.230 | 0.283 | 0.232 | 0.168 | 0.064 | 0.016 | 0.002 | 0.000 | 0.000 | 0.000 | 0.000 |
|     | 3   | 0.085 | 0.236 | 0.258 | 0.240 | 0.142 | 0.054 | 0.012 | 0.001 | 0.000 | 0.000 | 0.000 |
|     | 4   | 0.021 | 0.133 | 0.194 | 0.231 | 0.213 | 0.121 | 0.042 | 0.008 | 0.002 | 0.001 | 0.000 |
|     | 5   | 0.004 | 0.053 | 0.103 | 0.158 | 0.227 | 0.193 | 0.101 | 0.029 | 0.011 | 0.003 | 0.000 |
|     | 6   | 0.000 | 0.016 | 0.040 | 0.079 | 0.177 | 0.226 | 0.177 | 0.079 | 0.040 | 0.016 | 0.000 |
|     | 7   | 0.000 | 0.003 | 0.011 | 0.029 | 0.101 | 0.193 | 0.227 | 0.158 | 0.103 | 0.053 | 0.004 |
|     | 8   | 0.000 | 0.001 | 0.002 | 0.008 | 0.042 | 0.121 | 0.213 | 0.231 | 0.194 | 0.133 | 0.021 |
|     | 9   | 0.000 | 0.000 | 0.000 | 0.001 | 0.012 | 0.054 | 0.142 | 0.240 | 0.258 | 0.236 | 0.085 |
|     | 10  | 0.000 | 0.000 | 0.000 | 0.000 | 0.002 | 0.016 | 0.064 | 0.168 | 0.232 | 0.283 | 0.230 |
|     | 11  | 0.000 | 0.000 | 0.000 | 0.000 | 0.000 | 0.003 | 0.017 | 0.071 | 0.127 | 0.206 | 0.377 |
|     | 12  | 0.000 | 0.000 | 0.000 | 0.000 | 0.000 | 0.000 | 0.002 | 0.014 | 0.032 | 0.069 | 0.282 |
| 13  | 0   | 0.254 | 0.055 | 0.024 | 0.010 | 0.001 | 0.000 | 0.000 | 0.000 | 0.000 | 0.000 | 0.000 |
|     | 1   | 0.367 | 0.179 | 0.103 | 0.054 | 0.011 | 0.002 | 0.000 | 0.000 | 0.000 | 0.000 | 0.000 |
|     | 2   | 0.245 | 0.268 | 0.206 | 0.139 | 0.045 | 0.010 | 0.001 | 0.000 | 0.000 | 0.000 | 0.000 |
|     | 3   | 0.100 | 0.246 | 0.252 | 0.218 | 0.111 | 0.035 | 0.006 | 0.001 | 0.000 | 0.000 | 0.000 |
|     | 4   | 0.028 | 0.154 | 0.210 | 0.234 | 0.184 | 0.087 | 0.024 | 0.003 | 0.001 | 0.000 | 0.000 |
|     | 5   | 0.006 | 0.069 | 0.126 | 0.180 | 0.221 | 0.157 | 0.066 | 0.014 | 0.005 | 0.001 | 0.000 |
|     | 6   | 0.001 | 0.023 | 0.056 | 0.103 | 0.197 | 0.209 | 0.131 | 0.044 | 0.019 | 0.006 | 0.000 |
|     | 7   | 0.000 | 0.006 | 0.019 | 0.044 | 0.131 | 0.209 | 0.197 | 0.103 | 0.056 | 0.023 | 0.001 |
|     | 8   | 0.000 | 0.001 | 0.005 | 0.014 | 0.066 | 0.157 | 0.221 | 0.157 | 0.069 | 0.069 | 0.006 |
|     | 9   | 0.000 | 0.000 | 0.001 | 0.003 | 0.024 | 0.087 | 0.184 | 0.234 | 0.210 | 0.154 | 0.028 |
|     | 10  | 0.000 | 0.000 | 0.000 | 0.001 | 0.006 | 0.035 | 0.111 | 0.218 | 0.252 | 0.246 | 0.100 |
|     | 11  | 0.000 | 0.000 | 0.000 | 0.000 | 0.001 | 0.010 | 0.045 | 0.139 | 0.206 | 0.268 | 0.245 |
|     | 12  | 0.000 | 0.000 | 0.000 | 0.000 | 0.000 | 0.002 | 0.011 | 0.054 | 0.103 | 0.179 | 0.367 |
|     | 13  | 0.000 | 0.000 | 0.000 | 0.000 | 0.000 | 0.000 | 0.001 | 0.010 | 0.024 | 0.055 | 0.254 |
| 14  | 0   | 0.229 | 0.044 | 0.018 | 0.007 | 0.001 | 0.000 | 0.000 | 0.000 | 0.000 | 0.000 | 0.000 |
|     | 1   | 0.356 | 0.154 | 0.083 | 0.041 | 0.007 | 0.001 | 0.000 | 0.000 | 0.000 | 0.000 | 0.000 |
|     | 2   | 0.257 | 0.250 | 0.180 | 0.113 | 0.032 | 0.006 | 0.001 | 0.000 | 0.000 | 0.000 | 0.000 |
|     | 3   | 0.114 | 0.250 | 0.240 | 0.194 | 0.085 | 0.022 | 0.003 | 0.000 | 0.000 | 0.000 | 0.000 |
|     | 4   | 0.035 | 0.172 | 0.220 | 0.229 | 0.155 | 0.061 | 0.014 | 0.001 | 0.000 | 0.000 | 0.000 |
|     | 5   | 0.008 | 0.086 | 0.147 | 0.196 | 0.207 | 0.122 | 0.041 | 0.007 | 0.002 | 0.000 | 0.000 |
|     | 6   | 0.001 | 0.032 | 0.073 | 0.126 | 0.207 | 0.183 | 0.092 | 0.023 | 0.008 | 0.002 | 0.000 |
|     | 7   | 0.000 | 0.009 | 0.028 | 0.062 | 0.157 | 0.209 | 0.157 | 0.062 | 0.028 | 0.009 | 0.000 |
|     | 8   | 0.000 | 0.002 | 0.008 | 0.023 | 0.092 | 0.183 | 0.207 | 0.126 | 0.073 | 0.032 | 0.001 |
|     | 9   | 0.000 | 0.000 | 0.002 | 0.007 | 0.041 | 0.122 | 0.207 | 0.196 | 0.147 | 0.086 | 0.008 |
|     | 10  | 0.000 | 0.000 | 0.000 | 0.001 | 0.014 | 0.061 | 0.155 | 0.229 | 0.220 | 0.172 | 0.035 |
|     | 11  | 0.000 | 0.000 | 0.000 | 0.000 | 0.003 | 0.022 | 0.085 | 0.194 | 0.240 | 0.250 | 0.114 |
|     | 12  | 0.000 | 0.000 | 0.000 | 0.000 | 0.001 | 0.006 | 0.032 | 0.113 | 0.180 | 0.250 | 0.257 |
|     | 13  | 0.000 | 0.000 | 0.000 | 0.000 | 0.000 | 0.001 | 0.007 | 0.041 | 0.083 | 0.154 | 0.356 |
|     | 14  | 0.000 | 0.000 | 0.000 | 0.000 | 0.000 | 0.000 | 0.001 | 0.007 | 0.018 | 0.044 | 0.229 |

*(continued)*

## Table A-2 *(continued)*

Binomial probabilities:

$$\binom{n}{x} p^x (1-p)^{n-x}$$

| | | | | | | | | $p$ | | | | | |
|---|---|---|---|---|---|---|---|---|---|---|---|---|---|
| $n$ | $x$ | 0.1 | 0.2 | 0.25 | 0.3 | 0.4 | 0.5 | 0.6 | 0.7 | 0.75 | 0.8 | 0.9 |
| 15 | 0 | 0.206 | 0.035 | 0.013 | 0.005 | 0.000 | 0.000 | 0.000 | 0.000 | 0.000 | 0.000 | 0.000 |
| | 1 | 0.343 | 0.132 | 0.067 | 0.031 | 0.005 | 0.000 | 0.000 | 0.000 | 0.000 | 0.000 | 0.000 |
| | 2 | 0.267 | 0.231 | 0.156 | 0.092 | 0.022 | 0.003 | 0.000 | 0.000 | 0.000 | 0.000 | 0.000 |
| | 3 | 0.129 | 0.250 | 0.225 | 0.170 | 0.063 | 0.014 | 0.002 | 0.000 | 0.000 | 0.000 | 0.000 |
| | 4 | 0.043 | 0.188 | 0.225 | 0.219 | 0.127 | 0.042 | 0.007 | 0.001 | 0.000 | 0.000 | 0.000 |
| | 5 | 0.010 | 0.103 | 0.165 | 0.206 | 0.186 | 0.092 | 0.024 | 0.003 | 0.001 | 0.000 | 0.000 |
| | 6 | 0.002 | 0.043 | 0.092 | 0.147 | 0.207 | 0.153 | 0.061 | 0.012 | 0.003 | 0.001 | 0.000 |
| | 7 | 0.000 | 0.014 | 0.039 | 0.081 | 0.177 | 0.196 | 0.118 | 0.035 | 0.013 | 0.003 | 0.000 |
| | 8 | 0.000 | 0.003 | 0.013 | 0.035 | 0.118 | 0.196 | 0.177 | 0.081 | 0.039 | 0.014 | 0.000 |
| | 9 | 0.000 | 0.001 | 0.003 | 0.012 | 0.061 | 0.153 | 0.207 | 0.147 | 0.092 | 0.043 | 0.002 |
| | 10 | 0.000 | 0.000 | 0.001 | 0.003 | 0.024 | 0.092 | 0.186 | 0.206 | 0.165 | 0.103 | 0.010 |
| | 11 | 0.000 | 0.000 | 0.000 | 0.001 | 0.007 | 0.042 | 0.127 | 0.219 | 0.225 | 0.188 | 0.043 |
| | 12 | 0.000 | 0.000 | 0.000 | 0.000 | 0.002 | 0.014 | 0.063 | 0.170 | 0.225 | 0.250 | 0.129 |
| | 13 | 0.000 | 0.000 | 0.000 | 0.000 | 0.000 | 0.003 | 0.022 | 0.092 | 0.156 | 0.231 | 0.267 |
| | 14 | 0.000 | 0.000 | 0.000 | 0.000 | 0.000 | 0.000 | 0.005 | 0.031 | 0.067 | 0.132 | 0.343 |
| | 15 | 0.000 | 0.000 | 0.000 | 0.000 | 0.000 | 0.000 | 0.000 | 0.005 | 0.013 | 0.035 | 0.206 |
| 20 | 0 | 0.122 | 0.012 | 0.003 | 0.001 | 0.000 | 0.000 | 0.000 | 0.000 | 0.000 | 0.000 | 0.000 |
| | 1 | 0.270 | 0.058 | 0.021 | 0.007 | 0.000 | 0.000 | 0.000 | 0.000 | 0.000 | 0.000 | 0.000 |
| | 2 | 0.285 | 0.137 | 0.067 | 0.028 | 0.003 | 0.000 | 0.000 | 0.000 | 0.000 | 0.000 | 0.000 |
| | 3 | 0.190 | 0.205 | 0.134 | 0.072 | 0.012 | 0.001 | 0.000 | 0.000 | 0.000 | 0.000 | 0.000 |
| | 4 | 0.090 | 0.218 | 0.190 | 0.130 | 0.035 | 0.005 | 0.000 | 0.000 | 0.000 | 0.000 | 0.000 |
| | 5 | 0.032 | 0.175 | 0.202 | 0.179 | 0.075 | 0.015 | 0.001 | 0.000 | 0.000 | 0.000 | 0.000 |
| | 6 | 0.009 | 0.109 | 0.169 | 0.192 | 0.124 | 0.037 | 0.005 | 0.000 | 0.000 | 0.000 | 0.000 |
| | 7 | 0.002 | 0.055 | 0.112 | 0.164 | 0.166 | 0.074 | 0.015 | 0.001 | 0.000 | 0.000 | 0.000 |
| | 8 | 0.000 | 0.022 | 0.061 | 0.114 | 0.180 | 0.120 | 0.035 | 0.004 | 0.001 | 0.000 | 0.000 |
| | 9 | 0.000 | 0.007 | 0.027 | 0.065 | 0.160 | 0.160 | 0.071 | 0.012 | 0.003 | 0.000 | 0.000 |
| | 10 | 0.000 | 0.002 | 0.010 | 0.031 | 0.117 | 0.176 | 0.117 | 0.031 | 0.010 | 0.002 | 0.000 |
| | 11 | 0.000 | 0.000 | 0.003 | 0.012 | 0.071 | 0.160 | 0.160 | 0.065 | 0.027 | 0.007 | 0.007 |
| | 12 | 0.000 | 0.000 | 0.001 | 0.004 | 0.035 | 0.120 | 0.180 | 0.114 | 0.061 | 0.022 | 0.000 |
| | 13 | 0.000 | 0.000 | 0.000 | 0.001 | 0.015 | 0.074 | 0.166 | 0.164 | 0.112 | 0.055 | 0.002 |
| | 14 | 0.000 | 0.000 | 0.000 | 0.000 | 0.005 | 0.037 | 0.124 | 0.192 | 0.169 | 0.109 | 0.009 |
| | 15 | 0.000 | 0.000 | 0.000 | 0.000 | 0.001 | 0.015 | 0.075 | 0.179 | 0.202 | 0.175 | 0.032 |
| | 16 | 0.000 | 0.000 | 0.000 | 0.000 | 0.000 | 0.005 | 0.035 | 0.130 | 0.190 | 0.218 | 0.090 |
| | 17 | 0.000 | 0.000 | 0.000 | 0.000 | 0.000 | 0.001 | 0.012 | 0.072 | 0.134 | 0.205 | 0.190 |
| | 18 | 0.000 | 0.000 | 0.000 | 0.000 | 0.000 | 0.000 | 0.003 | 0.028 | 0.067 | 0.137 | 0.285 |
| | 19 | 0.000 | 0.000 | 0.000 | 0.000 | 0.000 | 0.000 | 0.000 | 0.007 | 0.021 | 0.058 | 0.270 |
| | 20 | 0.000 | 0.000 | 0.000 | 0.000 | 0.000 | 0.000 | 0.000 | 0.001 | 0.003 | 0.012 | 0.122 |

*(continued)*

# Chi-Square Table

Table A-3 shows right-tail probabilities for the Chi-square distribution (you can use Chapter 14 as a reference for the Chi-square test). To use Table A-3, you need three pieces of information from the particular problem you're working on:

- ✔ The sample size, $n$.

- ✔ The value of Chi-squared for which you want the right-tail probability.

- ✔ If you're working with a two-way table, you need $r$ = number of rows and $c$ = number of columns. If you're working with a goodness-of-fit test, you need $k - 1$, where $k$ is the number of categories.

The degrees of freedom for the Chi-square test statistic is $(r - 1) * (c - 1)$ if you're testing for an association between two variables, where $r$ and $c$ are the number of rows and columns in the two-way table, respectively. Or, the degrees of freedom is $k - 1$ in a goodness-of-fit test, where $k$ is the number of categories; see Chapter 15.

Go across the row for your degrees of freedom until you find the value in that row closest to your Chi-square test statistic. Look up at the number at the top of that column. That value is the area to the right of (beyond) that particular Chi-square statistic.

## Table A-3 — The Chi-Square Table

Numbers in the table represent Chi-square values whose area to the right equals $p$.

| df/p | 0.10 | 0.05 | 0.025 | 0.01 | 0.005 |
|---|---|---|---|---|---|
| 1 | 2.71 | 3.84 | 5.02 | 6.64 | 7.88 |
| 2 | 4.61 | 5.99 | 7.38 | 9.21 | 10.60 |
| 3 | 6.25 | 7.82 | 9.35 | 11.35 | 12.84 |
| 4 | 7.78 | 9.49 | 11.14 | 13.28 | 14.86 |
| 5 | 9.24 | 11.07 | 12.83 | 15.09 | 16.75 |
| 6 | 10.65 | 12.59 | 14.45 | 16.81 | 18.55 |
| 7 | 12.02 | 14.07 | 16.01 | 18.48 | 20.28 |
| 8 | 13.36 | 15.51 | 17.54 | 20.09 | 21.96 |
| 9 | 14.68 | 16.92 | 19.02 | 21.67 | 23.59 |
| 10 | 15.99 | 18.31 | 20.48 | 23.21 | 25.19 |
| 11 | 17.28 | 19.68 | 21.92 | 24.73 | 26.76 |
| 12 | 18.55 | 21.03 | 23.34 | 26.22 | 28.30 |
| 13 | 19.81 | 22.36 | 24.74 | 27.69 | 29.819 |
| 14 | 21.06 | 23.69 | 26.12 | 29.14 | 31.32 |
| 15 | 22.31 | 25.00 | 27.49 | 30.58 | 32.80 |
| 16 | 23.54 | 26.30 | 28.85 | 32.00 | 34.27 |
| 17 | 24.77 | 27.59 | 30.19 | 33.41 | 35.72 |
| 18 | 25.99 | 28.87 | 31.53 | 34.81 | 37.16 |
| 19 | 27.20 | 30.14 | 32.85 | 36.19 | 38.58 |
| 20 | 28.41 | 31.41 | 34.17 | 37.57 | 40.00 |
| 21 | 29.62 | 32.67 | 35.48 | 38.93 | 41.40 |
| 22 | 30.81 | 33.92 | 36.78 | 40.29 | 42.80 |
| 23 | 32.01 | 35.17 | 38.08 | 41.64 | 44.18 |
| 24 | 33.20 | 36.42 | 39.36 | 42.98 | 45.56 |
| 25 | 34.38 | 37.65 | 40.65 | 44.31 | 46.93 |
| 26 | 35.56 | 38.89 | 41.92 | 45.64 | 48.29 |
| 27 | 36.74 | 40.11 | 43.20 | 46.96 | 49.65 |
| 28 | 37.92 | 41.34 | 44.46 | 48.28 | 50.99 |
| 29 | 39.09 | 42.56 | 45.72 | 49.59 | 52.34 |
| 30 | 40.26 | 43.77 | 46.98 | 50.89 | 53.67 |
| 40 | 51.81 | 55.76 | 59.34 | 63.69 | 66.77 |
| 50 | 63.17 | 67.51 | 71.42 | 76.15 | 79.49 |

# Rank Sum Table

Tables A-4(a) and A-4(b) show the critical values for the rank sum test for $\alpha = 0.05$ and $\alpha = 0.10$, respectively; see Chapter 18 for more on this test. To use Table A-4, you need two pieces of information from the particular problem you're working on:

- The rank sum statistic, $T$
- The sample sizes of the two samples, $n_1$ and $n_2$

To find the critical value for your rank sum statistic using Table A-4, go to the column representing $n_1$ and the row representing $n_2$. Intersect the row and column to find the lower and upper critical values (denoted $T_L$ and $T_U$) for the rank sum test.

| Table A-4(a) | | | | | | | | | | | | | | | | |
|---|---|---|---|---|---|---|---|---|---|---|---|---|---|---|---|---|
| **The Rank Sum Table ($\alpha = 0.05$)** | | | | | | | | | | | | | | | | |
| $n_1$ / $n_2$ | **3** | | **4** | | **5** | | **6** | | **7** | | **8** | | **9** | | **10** | |
| | $T_L$ | $T_U$ | $T_L$ | $T_U$ | $T_L$ | $T_U$ | $T_L$ | $T_U$ | $T_L$ | $T_U$ | $T_L$ | $T_U$ | $T_L$ | $T_U$ | $T_L$ | $T_U$ |
| 3 | 5 | 16 | 6 | 18 | 6 | 21 | 7 | 23 | 7 | 26 | 8 | 28 | 8 | 31 | 9 | 33 |
| 4 | 6 | 18 | 11 | 25 | 12 | 28 | 12 | 32 | 13 | 35 | 14 | 38 | 15 | 41 | 16 | 44 |
| 5 | 6 | 21 | 12 | 28 | 18 | 37 | 19 | 41 | 20 | 45 | 21 | 49 | 22 | 53 | 24 | 56 |
| 6 | 7 | 23 | 12 | 32 | 19 | 41 | 26 | 52 | 28 | 56 | 29 | 61 | 31 | 65 | 32 | 70 |
| 7 | 7 | 26 | 13 | 35 | 20 | 45 | 28 | 56 | 37 | 68 | 39 | 73 | 41 | 78 | 43 | 83 |
| 8 | 8 | 28 | 14 | 38 | 21 | 49 | 29 | 61 | 39 | 73 | 49 | 87 | 53 | 93 | 54 | 98 |
| 9 | 8 | 31 | 15 | 41 | 22 | 53 | 31 | 65 | 41 | 78 | 51 | 93 | 63 | 108 | 66 | 114 |
| 10 | 9 | 33 | 16 | 44 | 24 | 56 | 32 | 70 | 43 | 83 | 54 | 98 | 66 | 114 | 79 | 131 |

| Table A-4(b) | | | | | | The Rank Sum Table ($\alpha = 0.10$) | | | | | | | | | |

| $n_2$ \ $n_1$ | 3 | | 4 | | 5 | | 6 | | 7 | | 8 | | 9 | | 10 | |
|---|---|---|---|---|---|---|---|---|---|---|---|---|---|---|---|---|
| | $T_L$ | $T_U$ | $T_L$ | $T_U$ | $T_L$ | $T_U$ | $T_L$ | $T_U$ | $T_L$ | $T_U$ | $T_L$ | $T_U$ | $T_L$ | $T_U$ | $T_L$ | $T_U$ |
| 3 | 6 | 15 | 7 | 17 | 7 | 20 | 8 | 22 | 9 | 24 | 9 | 27 | 10 | 29 | 11 | 31 |
| 4 | 7 | 17 | 12 | 24 | 13 | 27 | 14 | 30 | 15 | 33 | 16 | 36 | 17 | 39 | 18 | 42 |
| 5 | 7 | 20 | 13 | 37 | 19 | 36 | 20 | 40 | 22 | 43 | 24 | 46 | 25 | 50 | 26 | 54 |
| 6 | 8 | 22 | 14 | 30 | 20 | 40 | 28 | 50 | 30 | 54 | 32 | 58 | 33 | 63 | 35 | 67 |
| 7 | 9 | 24 | 15 | 33 | 22 | 43 | 30 | 54 | 39 | 66 | 41 | 71 | 43 | 76 | 46 | 80 |
| 8 | 9 | 27 | 16 | 36 | 24 | 46 | 32 | 58 | 41 | 71 | 52 | 84 | 54 | 90 | 57 | 95 |
| 9 | 10 | 29 | 17 | 39 | 25 | 50 | 33 | 63 | 43 | 76 | 54 | 90 | 66 | 105 | 69 | 111 |
| 10 | 11 | 31 | 18 | 42 | 26 | 54 | 35 | 67 | 46 | 80 | 57 | 95 | 69 | 111 | 83 | 127 |

# F-Table

Table A-5 shows the critical values on the *F*-distribution where $\alpha$ is equal to 0.05. (*Critical values* are those values that represent the boundary between rejecting Ho and not rejecting Ho; refer to Chapter 9.) To use Table A-5, you need three pieces of information from the particular problem you're working on:

- The sample size, *n*
- The number of populations (or treatments being compared), *k*
- The value of *F* for which you want the cumulative probability

To find the critical value for your *F*-test statistic using Table A-5, go to the column representing the degrees of freedom you need ($k - 1$ and $n - k$). Intersect the column degrees of freedom ($k - 1$) with the row degrees of freedom ($n - k$), and you find the critical value on the *F*-distribution.

## Table A-5   The F-Table ($\alpha = 0.05$)

$F_{(.05,\ df1,\ df2)}$

| df2/df1 | 1 | 2 | 3 | 4 | 5 | 6 | 7 | 8 | 9 | 10 | 12 | 15 | 20 | 24 | 30 | 40 | 60 | 120 |
|---|---|---|---|---|---|---|---|---|---|---|---|---|---|---|---|---|---|---|
| 1 | 161.4476 | 199.5000 | 215.7073 | 224.5832 | 230.1619 | 233.9860 | 236.7684 | 238.8827 | 240.5433 | 241.8817 | 243.9060 | 245.9499 | 248.0131 | 249.0518 | 250.0951 | 251.1432 | 252.1957 | 253.252 |
| 2 | 18.5128 | 19.0000 | 19.1643 | 19.2468 | 19.2964 | 19.3295 | 19.3532 | 19.3710 | 19.3848 | 19.3959 | 19.4125 | 19.4291 | 19.4458 | 19.4541 | 19.4624 | 19.4707 | 19.4791 | 19.487 |
| 3 | 10.1280 | 9.5521 | 9.2766 | 9.1172 | 9.0135 | 8.9406 | 8.8867 | 8.8452 | 8.8123 | 8.7855 | 8.7446 | 8.7029 | 8.6602 | 8.6385 | 8.6166 | 8.5944 | 8.5720 | 8.549 |
| 4 | 7.7086 | 6.9443 | 6.5914 | 6.3882 | 6.2561 | 6.1631 | 6.0942 | 6.0410 | 5.9988 | 5.9644 | 5.9117 | 5.8578 | 5.8025 | 5.7744 | 5.7459 | 5.7170 | 5.6877 | 5.658 |
| 5 | 6.6079 | 5.7861 | 5.4095 | 5.1922 | 5.0503 | 4.9503 | 4.8759 | 4.8183 | 4.7725 | 4.7351 | 4.6777 | 4.6188 | 4.5581 | 4.5272 | 4.4957 | 4.4638 | 4.4314 | 4.398 |
| 6 | 5.9874 | 5.1433 | 4.7571 | 4.5337 | 4.3874 | 4.2839 | 4.2067 | 4.1468 | 4.0990 | 4.0600 | 3.9999 | 3.9381 | 3.8742 | 3.8415 | 3.8082 | 3.7743 | 3.7398 | 3.704 |
| 7 | 5.5914 | 4.7374 | 4.3468 | 4.1203 | 3.9715 | 3.8660 | 3.7870 | 3.7257 | 3.6767 | 3.6365 | 3.5747 | 3.5107 | 3.4445 | 3.4105 | 3.3758 | 3.3404 | 3.3043 | 3.267 |
| 8 | 5.3177 | 4.4590 | 4.0662 | 3.8379 | 3.6875 | 3.5806 | 3.5005 | 3.4381 | 3.3881 | 3.3472 | 3.2839 | 3.2184 | 3.1503 | 3.1152 | 3.0794 | 3.0428 | 3.0053 | 2.966 |
| 9 | 5.1174 | 4.2565 | 3.8625 | 3.6331 | 3.4817 | 3.3738 | 3.2927 | 3.2296 | 3.1789 | 3.1373 | 3.0729 | 3.0061 | 2.9365 | 2.9005 | 2.8637 | 2.8259 | 2.7872 | 2.747 |
| 10 | 4.9646 | 4.1028 | 3.7083 | 3.4780 | 3.3258 | 3.2172 | 3.1355 | 3.0717 | 3.0204 | 2.9782 | 2.9130 | 2.8450 | 2.7740 | 2.7372 | 2.6996 | 2.6609 | 2.6211 | 2.580 |
| 11 | 4.8443 | 3.9823 | 3.5874 | 3.3567 | 3.2039 | 3.0946 | 3.0123 | 2.9480 | 2.8962 | 2.8536 | 2.7876 | 2.7186 | 2.6464 | 2.6090 | 2.5705 | 2.5309 | 2.4901 | 2.448 |
| 12 | 4.7472 | 3.8853 | 3.4903 | 3.2592 | 3.1059 | 2.9961 | 2.9134 | 2.8486 | 2.7964 | 2.7534 | 2.6866 | 2.6169 | 2.5436 | 2.5055 | 2.4663 | 2.4259 | 2.3842 | 2.341 |
| 13 | 4.6672 | 3.8056 | 3.4105 | 3.1791 | 3.0254 | 2.9153 | 2.8321 | 2.7669 | 2.7144 | 2.6710 | 2.6037 | 2.5331 | 2.4589 | 2.4202 | 2.3803 | 2.3392 | 2.2966 | 2.252 |
| 14 | 4.6001 | 3.7389 | 3.3439 | 3.1122 | 2.9582 | 2.8477 | 2.7642 | 2.6987 | 2.6458 | 2.6022 | 2.5342 | 2.4630 | 2.3879 | 2.3487 | 2.3082 | 2.2664 | 2.2229 | 2.177 |
| 15 | 4.5431 | 3.6823 | 3.2874 | 3.0556 | 2.9013 | 2.7905 | 2.7066 | 2.6408 | 2.5876 | 2.5437 | 2.4753 | 2.4034 | 2.3275 | 2.2878 | 2.2468 | 2.2043 | 2.1601 | 2.114 |
| 16 | 4.4940 | 3.6337 | 3.2389 | 3.0069 | 2.8524 | 2.7413 | 2.6572 | 2.5911 | 2.5377 | 2.4935 | 2.4247 | 2.3522 | 2.2756 | 2.2354 | 2.1938 | 2.1507 | 2.1058 | 2.058 |
| 17 | 4.4513 | 3.5915 | 3.1968 | 2.9647 | 2.8100 | 2.6987 | 2.6143 | 2.5480 | 2.4943 | 2.4499 | 2.3807 | 2.3077 | 2.2304 | 2.1898 | 2.1477 | 2.1040 | 2.0584 | 2.010 |
| 18 | 4.4139 | 3.5546 | 3.1599 | 2.9277 | 2.7729 | 2.6613 | 2.5767 | 2.5102 | 2.4563 | 2.4117 | 2.3421 | 2.2686 | 2.1906 | 2.1497 | 2.1071 | 2.0629 | 2.0166 | 1.968 |
| 19 | 4.3807 | 3.5219 | 3.1274 | 2.8951 | 2.7401 | 2.6283 | 2.5435 | 2.4768 | 2.4227 | 2.3779 | 2.3080 | 2.2341 | 2.1555 | 2.1141 | 2.0712 | 2.0264 | 1.9795 | 1.930 |
| 20 | 4.3512 | 3.4928 | 3.0984 | 2.8661 | 2.7109 | 2.5990 | 2.5140 | 2.4471 | 2.3928 | 2.3479 | 2.2776 | 2.2033 | 2.1242 | 2.0825 | 2.0391 | 1.9938 | 1.9464 | 1.896 |
| 21 | 4.3248 | 3.4668 | 3.0725 | 2.8401 | 2.6848 | 2.5727 | 2.4876 | 2.4205 | 2.3660 | 2.3210 | 2.2504 | 2.1757 | 2.0960 | 2.0540 | 2.0102 | 1.9645 | 1.9165 | 1.865 |
| 22 | 4.3009 | 3.4434 | 3.0491 | 2.8167 | 2.6613 | 2.5491 | 2.4638 | 2.3965 | 2.3419 | 2.2967 | 2.2258 | 2.1508 | 2.0707 | 2.0283 | 1.9842 | 1.9380 | 1.8894 | 1.838 |
| 23 | 4.2793 | 3.4221 | 3.0280 | 2.7955 | 2.6400 | 2.5277 | 2.4422 | 2.3748 | 2.3201 | 2.2747 | 2.2036 | 2.1282 | 2.0476 | 2.0050 | 1.9605 | 1.9139 | 1.8648 | 1.812 |
| 24 | 4.2597 | 3.4028 | 3.0088 | 2.7763 | 2.6207 | 2.5082 | 2.4226 | 2.3551 | 2.3002 | 2.2547 | 2.1834 | 2.1077 | 2.0267 | 1.9838 | 1.9390 | 1.8920 | 1.8424 | 1.789 |
| 25 | 4.2417 | 3.3852 | 2.9912 | 2.7587 | 2.6030 | 2.4904 | 2.4047 | 2.3371 | 2.2821 | 2.2365 | 2.1649 | 2.0889 | 2.0075 | 1.9643 | 1.9192 | 1.8718 | 1.8217 | 1.768 |
| 26 | 4.2252 | 3.3690 | 2.9752 | 2.7426 | 2.5868 | 2.4741 | 2.3883 | 2.3205 | 2.2655 | 2.2197 | 2.1479 | 2.0716 | 1.9898 | 1.9464 | 1.9010 | 1.8533 | 1.8027 | 1.748 |
| 27 | 4.2100 | 3.3541 | 2.9604 | 2.7278 | 2.5719 | 2.4591 | 2.3732 | 2.3053 | 2.2501 | 2.2043 | 2.1323 | 2.0558 | 1.9736 | 1.9299 | 1.8842 | 1.8361 | 1.7851 | 1.730 |
| 28 | 4.1960 | 3.3404 | 2.9467 | 2.7141 | 2.5581 | 2.4453 | 2.3593 | 2.2913 | 2.2360 | 2.1900 | 2.1179 | 2.0411 | 1.9586 | 1.9147 | 1.8687 | 1.8203 | 1.7689 | 1.713 |
| 29 | 4.1830 | 3.3277 | 2.9340 | 2.7014 | 2.5454 | 2.4324 | 2.3463 | 2.2783 | 2.2229 | 2.1768 | 2.1045 | 2.0275 | 1.9446 | 1.9005 | 1.8543 | 1.8055 | 1.7537 | 1.698 |
| 30 | 4.1709 | 3.3158 | 2.9223 | 2.6896 | 2.5336 | 2.4205 | 2.3343 | 2.2662 | 2.2107 | 2.1646 | 2.0921 | 2.0148 | 1.9317 | 1.8874 | 1.8409 | 1.7918 | 1.7396 | 1.683 |
| 40 | 4.0847 | 3.2317 | 2.8387 | 2.6060 | 2.4495 | 2.3359 | 2.2490 | 2.1802 | 2.1240 | 2.0772 | 2.0035 | 1.9245 | 1.8389 | 1.7929 | 1.7444 | 1.6928 | 1.6373 | 1.576 |
| 60 | 4.0012 | 3.1504 | 2.7581 | 2.5252 | 2.3683 | 2.2541 | 2.1665 | 2.0970 | 2.0401 | 1.9926 | 1.9174 | 1.8364 | 1.7480 | 1.7001 | 1.6491 | 1.5943 | 1.5343 | 1.467 |
| 120 | 3.9201 | 3.0718 | 2.6802 | 2.4472 | 2.2899 | 2.1750 | 2.0868 | 2.0164 | 1.9588 | 1.9105 | 1.8337 | 1.7505 | 1.6587 | 1.6084 | 1.5543 | 1.4952 | 1.4290 | 1.351 |

# Index

## BUSINESS, CAREERS & PERSONAL FINANCE

**Accounting For Dummies, 4th Edition***
978-0-470-24600-9

**Bookkeeping Workbook For Dummies†**
978-0-470-16983-4

**Commodities For Dummies**
978-0-470-04928-0

**Doing Business in China For Dummies**
978-0-470-04929-7

**E-Mail Marketing For Dummies**
978-0-470-19087-6

**Job Interviews For Dummies, 3rd Edition*†**
978-0-470-17748-8

**Personal Finance Workbook For Dummies*†**
978-0-470-09933-9

**Real Estate License Exams For Dummies**
978-0-7645-7623-2

**Six Sigma For Dummies**
978-0-7645-6798-8

**Small Business Kit For Dummies, 2nd Edition*†**
978-0-7645-5984-6

**Telephone Sales For Dummies**
978-0-470-16836-3

## BUSINESS PRODUCTIVITY & MICROSOFT OFFICE

**Access 2007 For Dummies**
978-0-470-03649-5

**Excel 2007 For Dummies**
978-0-470-03737-9

**Office 2007 For Dummies**
978-0-470-00923-9

**Outlook 2007 For Dummies**
978-0-470-03830-7

**PowerPoint 2007 For Dummies**
978-0-470-04059-1

**Project 2007 For Dummies**
978-0-470-03651-8

**QuickBooks 2008 For Dummies**
978-0-470-18470-7

**Quicken 2008 For Dummies**
978-0-470-17473-9

**Salesforce.com For Dummies, 2nd Edition**
978-0-470-04893-1

**Word 2007 For Dummies**
978-0-470-03658-7

## EDUCATION, HISTORY, REFERENCE & TEST PREPARATION

**African American History For Dummies**
978-0-7645-5469-8

**Algebra For Dummies**
978-0-7645-5325-7

**Algebra Workbook For Dummies**
978-0-7645-8467-1

**Art History For Dummies**
978-0-470-09910-0

**ASVAB For Dummies, 2nd Edition**
978-0-470-10671-6

**British Military History For Dummies**
978-0-470-03213-8

**Calculus For Dummies**
978-0-7645-2498-1

**Canadian History For Dummies, 2nd Edition**
978-0-470-83656-9

**Geometry Workbook For Dummies**
978-0-471-79940-5

**The SAT I For Dummies, 6th Edition**
978-0-7645-7193-0

**Series 7 Exam For Dummies**
978-0-470-09932-2

**World History For Dummies**
978-0-7645-5242-7

## FOOD, GARDEN, HOBBIES & HOME

**Bridge For Dummies, 2nd Edition**
978-0-471-92426-5

**Coin Collecting For Dummies, 2nd Edition**
978-0-470-22275-1

**Cooking Basics For Dummies, 3rd Edition**
978-0-7645-7206-7

**Drawing For Dummies**
978-0-7645-5476-6

**Etiquette For Dummies, 2nd Edition**
978-0-470-10672-3

**Gardening Basics For Dummies*†**
978-0-470-03749-2

**Knitting Patterns For Dummies**
978-0-470-04556-5

**Living Gluten-Free For Dummies†**
978-0-471-77383-2

**Painting Do-It-Yourself For Dummies**
978-0-470-17533-0

## HEALTH, SELF HELP, PARENTING & PETS

**Anger Management For Dummies**
978-0-470-03715-7

**Anxiety & Depression Workbook For Dummies**
978-0-7645-9793-0

**Dieting For Dummies, 2nd Edition**
978-0-7645-4149-0

**Dog Training For Dummies, 2nd Edition**
978-0-7645-8418-3

**Horseback Riding For Dummies**
978-0-470-09719-9

**Infertility For Dummies†**
978-0-470-11518-3

**Meditation For Dummies with CD-ROM, 2nd Edition**
978-0-471-77774-8

**Post-Traumatic Stress Disorder For Dummies**
978-0-470-04922-8

**Puppies For Dummies, 2nd Edition**
978-0-470-03717-1

**Thyroid For Dummies, 2nd Edition†**
978-0-471-78755-6

**Type 1 Diabetes For Dummies*†**
978-0-470-17811-9

* Separate Canadian edition also available
† Separate U.K. edition also available

Available wherever books are sold. For more information or to order direct: U.S. customers visit www.dummies.com or call 1-877-762-2974.
U.K. customers visit www.wileyeurope.com or call (0)1243 843291. Canadian customers visit www.wiley.ca or call 1-800-567-4797.

 WILEY

## INTERNET & DIGITAL MEDIA

**AdWords For Dummies**
978-0-470-15252-2

**Blogging For Dummies, 2nd Edition**
978-0-470-23017-6

**Digital Photography All-in-One Desk Reference For Dummies, 3rd Edition**
978-0-470-03743-0

**Digital Photography For Dummies, 5th Edition**
978-0-7645-9802-9

**Digital SLR Cameras & Photography For Dummies, 2nd Edition**
978-0-470-14927-0

**eBay Business All-in-One Desk Reference For Dummies**
978-0-7645-8438-1

**eBay For Dummies, 5th Edition***
978-0-470-04529-9

**eBay Listings That Sell For Dummies**
978-0-471-78912-3

**Facebook For Dummies**
978-0-470-26273-3

**The Internet For Dummies, 11th Edition**
978-0-470-12174-0

**Investing Online For Dummies, 5th Edition**
978-0-7645-8456-5

**iPod & iTunes For Dummies, 5th Edition**
978-0-470-17474-6

**MySpace For Dummies**
978-0-470-09529-4

**Podcasting For Dummies**
978-0-471-74898-4

**Search Engine Optimization For Dummies, 2nd Edition**
978-0-471-97998-2

**Second Life For Dummies**
978-0-470-18025-9

**Starting an eBay Business For Dummies, 3rd Edition†**
978-0-470-14924-9

## GRAPHICS, DESIGN & WEB DEVELOPMENT

**Adobe Creative Suite 3 Design Premium All-in-One Desk Reference For Dummies**
978-0-470-11724-8

**Adobe Web Suite CS3 All-in-One Desk Reference For Dummies**
978-0-470-12099-6

**AutoCAD 2008 For Dummies**
978-0-470-11650-0

**Building a Web Site For Dummies, 3rd Edition**
978-0-470-14928-7

**Creating Web Pages All-in-One Desk Reference For Dummies, 3rd Edition**
978-0-470-09629-1

**Creating Web Pages For Dummies, 8th Edition**
978-0-470-08030-6

**Dreamweaver CS3 For Dummies**
978-0-470-11490-2

**Flash CS3 For Dummies**
978-0-470-12100-9

**Google SketchUp For Dummies**
978-0-470-13744-4

**InDesign CS3 For Dummies**
978-0-470-11865-8

**Photoshop CS3 All-in-One Desk Reference For Dummies**
978-0-470-11195-6

**Photoshop CS3 For Dummies**
978-0-470-11193-2

**Photoshop Elements 5 For Dummies**
978-0-470-09810-3

**SolidWorks For Dummies**
978-0-7645-9555-4

**Visio 2007 For Dummies**
978-0-470-08983-5

**Web Design For Dummies, 2nd Edition**
978-0-471-78117-2

**Web Sites Do-It-Yourself For Dummies**
978-0-470-16903-2

**Web Stores Do-It-Yourself For Dummies**
978-0-470-17443-2

## LANGUAGES, RELIGION & SPIRITUALITY

**Arabic For Dummies**
978-0-471-77270-5

**Chinese For Dummies, Audio Set**
978-0-470-12766-7

**French For Dummies**
978-0-7645-5193-2

**German For Dummies**
978-0-7645-5195-6

**Hebrew For Dummies**
978-0-7645-5489-6

**Ingles Para Dummies**
978-0-7645-5427-8

**Italian For Dummies, Audio Set**
978-0-470-09586-7

**Italian Verbs For Dummies**
978-0-471-77389-4

**Japanese For Dummies**
978-0-7645-5429-2

**Latin For Dummies**
978-0-7645-5431-5

**Portuguese For Dummies**
978-0-471-78738-9

**Russian For Dummies**
978-0-471-78001-4

**Spanish Phrases For Dummies**
978-0-7645-7204-3

**Spanish For Dummies**
978-0-7645-5194-9

**Spanish For Dummies, Audio Set**
978-0-470-09585-0

**The Bible For Dummies**
978-0-7645-5296-0

**Catholicism For Dummies**
978-0-7645-5391-2

**The Historical Jesus For Dummies**
978-0-470-16785-4

**Islam For Dummies**
978-0-7645-5503-9

**Spirituality For Dummies, 2nd Edition**
978-0-470-19142-2

## NETWORKING AND PROGRAMMING

**ASP.NET 3.5 For Dummies**
978-0-470-19592-5

**C# 2008 For Dummies**
978-0-470-19109-5

**Hacking For Dummies, 2nd Edition**
978-0-470-05235-8

**Home Networking For Dummies, 4th Edition**
978-0-470-11806-1

**Java For Dummies, 4th Edition**
978-0-470-08716-9

**Microsoft® SQL Server™ 2008 All-in-One Desk Reference For Dummies**
978-0-470-17954-3

**Networking All-in-One Desk Reference For Dummies, 2nd Edition**
978-0-7645-9939-2

**Networking For Dummies, 8th Edition**
978-0-470-05620-2

**SharePoint 2007 For Dummies**
978-0-470-09941-4

**Wireless Home Networking For Dummies, 2nd Edition**
978-0-471-74940-0